Y0-AGX-015

Metallurgy and Materials Science

a series of student texts

General Editors:

Professor R W K Honeycombe
Professor P Hancock

Plastics

Microstructure, Properties and Applications

Second edition

N J Mills

Department of
Metallurgy and Materials,
University of Birmingham, UK

Halsted Press
an imprint of JOHN WILEY & SONS, INC.
NEW YORK TORONTO

First published in Great Britain 1986
Second edition 1993

Published in the United States and Canada by Halsted Press, an imprint of John Wiley & Sons, Inc.
605 Third Avenue, New York, NY 10158–0012

Library of Congress Cataloging-in-Publication

Mills, N. J. (Nigel J.)
Plastics : microstructure, properties and applications / N.J. Mills—2nd ed.
p. cm.
Includes bibliographical references and index.
ISBN 0–470–22132–1
1. Plastics. I. Title.
TA455.P5M515 1993 93–15986
668.4—dc20 CIP

Printed and bound in Great Britain

General editors' preface

Large textbooks with broad subject coverage still have their place in university teaching. However, staff and students alike are attracted to compact, cheaper books which cover well-defined parts of a subject up to and beyond final year undergraduate work. The aim of the series is to do just this in metallurgy and materials science.

Materials science as taught is now more integrated and less polarized towards metals or non-metallic materials, so we have planned the series to cover the basic aspects of the subject and to include the main groups of materials and processing routes of engineering interest.

The aim is not to deal with each subject in the greatest depth, but to provide the student with a compact treatment as a springboard to further detailed studies; and, through the texts, for the student to discover something of the real experience of research and development in the subject. Adequate general references are provided for further study. The books are aimed at materials scientists and metallurgists, and also towards engineers and scientists wishing to know more about the structure, processing and properties of engineering materials.

RWKH
PH

Preface

This book is intended for students on Engineering or Materials Science degree courses, and for scientists and engineers who require an introduction to the properties and applications of plastics. There is as much emphasis on the mechanical design of plastics products as there is on explaining physical properties in terms of the microstructure. When plastics are substituted for other materials, the mechanical properties of the rival materials must be compared. The responses of polymers to the chemical environment and the electrical properties differ from other materials, and this must also be considered. In order to take advantage of the low cost fabrication of plastics, the product design must be suitable for one of the existing processes. Therefore the merits and limitations of these processes must be understood. The processing has permanent consequences on the microstructure of the product. Hence the effects of processing must be anticipated, and used to advantage if possible. It is the interlocking nature of different aspects of plastics technology that provides a challenge to engineers.

The approach used is to emphasise concepts and to point out the links between the subject and other areas of science and technology. This is one reason for including the derivation of key equations; they do not have to be learnt, rather it is important to see the assumptions made in reaching the result. Catalogue treatments of the properties of polymers, and comprehensive accounts of the processing routes, should be turned to for reference if necessary. The aim here is to give the student confidence in the approach used in polymer engineering, and there are questions for each chapter to test this. There has been a re-ordering of material from the first edition to make the sections more manageable, and to emphasise the structural divisions of the topic areas.

In most materials science courses there is a considerable practical content, because there is great benefit in having to apply the principles. This necessitates experimental work on the microstructure of polymers, measurements of mechanical and physical properties and experience of the main types of process. The selection of which chapters to study in the progression through a degree course will be left to the academic course organiser. There is merit in covering the basics of microstructure and processing in the first year of a course. Detailed consideration of mechanical properties is best left until the properties of 'simpler' elastic materials have been studied. There is sufficient range of topics in Chapters 9 to 11 to suit option courses on electrical, chemical or optical properties. The subject of materials selection, covered in Chapter 12, necessitates access to a computer database. Some of these are available free of change from materials manufacturers, and the access to PCs should not be a problem. The design case studies covered in Chapter 13 are intended to illustrate the compromises needed in successful polymer products. Throughout there is reference to further reading from secondary sources. This is not to discourage the use of primary journals, appropriate when undertaking research projects, but critical reviews of subjects are the best starting point.

It is assumed that the reader has an elementary knowledge of the mechanics of materials: the elastic stress–strain relationships of isotropic materials, the stress analysis of beams and thin walled cylinders and some familiarity with mechanical tests. Two appendices provide the heat and fluid flow theory relevant to plastics, because it may not be readily available in student textbooks.

Contents

1

Molecular structures and manufacture of polymers

1.1 CATEGORIES OF POLYMERS

We live in an era when the sales of plastics are still expanding. Some of this growth is at the expense of traditional materials, and some is due to the development of new markets. When plastics are substituted for other materials there must be improvements in the performance of the product and/or reductions in the cost of manufacture. There usually needs to be a redesign of the product, because the properties of plastics differ markedly from those of metals, glass, wood or ceramics. In exploring the relation between the properties of plastics and their microstructure we begin to see the possibilities and the limitation of this class of materials. It is also vital to understand how plastics are processed. One of their main economic advantages is the ease and the low energy consumption of processing, but this is only achieved if the product design is suitable. An analysis of the heat transfer and fluid flow in the main processes brings out these points in later chapters.

Polymers can be subdivided into three main categories; *thermoplastics* consist of individual long chain molecules, and in principle any product can be reprocessed by chopping it up and feeding it back into the appropriate machine; *thermosets* contain an infinite three dimensional network which is only created when the product is in its final form, and cannot be broken down by reheating whilst *rubbers* contain looser three dimensional networks, where the chains are free to change their shapes. The relative importance of the individual members of these three classes can be judged by comparing their annual consumption. Table 1.1 shows that there are four thermoplastics which are referred to as *commodity thermoplastics*. A large number of manufacturers compete in supplying different versions of these, and the price range in 1991 was 1.0 to 1.5 Deutschmark per kg (1 DM per kg = £350 per tonne). The low density of these materials, ranging from 900 kg m^{-3} for polypropylene (PP) to 1400 kg m^{-3} for polyvinyl chloride (PVC), means that the material costs are low in volume terms. The remaining thermoplastics in Table 1.1 are called *engineering thermoplastics* because of their superior mechanical properties, but the distinction is a fine one. They are produced on

Table 1.1 Consumption of thermoplastics in Western Europe in 1991

Thermoplastic	Abbreviation	Form	% of	DM/kg
Polyethylene	HDPE	semi-crystalline	11	1.1–1.5
	LDPE	semi-crystalline	20	1.1–1.3
Polypropylene	PP	semi-crystalline	15	1.1–1.5
Polyvinyl chloride	PVC	glassy	20	1.0–1.2
Polystyrene	PS	glassy	8	1.7–1.9
Polyethylene terephthalate	PET	semi-crystalline	8	
Polyamide	PA	semi-crystalline	1.7	
Polyurethane	PU	rubber and semi-crystalline	2	~4
Acrylonitrile–butadiene–styrene	ABS	glassy and rubber	2.2	3.8–4.1
Polycarbonate	PC	glassy	0.7	
Polymethyl methacrylate	PMMA	glassy	1	~6
Polybutylene terephthalate	PBT	semi-crystalline	0.2	
Other			10	

Total 23 million tonnes. Courtesy of European Plastics News.

a smaller scale and have prices in the range 2 to 5+ DM/kg. Finally there are speciality plastics which only sell a few thousand tonnes per annum and may cost 10 DM/kg upwards. An example is polytetrafluoroethylene (PTFE) which has unique low friction properties.

The crosslinking reaction, that occurs in the production of thermosets, can be used to advantage to give good adhesion to other materials. Thus epoxy and polyester resins are used as the matrices for fibre reinforced composites, amino resins are used for bonding chipboard and phenolics are used for bonding fibres in brake pads and sand in metal casting moulds. Most of these products are specialised and do not fit in well with the discussion of thermoplastic properties in this book. The consumption of thermosets is almost static, which reflects a loss of some markets to thermoplastics with a high temperature resistance.

The consumption of rubbers is heavily dominated by the production of tyres. In this and many other markets (conveyor belts, pressure hoses) it is really a fibre reinforced rubber that is taking the main mechanical loads. The rubber allows flexibility in bending whereas the fabric reinforcement limits the in-plane stretching of the product. The major applications are dominated by natural rubber and styrene–butadiene copolymer rubber (SBR), whereas other rubbers have specialised properties of low air permeability (butyl rubbers), good oil resistance (nitrile rubbers) and high and low temperature resistance (silicone rubbers). Rubbers as such play a relatively small role in this book, but the rubbery behaviour of the armorphous phase in semi-crystalline thermoplastics is important.

In terms of microstructure thermoplastics can be divided into amorphous and semi-crystalline solids. The amorphous ones are glassy up to a temperature called the glass transition temperature T_g, whereupon they change into a rubbery liquid, the viscosity of which falls as the temperature is raised further. Semi-crystalline thermoplastics can be regarded as two phase materials, with an amorphous phase, and a crystalline phase with a melting temperature T_m. Table 1.2 lists the chemical structures and transition temperatures of several thermoplastics. These characteristic temperatures will be seen later to control the mechanical properties. The values are not as precise as the melting points of pure metals; the crystalline phase of polymers melts over a temperature range that ends at T_m, and the exact values of T_g and T_m depend on themolecular weight. For semi-crystalline polymers the percentage crystallinity is another important parameter; we will see in the next section that it is determined by the regularity of the chemical structure. We must also study the chemical structure to be able to specify precisely the polymer that is being discussed.

1.2 CHEMICAL CHARACTERISATION

1.2.1 Bonding in polymers

The most important type of bond is the single covalent bond, created by the sharing of an electron between the outer electron shells of two atoms. Carbon

Table 1.2 Structures and transition temperatures

Generic structure		Name	T_g (°C)	T_m (°C)
$-[CX_2-CX_2]-$	X = H	polyethylene	−120	140
	X = F	polytetrafluoroethylene	−113	327
$-[CH_2-CHX]-$	$X = CH_3$	polypropylene	−10	170
	X = Cl	polyvinylchloride	80	220
	$X = C_6H_6$	polystyrene	100	—
	X = CN	polyacrylonitrile	105	D
	$X = OCOCH_3$	polyvinylacetate	29	
$-[CH_2-CX_2]-$	X = Cl	polyvinylidene chloride	−18	205
	X = F	polyvinylidene fluoride	−45	210
	$X = CH_3$	polyisobutylene (butyl rubber)	−70	—
$-[CH_2-CXY]-$	$X = CH_3$, $Y = COOCH_3$	polymethyl methacrylate	105	—
$-[CH_2-HC=CX-CH_2]-$	X = H	polybutadiene	−85	11
	$X = CH_3$	polyisoprene (natural rubber)	−75	25
$-[O-(CH_2)_n]-$	$n = 1$	polyoxymethylene	−85	170
	$n = 2$	polyethylene oxide	−67	69
$-[O-(CH_2)_n-O-C(=O)-C_6H_4-C(=O)]-$	$n = 2$	polyethyleneterephthalate	69	265
	$n = 4$	polybutyleneterephthalate	80	232

Table 1.2 (*Continued*)

Generic structure		Name	T_g (°C)	T_m (°C)
$-[(CH_2)_n-C(=O)-NH]-$	$n = 5$	nylon 6	50	228
	$n = 10$	nylon 11	46	185
$-[(CH_2)_n-NH-C(=O)-(CH_2)_m-C(=O)-NH]-$	$n = 6$, $m = 4$	nylon 6,6	57	265
$-[C_6H_2(CH_3)_2-O]-$		polyphenylene oxide	209	(261)
$-[C_6H_4-S]-$		polyphenylene sulphide	90	290
$-[C_6H_4-O-C_6H_4-SO_2]-$		polyethersulphone	225	—
$-[C_6H_4-C(CH_3)_2-C_6H_4-O-C(=O)-O]-$		polycarbonate	145	(295)
$-[C_6H_4-O-C_6H_4-O-C_6H_4-C(=O)]-$		polyetheretherketone	143	343

Parentheses on melting point signify the polymer is usually amorphous.
'D' means that the polymer decomposes before melting.

has 4 electrons in its outer shell (quantum no. $n = 2$), which is full when it contains 8 electrons. Hydrogen has only 1 electron in its outer shell (quantum no. $n = 1$), which is filled by 2 electrons. In methane (CH_4) the carbon atom forms covalent bonds to 4 hydrogen atoms. As each atom now has a full outer shell of electrons, the energy is minimised. Covalent bonds indicated as C—H are directional. In the methane molecule the lines joining the centres of the H atoms to the C are at 109° 28′ to each other; geometrically the H atoms are at the corners of a tetrahedron with the C atom at the centre. This tetrahedral bonding geometry of carbon occurs in the diamond crystal lattice (Chapter 2), and to a good approximation in polymers.

Some polymers contain the double covalent bond, written as C=C. This indicates that two electrons are shared between the C atoms, but the two bonds are different in nature. The first is as described above (a σ bond) but the second is the less stable π bond. It has two special qualities: (a) it prevents the rotation of the C atoms relative to each other—the consequences of this will be seen when we consider the shapes of polymer molecules in Chapter 2, and (b) it is less stable than the σ bond. Monomers, that must react together to form polymers, usually contain π bonds.

The second most important bond in polymers is the van der Waals bond. It is a secondary form of atomic attraction; electron oscillations in one atom induce electron movement in neighbouring atoms and thereby attract them. This weak bond is never shown in diagrams of polymer structures but it is responsible for holding neighbouring polymer chains together. In solid methane it is the only type of bond holding the structure together. In any of the condensed polymer states (melt, glassy or crystalline) sections of polymer molecules are packed closely together. There are van der Waals attractive forces between polar or non-polar groups in neighbouring polymer molecules. The forces are easier to quantify in molecular solids than in polymers. For example the potential energy of two methane molecules, with their centres a distance R apart, is

$$\frac{E}{E_0} = \left(\frac{R_0}{R}\right)^{12} - 2\left(\frac{R_0}{R}\right)^{6} \tag{1.1}$$

where the constants $R_0 = 0.43$ nm and $E_0 = 0.0127$ eV. Fig. 1.1. shows the

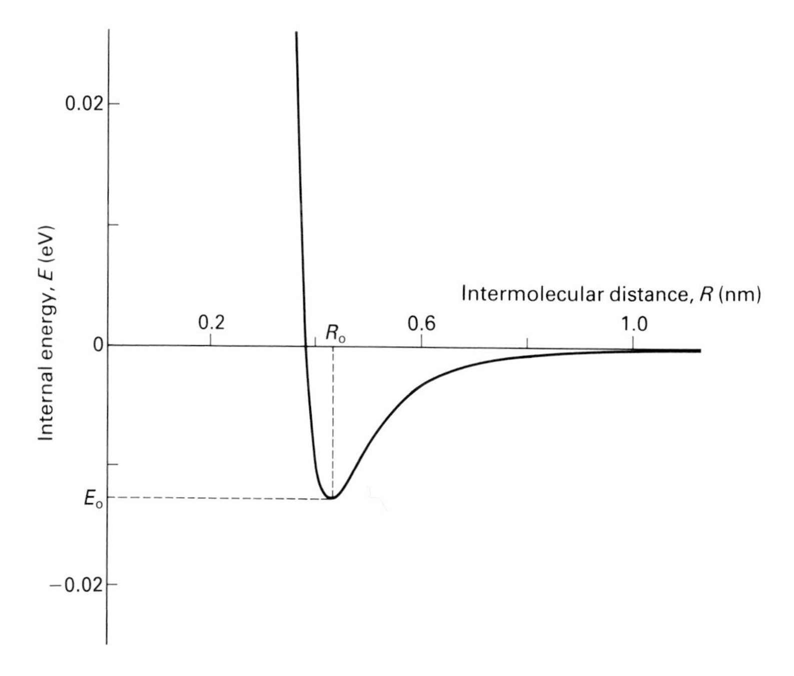

Fig. 1.1 Variation of the internal energy E of a pair of methane molecules with the intermolecular distance R

form of the potential energy versus distance graph. The R^{-12} term is a short range repulsion force, whereas the R^{-6} term is the attractive force. A potential energy minimum of depth E_0 occurs at R_0. We expect polymer molecules to have a similar shaped energy versus separation curve to Fig.1.1. At room temperature the thermal energy per carbon atom is of order kT, which is of the same order as the depth of the potential well. Consequently the occupied energy level will be close to the top of the potential well.

The overall energy E_0 needed to break the bond is only $\approx 1\%$ of that needed to break a covalent bond. Some idea of the relative strength of the bonds can be obtained by comparing the densities and melting points of the crystalline form:

Material	Bonding	Density ($kg\ m^{-3}$)	Melting point (K)
Diamond	Covalent	3510	>3000
Polyethylene	Covalent + van der Waals	1000	410
Methane	van der Waals	543	150

If a covalent bond is temporarily broken a free radical exists. This is shown as a dot as in $C^{\cdot}$. These bonds are extremely reactive and consequently shortlived; their lives are measured in milliseconds.

In certain polymers it is possible for hydrogen bonds to exist. These are intermediate in strength between covalent and van der Waals bonds. Apart from being responsible for the interesting properties of ice, they occur mainly in polyamides, where the hydrogen atom that is covalently bonded to nitrogen transfers part of this bond to the carbonyl group on the neighbouring polymer. The hydrogen bond will be shown by a series of dots as in

$$—N—H\cdots\cdots O{=}C—$$

Ionic bonds (in which electrons are donated or received from other atoms) occur rarely, and then only in the side groups in polymer chains. In contrast with the metallic bond, which allows some electrons to move freely throughout the crystal lattice, neither the covalent nor the van der Waals bond allows long range electron movement. We shall see later that this means that polymers are electrical insulators, and as a corollary are able to transmit light.

1.2.2 Addition polymers and their naming

The main types of polymerisation reaction can be classified as either *addition* or *step growth* polymerisations. The commodity plastics of Table 1.1 are all made by addition polymerisation, where a monomer is converted into the polymer with no by-products. For example the polymerisation of ethylene can be written

$$n\ CH_2{=}CH_2 \rightarrow \text{+}CH_2—CH_2\text{+}_n$$

where n is the degree of polymerisation. The name of the polymer is the prefix, poly-, plus the monomer name. The reaction needs to be initiated, usually by the thermal decomposition of an unstable initiator molecule, such as a peroxide, to produce free radicals. The initiator fragment with its free radical, shown as $I^\bullet$, attacks the covalent π bond in a monomer molecule, and leaves a free radical on the monomer.

$$I^\bullet + CH_2{=}CH_2 \rightarrow I{-}CH_2{-}CH_2^\bullet$$

The decomposition of the initiator is slow compared with the succeeding propagation step in which a succession of monomers undergo a chain reaction.

$$\sim\sim\sim CH_2{-}CH_2^\bullet + CH_2{=}CH_2 \rightarrow \sim\sim\sim CH_2{-}CH_2{-}CH_2{-}CH_2^\bullet$$

The growth of the chain is terminated by one of a number of reactions. These can either destroy the free radical, as in the termination reaction

$$\sim\sim CH_2{-}CH_2^\bullet + \sim\sim CH_2{-}CH_2^\bullet \rightarrow \sim\sim CH_2{-}CH_2{-}CH_2{-}CH_2\sim\sim$$

or allow it to survive to continue polymerisation as in the chain transfer reaction.

$$\sim\sim\sim CH_2{-}CH_2^\bullet + H_2 \rightarrow \sim\sim\sim CH_2{-}CH_3 + H^\bullet$$

There will be a mixture of monomer, completed polymer and a small proportion of growing chains in the polymerisation reactor at any time, because the propagation process is much faster than the initiation process. Therefore there is no need to take the reaction to completion; separation of the mixture at any time will produce some polymeric product. The polymerisation reaction is irreversible so no special precautions are necessary to prevent it reversing. Control of the degree of polymerisation is by the termination step. This may occur naturally as a result of side reactions or because of the amount of impurities in the monomer, or it may be necessary to add a specific reagent. Thiol compounds, containing the weak S—H bond, are highly effective chain transfer agents.

1.2.3 Condensation polymers and their naming

There is a limit to the complexity of the polymer structures that can be made by addition polymerisation. If it is wished to alternate two structures in the same chain the *condensation* polymerisation route is available. The name condensation polymerisation was used because there is usually a by-product of water or other small molecule. However there is not always a byproduct, so the name *step growth* polymerisation has been used to indicate that each step in the polymerisation is reversible. Each of the starting chemicals has a reactive group at each end. For example a diol can react with a dibasic acid.

$$HO{-}R{-}OH + HOOC{-}R'{-}COOH \rightleftharpoons HO{-}R{-}OCO{-}R'{-}COOH + H_2O$$

R and R′ represent unspecified chemical groups. This is an equilibrium reaction, so the water needs to be removed from the reactor to move the equilibrium to the right. The product still has a reactive group at each end, so a succession of further reaction steps leads to the production of a polyester.

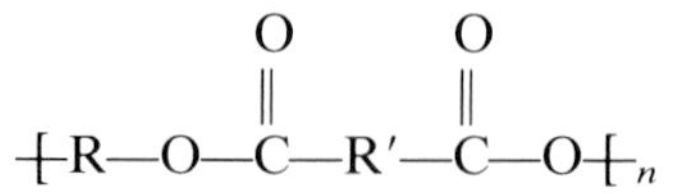

$$\left[R-O-\overset{\displaystyle O}{\overset{\|}{C}}-R'-\overset{\displaystyle O}{\overset{\|}{C}}-O \right]_n$$

Condensation polymers have generic names derived from the characteristic linking group that is formed in the polymerisation; the groups R and R′ need to be identified to specify a particular polyester. Table 1.3 lists the most common linking groups.

While the reaction is proceeding there is always an equilibrium between polymer molecules of different degrees of polymerisation. Those with $n = 1,2,3, \ldots, 10$ are referred to as oligomers. When a fraction p of the reactive groups have reacted, the mean degree of polymerisation is

$$\bar{n} = \frac{1}{1-p} \tag{1.2}$$

Consequently the reaction must be taken to a very high degree of completion, with $p > 0.999$, to obtain a useful high polymer. This means that the starting reagents must be very pure and that they must be present in exact stochiometric quantities. It may be necessary to prepare an intermediate monomer and

Table 1.3 Linking groups in step-growth polymerisation

Amide	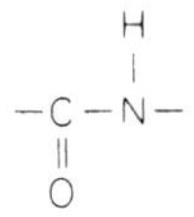 $-\underset{\underset{\displaystyle O}{\|}}{C}-\overset{\overset{\displaystyle H}{\vert}}{N}-$
Ester	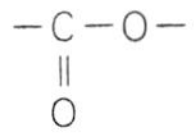 $-\underset{\underset{\displaystyle O}{\|}}{C}-O-$
Ether	$-O-$
Imide	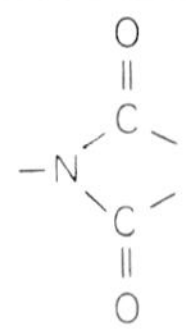 $-N\langle\begin{smallmatrix} C(=O) \\ C(=O) \end{smallmatrix}\rangle$
Sulphone	 $-\overset{\overset{\displaystyle O}{\|}}{\underset{\underset{\displaystyle O}{\|}}{S}}-$
Urethane	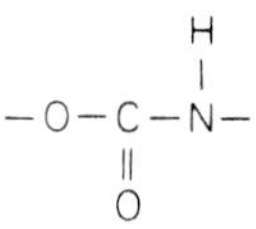 $-O-\underset{\underset{\displaystyle O}{\|}}{C}-\overset{\overset{\displaystyle H}{\vert}}{N}-$

purify it to allow the final reaction to proceed to a high polymer. The polymerisation takes place as a batch process over a period of hours. In contrast with addition polymerisations large amounts of heat are not evolved so the polymer does not have to be diluted with a heat transfer medium such as a gaseous monomer or a liquid. If there is no diluent the polymer must be in the melt state to allow the rapid diffusion of reactive groups towards each other; thus polymerisation may need to be completed at a high temperature.

1.2.4 Molecular weight distribution

The molecular weights are measured on the scale of atomic masses, so hydrogen = 1 unit, carbon = 12 units etc. The polymer molecular weight M is related to the *degree of polymerisation* n and the repeat unit molecular weight M_r by

$$M = nM_r \tag{1.3}$$

It is impossible to manufacture a truly monodisperse polymer in which every molecule has the same value of M. Hence a statistical average of the *molecular weight distribution* (MWD) must be quoted to specify the polymer under study. Usually the shape of the MWD is constant for a polymer made by a particular route, so the specification of a single average may be adequate. If f_i is the frequency of molecules that have degree of polymerisation i, then the mean degree of polymerisation n is given by

$$\bar{n} = \sum_{i=1}^{\infty} f_i i \Big/ \sum_{i=1}^{\infty} f_i \tag{1.4}$$

This familiar statistical measure is converted into the *number average molecular weight* M_N by multiplying by the repeat unit molecular weight M_r

$$M_N = \bar{n}M_r \tag{1.5}$$

There are properties of dilute polymer solutions, such as osmotic pressure, that are proportional to the number of molecules present per unit volume. Hence the ratio of osmotic pressure to the solution concentration is inversely proportional to M_N. If a polyethylene has $n = 400$, then as $M_r = 28$, $M_N = 11\,200$. The moderate value of n may give a false impression that there are no really large molecules present.

The *weight average molecular weight* M_W defined by

$$\bar{M}_W = M_r \sum_{i=1}^{\infty} f_i i^2 \Big/ \sum_{i=1}^{\infty} f_i i \tag{1.6}$$

can be measured by measuring the amount of light scattered from a dilute polymer solution. Other information such as the radius of gyration (Section 2.3.2) can be measured from the angular spread of the scattered light.

The standard deviation, the most common measure of the spread of a statistical distribution, is not used to quantify the spread of the MWD. This is partly because of the skew shape of the MWD, and partly because the ratio

M_W/M_N is used instead. Samples of so-called 'monodisperse' polystyrene, having $M_W/M_N = 1.05$, can be used to calibrate gel permeation chromatography (GPC) the commonly used *indirect* method of determining the MWD. In this method a constant flow rate of solvent passes through columns of swollen gel of crosslinked polymer, and the sample to be characterised is injected into the flow stream at the start of the columns. Columns containing different pore sizes are placed in series so that GPC can separate polymers over a wide molecular weight range. The polymer concentration at the far end of the columns is detected by refractive index measurement and is plotted against the elution volume. The largest molecules pass through the columns quickest, because they are unable to diffuse into the smaller passages in the gel. Once a GPC instrument has been calibrated with monodisperse polystyrenes, a molecular weight distribution can be plotted for the polymer of interest. Figure 1.2 shows that the range of molecular weights in a polyethylene is extremely wide, extending from oligomers with $n < 10$ to very large molecules with $M > 10^6$. For this polyethylene the distribution appears to be close to the Gaussian or normal distribution, but the molecular weight scale is logarithmic. Hence such distribution is referred to as a log-normal distribution.

For a polymer produced by a particular polymerisation route the MWD often has the same relative width, so the ratios M_W/M_N and M_Z/M_W remain constant [M_Z is an average of i^3 calculated by an equation similar to (1.6)]. For example, most commercial PVCs have $M_W/M_N = 2$. In such a case the measurement of a single molecular weight average is enough to specify the whole MWD.

It is possible to calculate the theoretical form of the MWD for a step growth polymerisation that is in equilibrium. In such a case, the number fraction f_i of molecules with i repeat units is equal to the probability that a molecule chosen at random has i repeat units. Since each step in forming a chain is mutually exclusive, we can multiply the probabilities that the first unit is polymerised, that the second is polymerised and so on. These probabilities are equal to p,

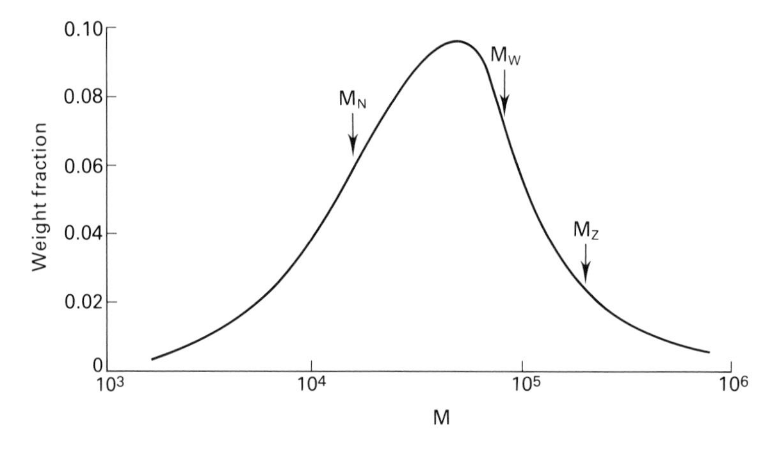

Fig. 1.2 The molecular weight distribution of a polyethylene, determined using GPC. The various molecular weight averages are shown

the average extent of reaction, except the probability that the ith unit is not polymerised is $(1-p)$. Consequently

$$f_i = p^{i-1}(1-p) \tag{1.7}$$

When this theoretical distribution is substituted in equations (1.4) and (1.6), we find that

$$M_N = \frac{M_r}{1-p} \quad \text{and} \quad M_W = M_r\left(\frac{1+p}{1-p}\right) \tag{1.8}$$

Therefore as commercially available polymers have $p \approx 1$ the ratio M_W/M_N is equal to 2. There is no such simple theory for addition polymerisation. If there are a number of different types of polymerisation sites on a catalyst used for free radical addition polymerisation then the MWD can be very broad.

We shall see later that desirable mechanical properties, such as resistance to cracking, improve as M_N increases. On the other hand the ease of fabrication of polymers by melt processing decreases rapidly as M_W increases. Commercial polymers therefore have MWDs that are the best compromise for a particular process and application area. There is a trend to manufacture polymers of narrower molecular weight distributions as the mechanical properties are better without sacrificing processability.

For *quality control* purposes it is not even necessary to measure a molecular weight; instead a variable that correlates with molecular weight will do. Examples are the dilute solution intrinsic viscosity, and the melt viscosity measured under specified conditions, such as the melt flow index (see Section 1.4.1).

1.2.5 Stereoregularity

If a polymer chain is to crystallise it must have a regular molecular structure. By regular is meant that the shape repeats itself at regular intervals. The reason for this is that there must be a repeating group of atoms (a motif) that is repeated through the crystal lattice by symmetry operators such as rotation, reflection and translation. In this and the next section we explore two measures of regularity. The monomer units for most addition polymers, with the exception of polyethylene, have asymmetric side groups. Let us consider the polymerisation of the vinyl monomer $H_2C{=}CHX$ where the side group X can represent Cl, CH_3 etc. We assume that the monomer units add head to tail, so that in the polymer there are X side groups on alternate C atoms. Since the C—C—C bond angle is 112° there is only one way in which the polymer molecule can be fully extended, with C atoms lying in a plane (Fig. 1.3). In a plan view the side groups appear to be on one side of the chain or the other. When a monomer unit adds itself to the end of a growing chain its side group can be on the same side as the last one—this is a meso (m) placement—or on the opposite side as the last side group—this is a racemic (r) placement. It is usual to idealise the stereoregularity of complete chains as being one of three types; the side groups X can all be on the same side of the chain, they can alternate, or their positions can be random. The word *isotactic*

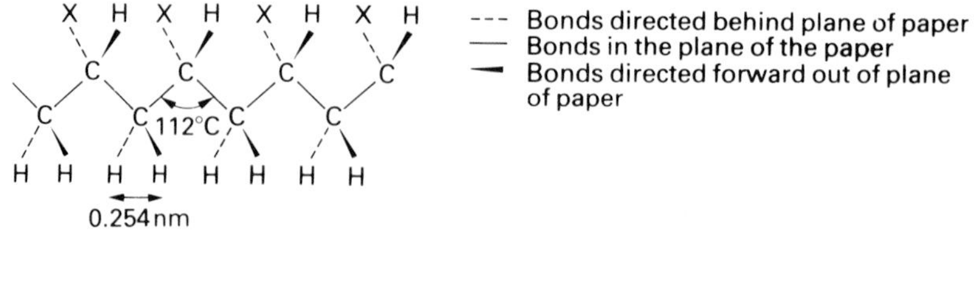

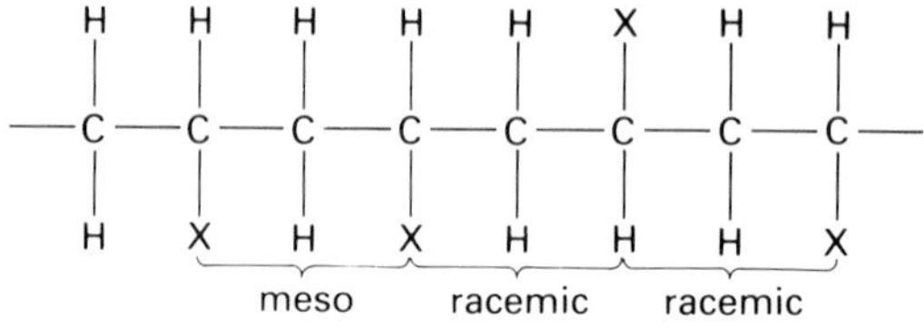

Fig. 1.3 Views of a fully extended vinyl polymer chain, showing the various stereo-isomers

(iso = same, tactos = form) was coined for the first possibility, and the words *syndiotactic* and *atactic* for the others. The section in Fig. 1.3 is part of an atactic chain. Special catalysts are necessary to produce polypropylene in the isotactic form, which is highly crystalline, whereas polystyrene is produced in the atactic form without the use of special catalysts. It is possible to produce isotactic polystyrene, but there is no commercial market for it.

A more accurate picture of stereoregularity involves the statistical distribution of regular sequences in the chain. In the simplest form of chain growth statistics (Bernoullian) the probability of an m placement, α, is independent of the previous monomer unit placements. Consequently the probability of two consecutive m placements, or mm, is α^2, and of an r placement is $(1-\alpha)$. One method of determining stereoregularity is nuclear magnetic resonance (NMR), which can determine the proportions of different placements in triads of monomer units. The possible triad conformations are shown in Table 1.4. When such measurements are made for commercial PVC it is found that the triad populations obey Bernoullian statistics and that the fraction of syndiotactic triads $S = 0.51 \pm 0.02$. Consequently it is possible to calculate the probability of n successive racemic placements as 0.71^n. There is a probability of 0.06 that a sequence of 8 racemic units occur in a row. Since the crystallinity of PVC is about 10% it can be inferred that such sequences are regular enough to fit into a crystal lattice.

Polypropylene is the commodity polymer whose properties depend to the

Table 1.4 Possible triad combinations

Conformation	Name	Bernoulian probability
mm	isotactic I	α
rr	syndiotactic S	$(1-\alpha)$
mr, rm	heterotactic H	$2\alpha(1-\alpha)$

greatest extent on its stereoregularity. Until organometallic catalysts that would produce isotactic polymers were found, there had been very little market for the irregular or atactic form, which is a rubbery liquid at room temperature. Different catalysts and polymerisation conditions produce polypropylenes of different triad proportions; a typical ratio of I:S:H triads is 0.95:0.01:0.04 for crystalline isotactic polypropylene and 0.35:0.36:0.30 for amorphous polypropylene. The triad proportions do not obey Bernoullian statistics, and the probability of an m placement must depend on whether the end of the growing chain has an m or r conformation.

1.2.6 Copolymerisation and chain branching

When two monomers are copolymerised (given that this is possible) the simplest classification of the products is into random copolymers and block copolymers (Fig. 1.4). Block copolymers are more difficult to make, requiring special catalysts, and the sequential addition of the monomers. To distinguish them the nomenclature 'poly (M_1—b—M_2)' is used where M_1 and M_2 are the monomer names, for example poly(styrene-b-butadiene).

Figure 1.4 suggests that the composition of a random copolymer remains the same as that of the monomer mixture. However, it is rare that two monomers add to the end of a growing chain at the same rate. Let us consider the free radical addition of monomer 1 or 2 to the end of a chain that terminates in an M_1 or M_2 unit

$$—M_1^{\bullet} + M_1 \xrightarrow{k_{11}} —M_1M_1^{\bullet}$$

where k_{11} is the appropriate rate constant. There are three similar equations for adding M_2 to $M_1^{\bullet}$ etc. The rate at which M_1 is added to a growing chain is equal to the negative rate of change of the molar concentration $[M_1]$, which is

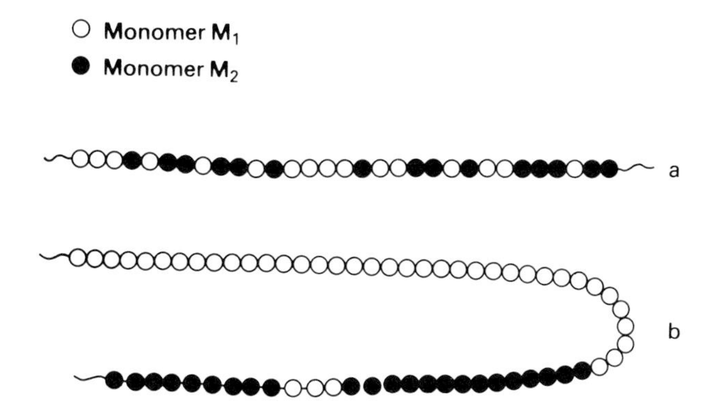

Fig. 1.4 Copolymers of two monomers. (a) Part of a random copolymer containing 60% M_1. (b) Part of a block copolymer in which there is a 90% probability that each monomer is joined to another of the same kind

$$-\frac{d[M_1]}{dt} = k_{11}[M_1^\cdot][M_1] + k_{21}[M_2^\cdot][M_1]$$

Similarly

$$-\frac{d[M_2]}{dt} = k_{22}[M_2^\cdot][M_2] + k_{12}[M_1^\cdot][M_2]$$

Therefore the ratio of the mole fractions F_1 and F_2 of M_1 and M_2 in the polymer being produced is

$$\frac{F_1}{F_2} = \frac{k_{11}[M_1^\cdot][M_1] + k_{21}[M_2^\cdot][M_1]}{k_{22}[M_2^\cdot][M_2] + k_{12}[M_1^\cdot][M_2]} \tag{1.9}$$

This expression can be simplified by using the fact that the concentration of free radicals rapidly reaches an equilibrium value. The second terms in the numerator and denominator of equation (1.9) represent the rates of $[M_2^\cdot]$ radicals becoming $[M_1^\cdot]$ radicals, and vice versa, so they must be equal. Therefore the right hand side of the equation can be divided through by either term to yield

$$\frac{F_1}{F_2} = \frac{[M_1]r_1[M_1] + [M_2]}{[M_2][M_1] + r_2[M_2]} \tag{1.10}$$

where the reactivity ratios are

$$r_1 = k_{11}/k_{12} \text{ and } r_2 = k_{22}/k_{12}$$

Equation (1.10) shows that unless $r_1 = r_2$ the composition of the polymer will be different from that of the monomer mixture. Consequently in a batch copolymerisation the monomer ratio will drift as the polymerisation proceeds, and hence the polymer formed at the end of the polymerisation can differ markedly in composition from that at the start. Some values of the reactivity ratios are given in Table 1.5 for common copolymer systems. With acrylonitrile–styrene copolymer the A units almost invariably occur singly but the S, SS, SSS units occur in the proportion 0.4:0.16:0.06. Consequently styrene will be used up faster. If the reaction is taken to completion without the addition of further supplies of one monomer there will be a drift in the composition with acrylonitrile rich copolymer being produced at later stages of the reaction. Ethylene–propylene copolymers are likely to contain long sequences of ethylene units, because of the high r_1 value.

Table 1.5 Reactivity ratios for random copolymers

Monomer 1	Monomer 2	r_1	r_2	r_1r_2
Acrylonitrile	Styrene	0.01	0.40	0.004
Butadiene	Styrene	1.40	0.78	1.1
Maleic anhydride	Styrene	0.015	0.04	0.0006
Ethylene	Propylene	17.8	0.065	1.16
Vinyl chloride	Vinyl acetate	1.35	0.65	0.88

The product r_1r_2 in Table 1.5 indicated the type of structure that will form. If $r_1r_2 = 1$ then the copolymer is referred to as ideal; if $r_1r_2 \gg 1$ then there is a tendency for block formation to occur; and if $r_1r_2 \ll 1$ then there is a tendency for alternating structures to form. Hence there tend to be alternating structures in acrylonitrile styrene copolymers.

The problem with making block copolymers is that the termination and transfer processes in addition polymerisation limit the lifetime of a growing polymer chain. Unless the chain remains alive while the first monomer is being replaced by the second, the result will be a mixture of homopolymers rather than a block copolymer. To overcome this, some special ionic polymerisation catalysts have been developed to make precise block copolymer structures. With the so-called 'living polymers' a fixed number of ions are introduced into an inert solvent. A dianion such as $^{-}[C_6H_5CHCH_2CH_2CHC_6H_5]^{-}$ will propagate from both ends if a suitable monomer is introduced. As there are no termination or transfer reactions, once the monomer has been used up a second monomer can be introduced to produce a triblock copolymer {2}{1}{2}. In this way {styrene}{butadiene}{styrene} block copolymers can be made in which each block has a precisely defined molecular weight (about 12 000 for the styrene and 60 000 for the butadiene). These materials undergo phase separation (see Chapter 3) and act as thermoplastic rubbers; that is they can be processed as thermoplastics but act like rubbers at room temperature.

When ethylene (ethene) is copolymerised with small proportions of higher alkenes (olefins), the resulting short chain branches modify the polymer crystallinity (see Section 1.4.1). We will limit our discussion to long chain branched molecules (Fig. 1.5). These can either occur as a result of a side reaction; for example when a propagating polyethylene molecule abstracts a H atom from a dead polyethylene molecule

$$—CH_2—CH_2^{\bullet} + —CH_2—CH_2— \rightarrow$$

$$—CH_2—CH_3 + —CH^{\bullet}—CH_2—$$

and the side chain then continues to propagate. Alternatively if a small proportion of a tri- or tetra-functional monomer is used in a step growth polymerisation, this will produce single or double branches in the resulting polymer. Care must be taken to limit this process or otherwise an infinite network molecule is formed.

Graft copolymers can be made in which the polymer backbone consist of one monomer and the branches of another. For example polybutadiene contains carbon double bonds that can be attacked by a free radical initiator

$$I^{\bullet} + \!\!+\!\! CH_2—CH{=}CH—CH_2 \!\!+\!\! \rightarrow$$

$$+\!\!CH_2—CH^{\bullet}—CHI—CH_2\!\!+$$

In this way styrene can be grafted on to butadiene. The efficiency of the process will not be high as polystyrene will also be formed. Once the polystyrene concentration reaches 2%, phase separation will occur, with spheres of polystyrene forming in the polybutadiene matrix. Further grafting will then only be possible in the monomer-swollen polybutadiene phase.

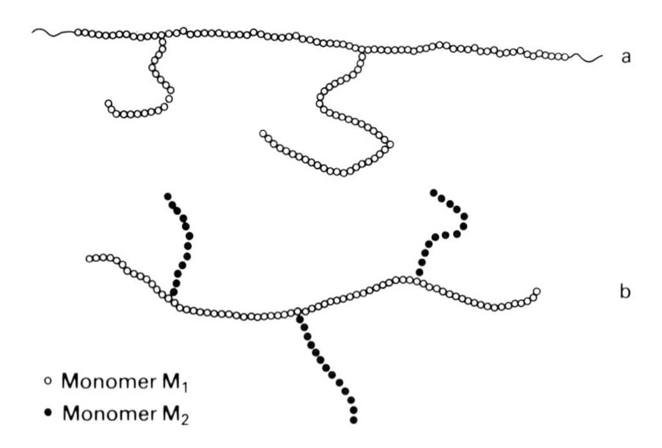

Fig. 1.5 (a) Long chain branching as found in LDPE polymerised at high pressures. (b) Graft copolymerisation of a monomer M_2 on to a backbone of monomer M_1

Although the concentration of true graft copolymer is not high, it is concentrated at the polystyrene–polybutadiene phase boundaries where it aids the mechanical properties.

1.2.7 Thermosets

If the relative number of branch points is increased in a polymer system there is a progression, from a number of branched molecules, to a single infinite tree molecule that contains no closed rings (Fig. 1.6a), to a three-dimensional network molecule (Fig. 1.6b). When a single tree molecule has been formed the *gel point* is said to be reached as at this point the majority of the polymer is no longer soluble, rather it forms a swollen gel if a solvent is added. Both thermosets and rubbers are examples of infinite three-dimensional network molecules. The chemical structure can be illustrated by the epoxy thermoset system. There are two components, a prepolymer of molecular weight 1000 to 2000 with reactive epoxy groups at each end, and a multifunctional amine 'hardener'. In the crosslinking reaction

$$-\overset{\mathrm{O}}{\overbrace{\underset{\mathrm{H}}{\mathrm{C}}-\mathrm{CH_2}}} + \underset{\mathrm{H}}{\overset{\mathrm{H}}{\mathrm{N}}}-\mathrm{R}-\underset{\mathrm{H}}{\overset{\mathrm{H}}{\mathrm{N}}} \longrightarrow -\underset{\mathrm{H}}{\overset{\overset{\mathrm{H}}{\mathrm{O}}}{\mathrm{C}}}-\underset{\mathrm{H}}{\overset{\mathrm{H}}{\mathrm{C}}}-\underset{\mathrm{H}}{\mathrm{N}}-\mathrm{R}-\underset{\mathrm{H}}{\overset{\mathrm{H}}{\mathrm{N}}}$$

the epoxy ring is opened without any by-product being produced. Figure 1.6b shows that if stochiometric quantities of amine hardener are used, each of these molecules is linked to 4 others by a network chain of the prepolymer molecular weight. Other thermoset systems produce less well defined networks. For example the polyester system starts with a partly unsaturated linear polyester from the step-growth polymerisation of propylene glycol, phthalic anhydride and maleic anhydride

$HO-CH(CH_3)-CH_2-OH$
propylene glycol

phthalic anhydride

$O=C-O-C=O$ / $HC=CH$
maleic anhydride

The proportion of maleic to phthalic anhydride determines the proportion of C=C bonds in the polyester. These can then react with styrene in the curing stage to produce links of one to three styrene units (the reactivity ratios of Table 1.5 show that an alternating copolymer is formed). The crosslinking reaction increases the glass transition temperature of the thermoset. There is an upper limit to the T_g of a particular thermoset (typically 145 °C for an

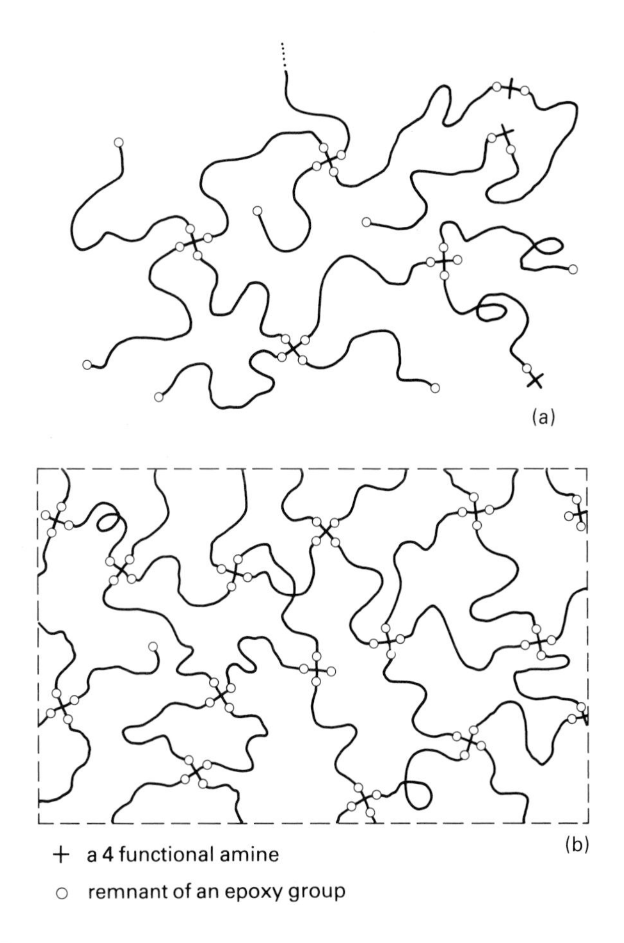

Fig. 1.6 (a) Part of an infinite tree molecule that forms during the crosslinking of a thermoset. (b) Part of an infinite three-dimensional network molecule

epoxy thermoset). If the thermoset is cured at a temperature that is below this limiting temperature, the reaction stops when the T_g reaches the curing temperature, because when the polymer becomes glassy there is no longer enough molecular mobility for the reaction to proceed any further. Consequently to obtain the maximum degree of crosslinking requires a curing temperature above the limiting T_g of the thermoset system.

1.2.8 Rubbers

Rubbers are crosslinked polymers that happen to be above their glass transition temperatures (T_g) or melting temperatures (T_m) at room temperature. In contrast with the thermoset systems it is usual to prepare a high molecular weight polymer and then crosslink it (although polyurethane rubbers are an exception to this rule). Table 1.6 lists some of the structures of common rubbers. The monomer structures do not contain rigid phenyl rings either in the chain, or as side groups where they would hinder chain rotation. There are no polar groups present either, which would increase the strength of the intermolecular forces. In order that the chains can easily be crosslinked, the polymers must contain unsaturated carbon bonds. If the repeat unit is apparently saturated, then it will be copolymerised with 1 to 2% of a monomer that produces unsaturated groups.

Table 1.6 Rubber structures

Rubber	Polymer	Structure	T_g (°C)	T_m (°C)
Natural	polyisoprene	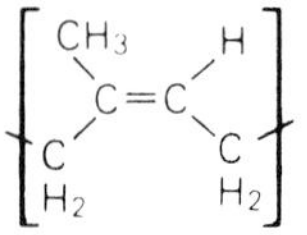 $[-CH_2-C(CH_3)=CH-CH_2-]$	−73	25
	polybutadiene	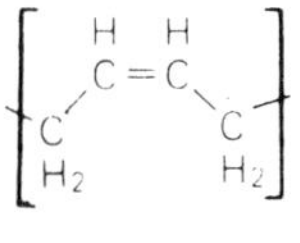 $[-CH_2-CH=CH-CH_2-]$	−55	
Butyl	polyisobutylene	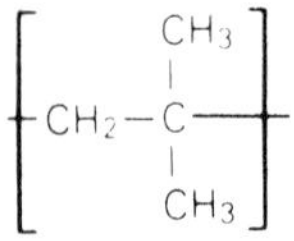 $[-CH_2-C(CH_3)_2-]$	−70	0
Silicone	polydimethyl siloxane	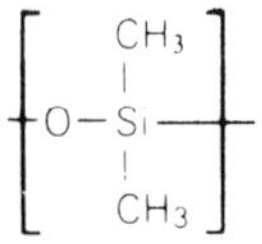 $[-O-Si(CH_3)_2-]$	−123	−70

The original crosslinking system for natural rubber, called *vulcanisation*, involves mixing in 2 to 3% of sulphur plus an accelerator. On heating to 140 °C the sulphur reacts with C=C bonds on neighbouring chains to form sulphur crosslinks C—$(S)_n$—C. Typically 15% of the crosslinks are monosulphide ($n = 1$), 15% are disulphide, and the rest are polysulphide with $n>2$. The polysulphide crosslinks are partially labile, which means that they can break and reform with other broken crosslinks when the applied stresses are high. This leads to permanent creep in compressed rubber blocks. To avoid such *permanent set* the so-called 'efficient' vulcanisation systems have been developed that produce only monosulphide crosslinks.

The rubber network can be characterised by the network chain molecular weight M_c. This is the number average molecular weight of the lengths of chain between neighbouring crosslinks. M_c g of rubber contains 1 mole of network chains i.e. it contains Avogadro's number N_A of network chains. Alternatively 1 m^3 of rubber has a mass 1000ρ g where ρ is the density of the rubber in kg m^{-3}, and contains N network chains, where N is the density of network chains. Consequently the network chain density is related to the network chain molecular weight by

$$\frac{M_c}{1000\rho} = \frac{N_A}{N} \tag{1.11}$$

If the crosslinks are all connected to 4 network chains, then every additional crosslink increases the number of network chains by two, and so the density of crosslinks is $N/2$. Hence the network chain molecular weight is inversely proportional to the density of crosslinks.

1.3 THE TECHNOLOGY AND ECONOMICS OF MANUFACTURE

1.3.1 Manufacture of monomers

The prices of monomers are determined by the number of stages in their manufacture, the raw material and energy costs and the scale on which they are made. The main starting material for manufacture is the naphtha fraction from the distillation of crude oil; it contains hydrocarbon molecules with 4 to 12 C atoms and boils in the range 20 °C to 200 °C. The *cracking* of a mixture of naphtha and steam at a temperature of about 850 °C for 0.5 s produces a complex mixture of products, only some of which are used for plastics manufacture. Figure 1.7 shows in outline how the products are separated. Liquid mixtures can be separated by distillation, and temperatures of down to −140 °C and pressures of up to 40 bar are required for liquification. A series of fractionation towers are required to strip off the products in turn. Roughly 30% of ethylene (ethene) and 15% of propylene (propene) are produced, with 20% fuel gas, 20% gasoline and 9% of 4 carbon atom hydrocarbons. A separate part of the petrochemical complex produces the aromatic compounds (e.g. benzene, toluene, xylene) that are also used in monomer production.

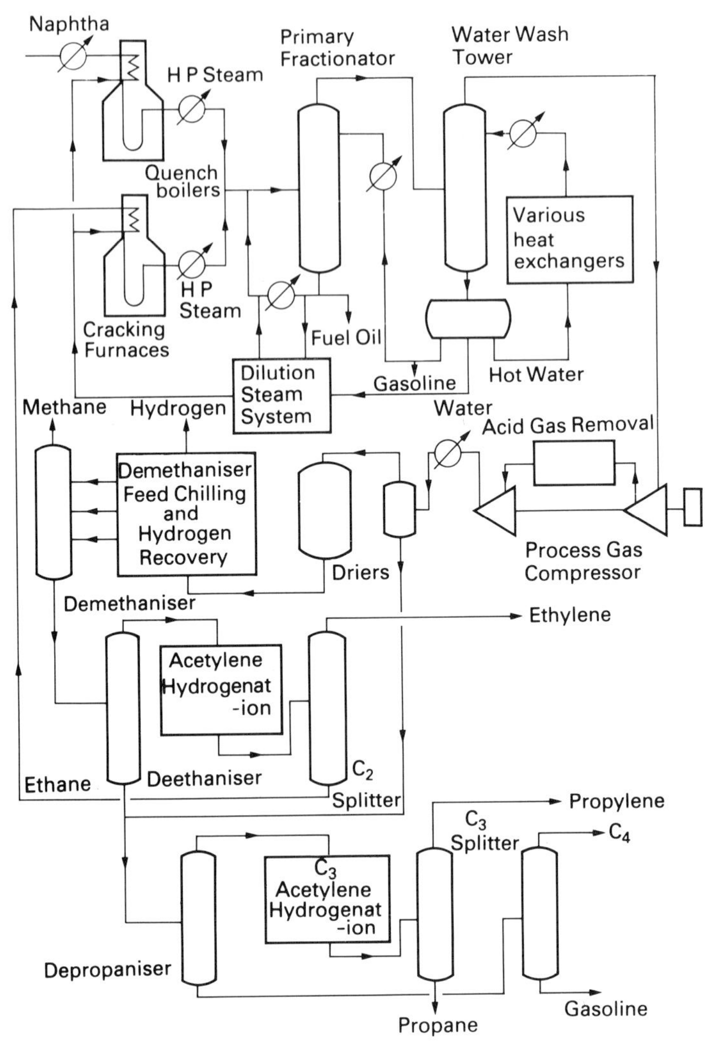

Fig. 1.7 Flow chart for the separation of the products produced by cracking naphtha

In contrast with the direct production of the ethylene (ethene) monomer there are many stages in the production of the monomers adipic acid and hexamethylenediamine used for the production of nylon 6,6 (Fig. 1.8). It is not surprising that the cost per tonne of nylon 6,6 is 4 or 5 times that of polyethylene since more energy has to be consumed at the different reaction stages and the capital cost of the different reactors is much higher. Moreover the scale of production is smaller by a factor of 50; the implications of this are explored later.

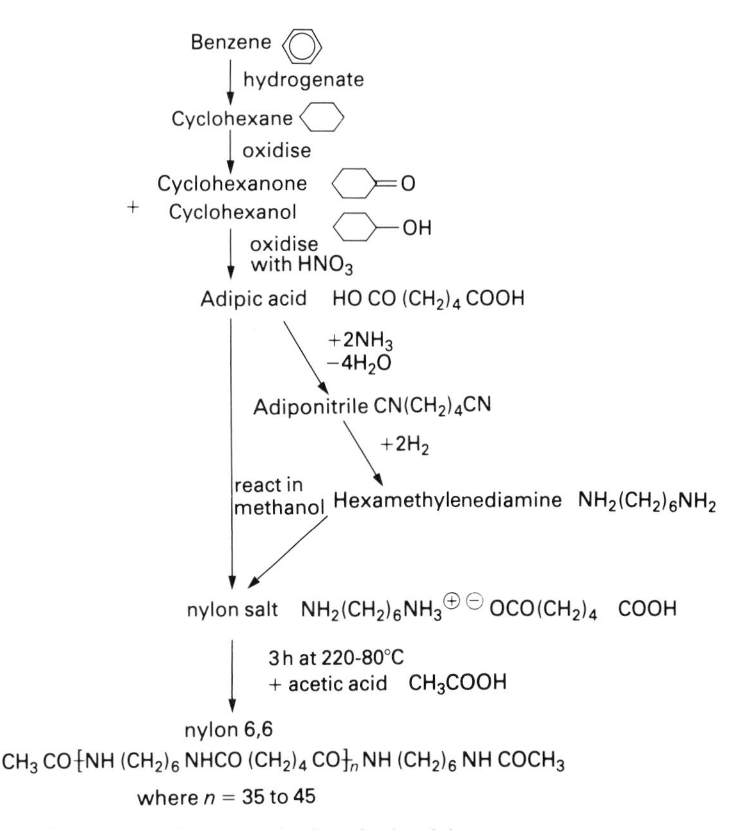

Fig. 1.8 A chemical route for the production of nylon 6,6

1.3.2 Polymerisation processes

One polymerisation process for a commodity plastic is used to illustrate some of the technology of large scale polymerisations. Generalisations should not be drawn from this because other processes operate in the batch rather than the continuous mode, and involve liquids or liquid/solid slurries rather than gases. The process chosen is the gas phase polymerisation of ethylene, developed by Union Carbide. In this the ethylene gas is used as the heat transfer medium to remove the heat of reaction, and solid polyethylene powder forms in a fluidised bed (Fig. 1.9).

A catalyst is needed for the polymerisation, and one objective is to make this so efficient that the catalyst does not need to be removed from the polymer (certain catalysts speed up polymer degradation or are toxic and need to be removed). Chromium compounds supported on finely divided silica are used as catalysts, and the chromium content of the polymer is about 1 part per million. Ethylene gas at about 20 bar pressure is introduced at the bottom of the fluidised bed reactor which can be 2.5 m in diameter and 12 m high. The temperature is maintained in the range 85 °C to 100 °C so that the

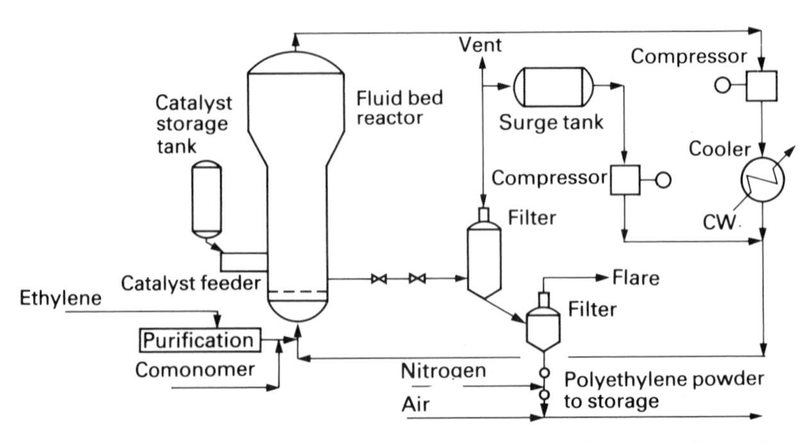

Fig. 1.9 Fluidised bed reactor used in the Union Carbide process for the gas phase polymerisation of ethylene (from *Chemical Engineering Symposium Series*, 1977, **50**, Institute of Chemical Engineers)

polyethylene particles are solid and hence do not stick together. There is continuous injection of catalyst into the pressure vessel. Only about 2% of the ethylene is polymerised each time it passes through the reactor; the remaining monomer is compressed and cooled before being returned to the reactor. The solid particles of polyethylene sink down the reactor as they grow in size; after 3 to 5 hours they are about 0.5 mm in diameter and can be removed from the reactor. The polymer needs to be purged with nitrogen before being conveyed to storage in an inert gas stream.

The reactor can produce polyethylene homopolymer, or copolymers (usually with butene), so the overall crystallinity of the product can be controlled. The molecular weight is controlled by additions of hydrogen, and the width of the molecular weight distribution can be changed by modifying the catalyst. The material and energy inputs to produce 1 tonne of homopolymer are given in Table 1.7. Slightly more power is required to extrude the powder into granules than is used in the polymerisation stage.

1.3.3 The economics of scale

The polyethylene process just described was originally operated as a pilot

Table 1.7 Inputs for 1 tonne of polyethylene homopolymer

Ethylene	1020 kg	to produce powder
Power	250 kWh	
Cooling water	110 m^3	
Nitrogen	6 m^3	
Catalyst	0.01 kg	
Power	300–450 kWh	to compound the powder into granules
Steam	200 kg	
Cooling water	40 m^3	

plant with a reactor 0.1 m in diameter, producing 50 tonnes per year. The largest reactor operating has a 4.5 m diameter reactor and is capable of producing 100 000 tonnes per year. As a result of this scaling up, the production cost per tonne of polymer is decreased considerably. The simplest type of analysis divides the production costs into 3 main elements:

$$\text{cost} = \text{monomer} + \text{energy} + \text{share of capital cost} \quad (1.12)$$

with the capital cost of constructing the plant and services being depreciated over a fixed period of say 5 years. If the plant is taken to be constructed of steel pipes and pressure vessels of a characteristic diameter D, then the capital cost £C is proportional to the surface area of steel used, i.e. to D^2. On the other hand the annual capacity Q tonnes increases in proportion to the volume of the reactors, or D^3. Consequently the relationship between capital cost and capacity is

$$C = AQ^{0.66} \quad (1.13)$$

where A is a constant. On this basis polyethylene produced in a plant with a 100 000 tonne annual capacity has a lower manufacturing cost than that made in a plant using the same process with a 10 000 tonne capacity. This simplistic analysis does not take into account the problems of overcapacity for certain commodity plastics, which can severely depress the selling price. Nor does it address the problem of plant reliability; it may well be better to have 2 or 3 units operating in parallel than to rely on a single very large reactor. Nevertheless it does indicate the cost advantage that the commodity plastics have. They are produced on a scale that is 10 to 30 times as large as the engineering plastics, and they have much simpler production processes. For a new polymer to establish itself on any scale it must be able to offer some specific property advantages. In spite of these advantages it may still face very strong competition from upgraded forms of commodity plastics which are improved by adding reinforcing fillers and toughened by rubbery additions. Figure 1.10 shows that the lower the selling price of a plastic the greater are its annual sales.

1.4 GRADES AND APPLICATIONS OF COMMODITY PLASTICS

Each of the commodity plastics consists of a family of closely related grades. The major variants need to be described and some of the specialised terminology explained. A few of the more important applications will be described because it is often the case that a particular type of polymer has been developed to suit a particular application.

1.4.1 Polyethylenes

The original ICI process for polymerising ethylene was developed in the 1940s and produces low density polyethylene (LDPE) with a density in the range 910 to 935 kg m^{-3}. It involves compressing ethylene to very high pressures

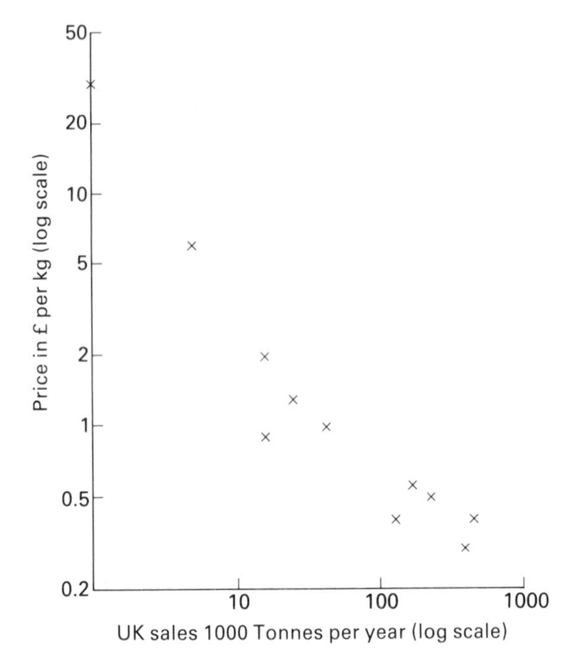

Fig. 1.10 Correlation between the selling price of plastics and the annual sales in the UK

(1400 to 2400 bar) and polymerising it at a high temperature (200 to 250 °C) using free radical catalysts. There are both short and long chain branches on the molecules because of side reactions in the process. For example between 1.6% and 3% of the C atoms in the chain have ethyl ($—CH_2CH_3$) or *n*-butyl ($—CH_2CH_2CH_2CH_3$) branches replacing a hydrogen atom. There are far fewer of the long chain branches but these have the effect of making the molecule into a star or comb shaped molecule (Fig. 1.5). If the polymerisation pressure is increased the molecular weight increases and the number of branches decreases.

From 1955 onwards a number of processes for producing high density polyethylene (HDPE), with densities in the range 955 to 970 kg m^{-3}, were introduced. These used organometallic catalysts which allowed polymerisation at low to medium pressures of 1 to 200 bar. There are less side reactions and hence fewer short chain branches (0.5 to 1% of the C atoms) and no long chain branching. The range of polymer densities produced by low pressure processes has been extended by copolymerising ethylene with butene or hexene. When these units are incorporated in the polymer they produce ethyl and butyl short chain branches. The copolymers produced in this way are often referred to as medium density polyethylenes (MDPE) of density about 940 kg m^{-3}, and linear low density polyethylenes (LLDPE) of densities 920–30 kg m^{-3}.

The short chain branches cannot easily be accommodated in the polymer crystals (see Chapter 2) so the percentage crystallinity decreases as the amount of comonomer is increased (Fig. 1.11). It is customary to measure

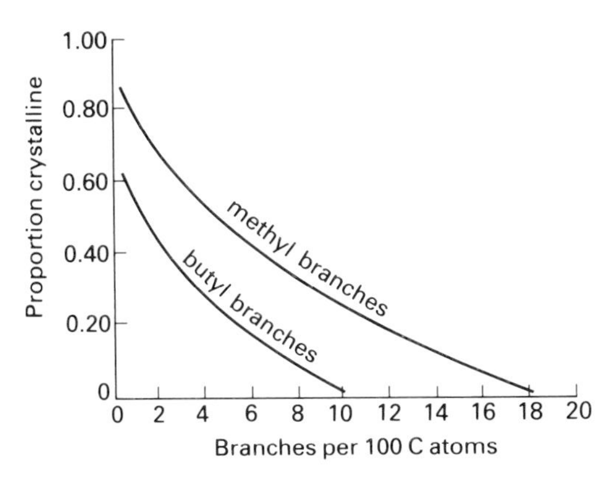

Fig. 1.11 Variation of the crystallinity of polyethylene copolymers with the number of short branches per 100 C atoms

density using density gradient columns to obtain the volume fraction crystallinity as this is quicker than using X-ray methods. It is assumed that the density of the crystalline phase does not change as the % crystallinity varies, i.e. that there is insignificant defect population inside the crystals that would reduce the density. This assumption is not completely valid, so every method of measuring crystallinity gives a slightly different value. The mass of crystals in 1 m^3 of polymer is the product of the volume fraction crystallinity V and the crystal density ρ_c, so the polymer density ρ is linearly related to V by

$$\rho = V\rho_c + (1 - V)\rho_a \tag{1.16}$$

where the crystal density ρ_c at 25 °C = 1000 kg m^{-3} and the amorphous density ρ_a = 854 kg m^{-3}.

The melt flow index (MFI) is used as a production control assessment of the average molecular weight. The MFI is an inverse measure of melt viscosity, so, as the MFI increases, the polymer flows more easily into moulds. It is measured by heating the polyethylene to 190 °C and applying a fixed pressure of 3.0 bar to the melt via a piston and load of total mass 2.16 kg (Fig. 1.12). The MFI is defined as the output rate, in g per 10 min from a standard die of 2.1 mm diameter and 8.0 mm length. The highest MFI grade polyethylenes are used for injection moulding whereas the lower MFI grade polyethylenes are for extrusion and blow moulding. Ultra high molecular weight polyethylene (UHMWPE) with M_W of 2 or 3×10^6 is so viscous that its MFI cannot be measured under the standard conditions. It is used for large blow mouldings or other applications demanding high toughness.

About 80% of LDPE is used as film, mainly for packaging. Minor percentages are used for cable insulation, injection moulding and the extrusion coating of cardboard or other materials. In contrast the more rigid HDPE is used for the blow moulding of containers, and for the extrusion of pipe for gas and water supplies. A proportion is used for film either on its own or blended with LDPE. Blends of such similar polymers are compatible, but

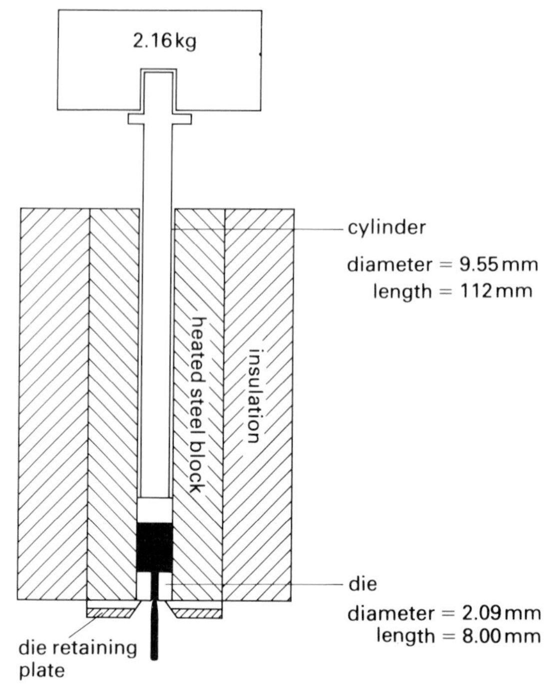

Fig. 1.12 Melt flow indexer used for the quality control of the molecular weight of polyolefins

when most other pairs of polymers are blended they separate into two distinct phases, often with a weak interfacial bond.

1.4.2 Polypropylene

Propylene is invariably polymerised with organometallic catalysts, because of the necessity of achieving a high isotactic content. A typical catalyst used is a mixture of titanium chloride and an aluminium alkyl such as $Al(C_2H_5)_2Cl$. The process has more stages than the polymerisation of ethylene in Fig. 1.9, because of the need to remove the catalyst by solvent treatment and centrifuging, and because there is a certain amount of atactic polypropylene by-product that is extracted using *n*-heptane as a solvent. The homopolymer is more rigid than HDPE and therefore is preferred for the injection moulding of thin walled products. However, it becomes brittle at about 0 °C due to the amorphous fraction becoming glassy. To overcome this propylene is copolymerised, mainly with ethylene. If random copolymers are made with 5 to 15% of ethylene, the crystallinity is reduced markedly from the usual 60 to 70%, so the stiffness and strength are low. To improve the toughness without losing rigidity, 'block copolymers' of 5 to 15% ethylene are produced. One method of producing these is to polymerise propylene, then purge the reactor of propylene using an inert gas, and introduce ethylene. In principle the lifetime of the growing chains is long enough for a true block copolymer to form, but in practice the product is a mixture of polypropylene, some polyethylene, plus a little block copolymer that aids the bonding between the two phases. If, instead of ethylene in the second stage, a mixture of 60% ethylene and 40% propylene by weight is used then a rubbery copolymer phase with a glass transition of −60 °C is formed. This forms spherical particles in the polypropylene matrix and aids its toughness. There have been significant increases in the toughness of polypropylenes for automotive applications in recent years.

The higher melting point of polypropylene (170 °C) compared with HDPE (135 °C) makes it more suitable for fibre applications, and 30% of the production goes into this market. About 20% is used in the form of oriented polypropylene film for wrapping crisps, cigarettes etc. where good clarity is required.

1.4.3 Polyvinyl chloride

Most of the PVC intended for melt processing is made by the suspension polymerisation process. A suspension of droplets of vinyl chloride monomer (VCM) in water is polymerised in a 10 m^3 autoclave at a temperature in the range 50 to 70 °C. The droplets of size 30 to 150 μm are formed by agitation with a stirrer and are stabilised by a colloidal layer of partially hydrolysed polyvinyl acetate, or other water soluble polymer. The colloid also has the role of controlling the porosity of the PVC grain that forms; there is a considerable contraction in volume from VCM of density 850 kg m^{-3} to PVC of density 1400 kg m^{-3}, and a porous grain is required to allow the rapid absorption of liquid additives at a later stage of processing. When the polymerisation is initiated PVC molecules form in the droplet of VCM. They are insoluble in VCM so they precipitate in the form of primary particles that are initially 0.1 to 0.2 μm in diameter. As the percentage conversion increases over a 5 to 6 hour period the number of primary particles per droplet increases and the particles coalesce into aggregates (Fig. 1.13a). There is also some coalescence of the droplets to form irregular shaped grains. The reaction is stopped before 90% conversion and the remaining VCM removed before the autoclave is discharged. When the powder is examined the grains are found to be roughly spherical, with an internal porosity (Fig. 1.13b). The porosity is utilised in the dry blending stage of processing, when stabilisers, lubricants and plasticisers are added.

It is also possible to polymerise an emulsion polymer with a smaller particle size. While the polymer is still an emulsion in water the spherical particles have diameters of 0.1 to 1 μm, but on drying these agglomerate to grains of mean size 30 to 60 μm. Therefore the emulsion polymer has a smaller particle size than the 100–160 μm grain size of suspension PVC. It is not possible to measure an unambiguous melt viscosity for PVC because the result depends on the degree to which the particle structure is retained. Consequently solution viscosities are measured to assess the molecular weight of the product. European practice is to measure '*K*-values' from the viscosity of a 0.5 g per decilitre solution in cyclohexanone at 25 °C. M_w increases with the K value, and the grade is selected to match the ease of the processing application. Thus PVC with K values of 55 to 62 is used in injection moulding

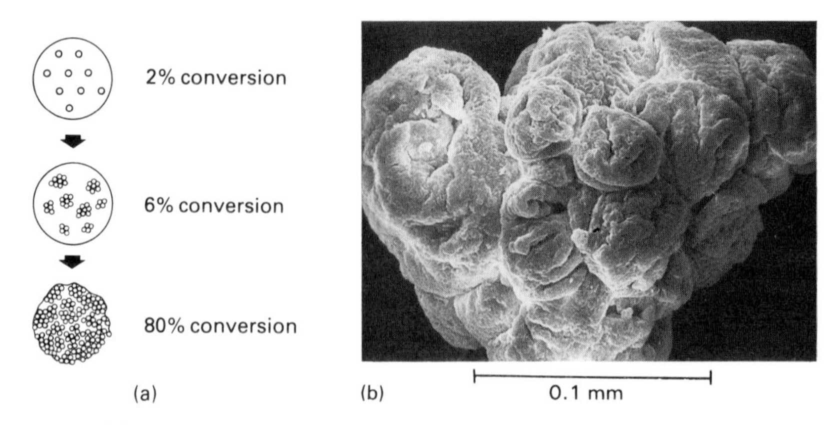

Fig. 1.13 (a) Precipitation of PVC particles inside a VCM droplet during suspension polymerisation. (b) SEM of the dried suspension powder

and the extrusion of thin foil, whereas large diameter pipe, which requires a maximum toughness, uses PVC with K values of 66 to 70.

PVC is unique among the commodity plastics in that about 50% of the production is sold with plasticiser incorporated. The plasticisers are high boiling point liquids which swell the PVC and reduce the glass transition temperature. Figure 1.14 shows how the shear modulus, measured by the twisting of a strip, changes with temperature at different plasticiser contents. The glass transition temperature, at which the modulus falls most rapidly, is reduced to below room temperature at a 40% plasticiser content. The 10% crystallinity of PVC prevents such a material being a sticky liquid at room temperature; instead it is a rubbery solid. For a plasticiser to be compatible, so that large amounts dissolve in PVC, it should contain polar ester groups or polarisable benzene groups. It must also have a low vapour pressure at room temperature and not diffuse out of the polymer rapidly. Most PVC plasticisers are esters which are the products of reactions between long chain alcohols with aromatics such as phthalic anhydride or dibasic acids such as adipic acid.

$$\text{alcohol} + \text{dibasic acid} \rightarrow \text{ester}$$

e.g.

$$\text{butanol} + \text{phthalic anhydride} \rightarrow \text{dibutyl phthalate}$$
$$\text{2-ethylhexanol} + \text{phthalic anhydride} \rightarrow \text{2-diethylhexyl phthalate}$$
$$\text{2-ethylhexanol} + \text{adipic acid} \rightarrow \text{2-diethylhexyl adipate}$$
$$\text{2-ethylhexanol} + \text{sebacic acid} \rightarrow \text{2-diethylhexyl sebacate}$$

Plasticised suspension PVC can be compounded in the melt and extrusion coated onto copper wire for electrical wiring; alternatively it can be calendered between parallel rolls to produce flexible sheeting for flooring and other applications. Emulsion PVC can be converted into a 'plastisol' by mixing it with a high proportion (50 to 70%) of a plasticiser. In this state it is a medium viscosity liquid at room temperature with a viscosity of 20 to $50\ \text{N s m}^{-2}$. In this plastisol form, of swollen 1 μm particles suspended in the plasticiser, it can be rotationally cast in moulds, coated onto wallpaper or

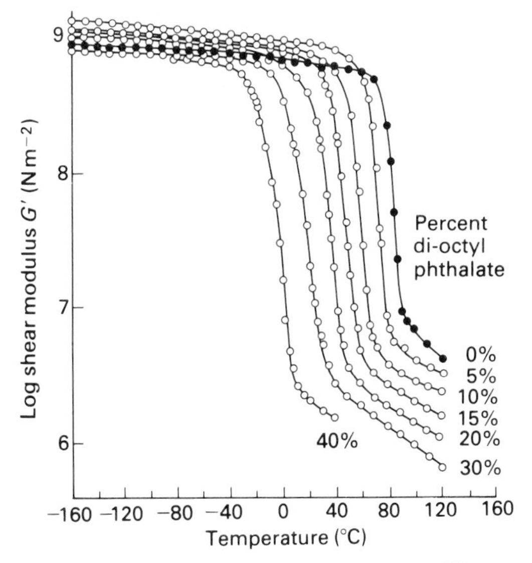

Fig. 1.14 The shear modulus of PVC versus temperature at different contents of dioctyl phthalate plasticiser (from Koleske and Wartman, *Polyvinylchloride*, 1967, Macdonald and Co.)

cloth or used to dip-coat metal products. A subsequent heating to between 150 °C and 175 °C causes the rest of the plasticiser to be absorbed in the PVC and the particles to fuse into a homogeneous solid.

1.4.4 Polystyrene and toughened derivatives

Styrene homopolymers are produced by a free radical polymerisation that proceeds to completion as the styrene/polystyrene mixture is taken through a series of gradually hotter reactors, starting at 110 °C and finishing at about 200 °C. Unlike PVC the polymer is soluble in the monomer at all concentrations, and the product is atactic, so no particle structure develops. It is usual to dilute the system with 3 to 12% of ethylbenzene solvent to reduce the melt viscosity at the later stages. Although the unreacted monomer and the ethylbenzene are flashed off under reduced pressure at the end of the process some of these materials remain in the polymer. The molecular weight of the product is varied from 100 000 to 400 000 for different applications. For injection moulding the‘ melt flow can be improved by adding about 1% of a lubricant such as butyl stearate and 0.3% of a mould release agent (a wax or zinc stearate).

Polystyrene homopolymer is an optically clear glass, and in an oriented film form can be used for the electrical insulation of capacitors. When converted into a low density foam it is used for building insulation and shock absorbing packaging. However, it is relatively brittle and weathers badly out of doors. Consequently several toughened versions of polystyrene have been developed, and these are described in Chapter 3.

1.5 ADDITIVES

A plastic consists of a polymer plus all the additives necessary for its performance. The additives are of many kinds, and will be discussed where they are appropriate to the properties described in other chapters. Fibres or a rubber can be added to modify the mechanical properties (Chapter 3), or plasticisers can be added to change the state of PVC from a glass to a rubber. Additives can be used to make melt processing easier; this is particularly necessary for difficult to process polymers like PVC. Then there are the additives to delay the various types of degradation that polymers suffer (Chapter 9). The optical properties discussed in Chapter 10 will be modified by the addition of inorganic pigments or organic dyes. There is a degree of interaction between these effects; for example the addition of carbon black may increase the tensile modulus of rubbery polymers, it certainly changes the colour, and it will, by absorbing u.v. radiation, improve the outdoor weathering behaviour.

The physical form of a polymer affects the ease with which additives can be dispersed. Powder blending is the easiest and lowest energy consumption process, so polymers that need large proportions of additives like PVC are sold in a porous powder form. Pigments or stabiliser that are needed in smaller proportions of 1% or less are often dispersed by the mixing action of the extruder screw in the process used. They may, for ease of handling, be provided as granules of a masterbatch, which contains perhaps 50% of the additive mixed into the relevant polymer. Universal masterbatches are becoming more common, in which the host polymer is compatible in small amounts in most of the major plastics.

2

Microstructure

2.1 INTRODUCTION

The microstructure of the main types of polymer will be described in this chapter, paying particular attention to features which will be used later to explain the physical properties of polymers. For example in explaining the mechanical properties we need to know how forces are transmitted from one part of the microstructure to another. To economise on the description, the microstructure of the main states in which polymers can exist will be described, without details for specific polymers. This is because, for any particular polymer, the microstructure will pass through two or more of the main states as the temperature changes. Fig. 2.1 shows what happens for five typical polymers, chosen to be in different states at 20 °C. The main transition temperatures are T_g, the glass transition temperature, and T_m the melting point of the crystalline phase. The first three examples are linear polymers, and the others are network polymers. Before considering the details of the microstructural states, the consequences of the covalent bonding of carbon on the shape of polymer molecules will be explored.

2.2 MODELLING THE SHAPE OF A POLYMER MOLECULE

2.2.1 Conformations of the C—C bond

When a carbon atom forms four covalent single bonds, these bonds point towards the corners of a tetrahedron, with the carbon atom at the centre. The angle between any two of the bonds is 109.5°, and the C—C interatomic distance is 0.154 nm. In a polymer chain consisting of carbon atoms it appears at first that every C—C bond is free to rotate about its axis. Fig. 2.2 shows how the rotation of bond C_2—C_3 is defined by the relative positions of bonds C_1—C_2 and C_3—C_4. If the rotation angle $\phi = 0$ then the atoms that are covalently bonded to C_1 and to C_4 come into such close contact that the potential energy of that shape or *conformation* is increased. On the other hand if $\phi = 180°$, the four atoms again lie in a plane but the potential energy has a minimum value. This is referred to as the trans (t) conformation, and

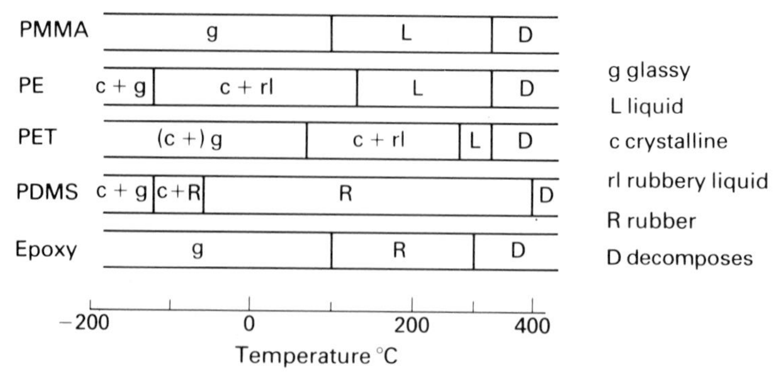

Fig. 2.1 Changes in the state of 5 typical polymers with temperature

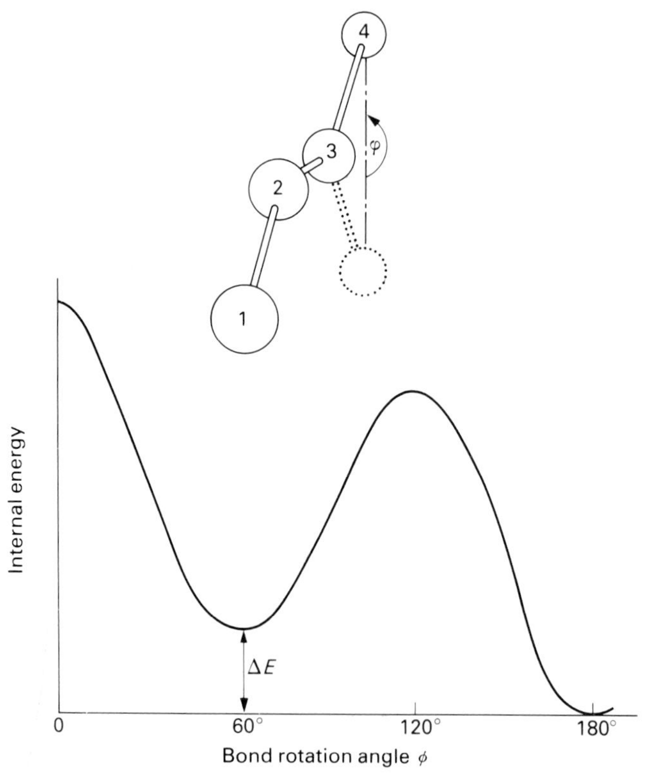

Fig. 2.2 Rotation of a C—C blond in a polyethylene molecule, and the variation of the internal energy of that bond with rotation angle

further subminima at $\phi = \pm 60°$ are referred to as gauche$^+$ (g^+) and gauche$^-$ (g^-) conformations.

In the *rotational isomeric approximation* each C—C bond is assumed to be in either gauche or trans conformations. A chain of 4 carbon atoms has 3 rotational isomers, one of 5 carbon atoms has 3^2 rotational isomers, so a polymer chain of n carbon atoms has 3^{n-3} rotational isomers. Of these only a very few will contain regularly repeating sets of bond rotations, such as the all-trans conformation that occurs in the polyethylene crystal, and the sequence $tg^+ tg^+ tg^+ \ldots$ that occurs in the monoclinic crystalline form of isotactic polypropylene (see Fig. 2.19). The overwhelming majority of conformations are irregular shapes, and we now consider how to generate typical members of this set.

In the liquid state the C—C bonds transform from one rotational isomeric state to another, and the lifetime of a rotational isomer is about 10^{-10} s. For the polyethylene chain the potential energy ΔE of the gauche isomers is about 2 kJ mol^{-1} higher than that of the trans isomers. The relative numbers $n(g^+)$ and $n(t)$ of bonds in the isomeric states can be calculated by using Boltzmann statistics as

$$n(g^+) = n(t) \exp(-\Delta E/RT) \tag{2.1}$$

where R is the gas constant and T the absolute temperature. This gives at $T = 410$ K that $n(g^+) = 0.26\, n(t)$ so the proportions of three isomers are roughly 1:1:4. These proportions can then be used in calculating the shape and size of the polyethylene chain.

2.2.2 Walks on a diamond lattice

One method of modelling the shape of a polyethylene molecule is as a random walk on the diamond crystal lattice (Fig. 2.3). The C—C bonds in the lattice have the correct orientations to produce the trans and gauche rotational isomers. Each step in the walk has components $\pm a/4$, $\pm a/4$, $\pm a/4$, where a is the side of the face centred cubic unit cell. The next step in the walk is produced by changing the sign of one of the components of the current step. Trans conformations occur when the sign of the same component is reversed in successive steps; this can be allowed to happen with a probability of $n(t)$.

In terms of the Miller index notation for directions in crystals, the bond directions are of the $\langle 111 \rangle$ type. If when modelling the shape of a polymer chain the position of a C atom in the diamond lattice is approached along a [111] direction, the possible next steps are $[\bar{1}11]$, $[1\bar{1}1]$ and $[11\bar{1}]$. However to generate a trans isomer requires two successive sign change of the same Miller index, e.g. $[11\bar{1}]$, [111], $[11\bar{1}]$. To generate a g^+ isomer requires changing the sign of progressively greater Miller indices, so if the sign of the 2nd index is changed in the first step and the sign of the 3rd index is changed in the second step. Hence the steps $[\bar{1}\bar{1}1]$, $[\bar{1}11]$ and $[\bar{1}1\bar{1}]$ generate a g^+ conformation. If the sign of the 3rd index is changed in the first step the sign of the 1st index must be changed in the second step—this is adding 1 in modulo 3 arithmetic.

Closed walks occur frequently, the smallest one being the 6 step closed walk that models the shape of the cyclohexane molecule. Since we are

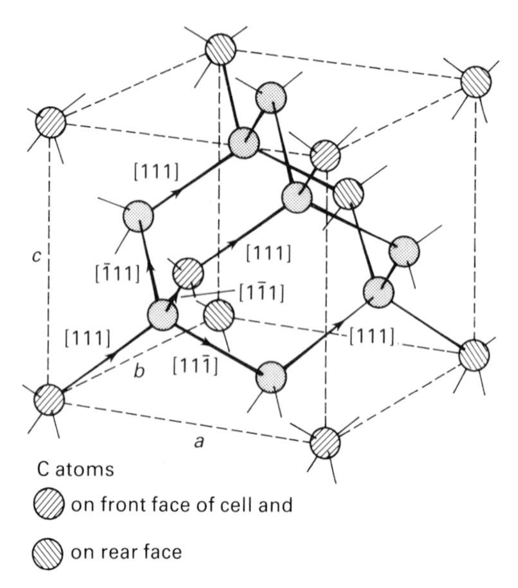

Fig. 2.3 The crystal unit cell of diamond and the Miller indices of the directions of the C—C bonds leading from the C atom at the cell origin

modelling a linear polymer molecule, the walk must be a self-avoiding. In a computer simulation of the walk on the diamond lattice the current position (in units of $a/4$) is the sum of the last position and the Miller indices of the last step. The current position must be compared with all previous positions to check that there is no duplication.

Figure 2.4 shows a typical simulation of a 400 carbon atom polyethylene chain; the coiled nature is readily apparent. The statistical nature of random walks is such that the length of the end to end vector **r** has a distribution of the form shown in Fig. 2.7. There is a range of chain end to end lengths, and the mean value of this can be found by repeating the computer simulation a large number of times (the Monte Carlo method!).

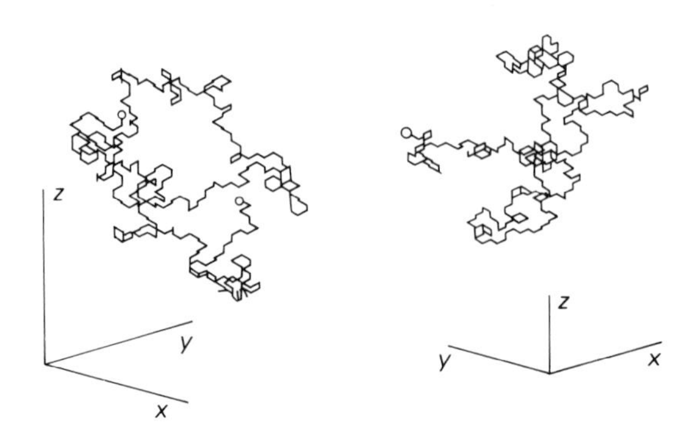

Fig. 2.4 Two projections of a 400 step self-avoiding random walk on a diamond lattice

2.3 NON-CRYSTALLINE FORMS

2.3.1 The four main forms

Before describing the details of the various non-crystalline forms it is useful to have an overview. Fig. 2.5 shows that there are 4 main forms, related to each other by temperature and crosslinking. The crosslinking reaction is irreversible, and it can only occur when the chains are above T_g, so their shapes can change. The degree of crosslinking in a typical thermoset is higher than in a typical rubber, but otherwise their structures are identical.

Crosslinked polymers are called *thermosets* if T_g is above room temperature, or *rubbers* if T_g is below room temperature. In order to deal with these 4 forms in a unified way we define the *network chain* in a crosslinked system, as the section of network between neighbouring crosslinks (Fig. 2.6). The shape and behaviour of a network chain in a rubber and of a molecule in a polymer melt are very similar.

Before considering the shape of a polymer chain the effect of stress on that shape must be mentioned. The shape of a network chain in a rubber and of a molecule in a melt can be changed dramatically by stress. However when the polymer is cooled below T_g the elastic strains are limited to a few % (unless a glassy polymer yields). If the melt or rubber was under stress while it was cooled into the glassy state, the molecular shape in the glass is *not* the equilibrium one. This *molecular orientation* will be discussed later; it may be deliberate, as in biaxially stretched polymethyl methacrylate used in aircraft

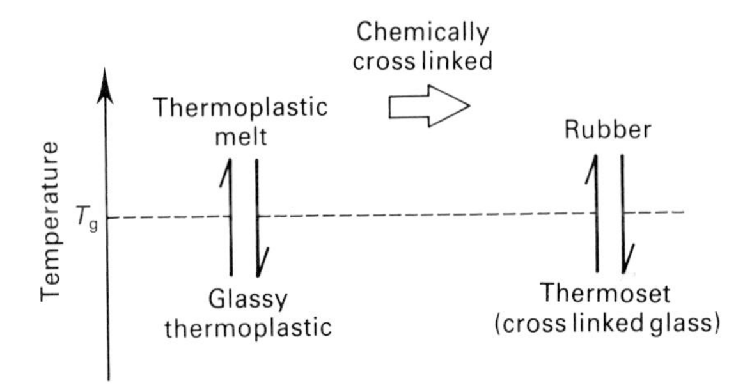

Fig. 2.5 The four forms of non-crystalline polymer on a graph with temperature and degree of crosslinking as axes

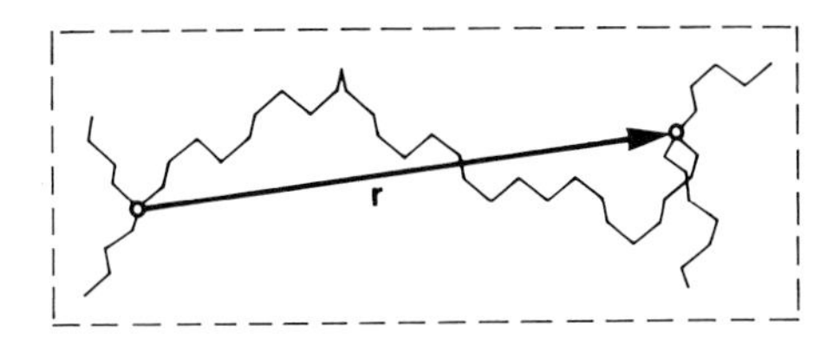

Fig. 2.6 Planar projection of a network chain in a three-dimensional rubber network

windows, or merely a byproduct of processing, as the oriented skin on a polystyrene injection moulding.

2.3.2 The effect of molecular weight on the size of an amorphous molecule

The molecule will be assumed to be unstressed, so that it is in its equilibrium shape. The most common way to specify the size of the molecule is by the length of its *end to end vector* **r** (Fig. 2.6). In a melt this vector runs from the start to the end of the chain, whereas for a network chain it runs from between the crosslinks at the two ends. The calculation will be of the *average length* of the end to end vector for a chain of a particular molecular weight. The average is over a large number of similar chains, and over a period of time—experimental measurements have long exposure times and involve grams of material. For the equilibrium shape of the molecule the end to end vector is denoted $\mathbf{r}_0$.

The model used here is walk on a diamond lattice, in which the gauche and trans rotations are chosen with equal probabilities of 1/3. It only applies to polymers with C—C single bond backbones. For the simplest result the self-avoiding condition is relaxed so the walks are allowed to intersect with themselves. The end to end vector **r** is the sum of the link vectors $\mathbf{l}_i$. The mean square length of a walk on the diamond lattice can be calculated when r^2 is calculated from the scalar product of **r** with itself, and the terms in the expansion of r^2 can be grouped according to the distance between the pairs of links

$$\begin{aligned} r_0{}^2 &= \mathbf{r}\cdot\mathbf{r} = (\mathbf{l}_1+\mathbf{l}_2+\ldots\,.+\mathbf{l}_n)\cdot(\mathbf{l}_1+\mathbf{l}_2+\ldots\,.+\mathbf{l}_n) \\ &= \Sigma\mathbf{l}_i\cdot\mathbf{l}_i + 2\Sigma\mathbf{l}_i\cdot\mathbf{l}_{i+1} + 2\Sigma\mathbf{l}_i\cdot\mathbf{l}_{i+2} + \ldots\,. \\ &= nl^2 + 2l^2\Sigma\cos\theta_{i,i+1} + 2l^2\Sigma\cos\theta_{i,i+2} + \ldots\,. \end{aligned}$$

where the summations are for $i = 1$ to n, and $\theta_{i,j}$ is the angle between link i and link j. The geometry of the polyethylene chain (or the diamond lattice) is such that $\overline{\cos\theta_{i,i+1}} = 1/3$. The correlation in direction between two links decreases as the number of links between them increases. In the equilibrium shape chain the probability of trans and either gauche conformation is equal, and it can be shown that the average value $\overline{\cos\theta_{i,i+n}}$ is equal to $(1/3)^n$. Consequently the series for the mean square length can be summed to give

$$\overline{r_0^2} = 2nl^2 \qquad (2.2)$$

The Monte Carlo method confirms this result; the root mean square length of the 100 step walks in Fig. 2.7 is 14.135. The theoretical distribution in the figure is derived in the analysis of rubber elasticity in Section 2.4. It is the product of a $4\pi r^2$ term (the surface area of a sphere of radius r on which the chain end lies) and a Gaussian distribution [see equation (2.13) in Section 2.4].

Equation (2.2) needs to be re-expressed in terms of measurable quantities if it is to be experimentally verified. It is possible to measure the radius of gyration r_g of the molecule by light or neutron scattering experiments. The

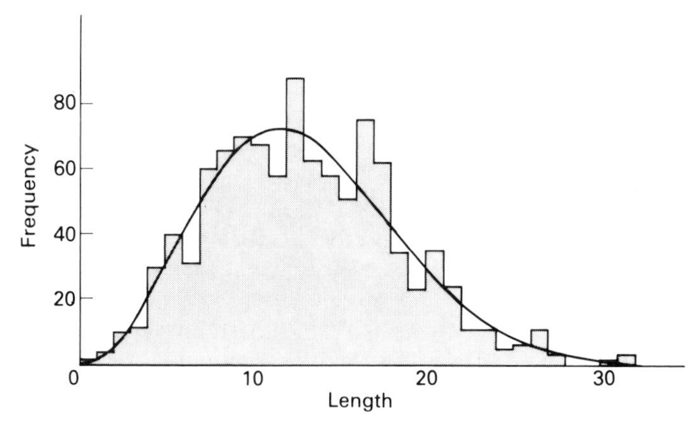

Fig. 2.7 Histogram of the frequency of different lengths of the end to end vectors for 1000 random walks of 100 steps on a diamond lattice. The curve is the theoretical distribution in equation (2.13).

radius of gyration is defined (as in mechanics) in terms of the mass $m(R)$ in a volume element $\mathrm{d}V$ at a radial distance R from the centre of mass, using

$$r_{\mathrm{g}}^2 = \frac{\int R^2 m(R)\,\mathrm{d}V}{\int m(R)\,\mathrm{d}V} \tag{2.3}$$

It is clear that this will be smaller than the mean square end to end length but related to it. It is shown by Shultz (1974) that

$$\overline{r_{\mathrm{g}}^2} = \frac{1}{6}\overline{r_0^2} \tag{2.4}$$

The number of links n in an addition polymer is equal to twice the molecular weight M divided by M_{r}, the repeat unit mass. Hence the radius of gyration of monodisperse polymers, that have $n(\mathrm{t}) = n(\mathrm{g}^+)$, should be related to M by

$$\sqrt{\overline{r_{\mathrm{g}}^2}} = 2l\sqrt{\frac{2M}{3M_{\mathrm{r}}}} \tag{2.5}$$

assuming that the fraction of trans isomers $n(\mathrm{t}) = 1/3$. The molecular weight dependence of r_{g} for PMMA, both in the glassy state and in dilute solution, agrees with equation (2.5) (Fig. 2.8), but the experimental values are about twice the theoretical values. The difference is due to a greater population of trans rotational isomers than that assumed. The solution values were obtained by light scattering measurements, but for the glassy state measurements the PMMA was deuterated (replacing the hydrogen atoms by deuterium atoms) so that it had a greater neutron scattering cross section. About 1% of the ordinary polymer was dispersed in the deuterated glass so that the ordinary molecules were separated from each other. The angular distribution of neutron scattering was then analysed to find the radius of gyration of the molecules.

The value of the mean square end to end distance divided by nl^2 can be

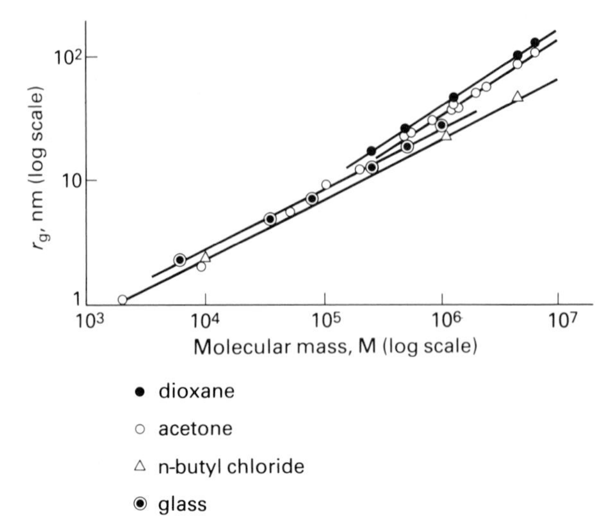

Fig. 2.8 Comparison for PMMA in the glassy state and in dilute solution, of the radius of gyration versus the weight average molecular weight (from Frishart, *The Physics of Non-crystalline Solids*, Transtech, Switzerland, 1977)

used to quantify the relative size of polymer molecules. Table 2.1 lists these values for a variety of polymers. The higher value for polystyrene reflects the expansion in the chain size necessary to accommodate the large phenyl side groups. The values for PC and PET, which have benzene rings in the main chain, are not calculated on the same basis; rather the link length l is taken as the length of the in-chain rigid unit.

2.3.3 Entanglements in polymer melts

The molecular weight variation of the radius of gyration of polyethylene molecules, measured in the melt, was found to agree with equation (2.5),

Table 2.1 Chain size and entanglement data

Polymer	r_0^2/nl^2	Entanglement molecular weight (g mol^{-1})	Entanglement density (10^{-5} mol m^{-3})
PS	10.8	18700	6
SAN	10.6	11600	9
PMMA	8.2	9200	13
PVC	7.6	5560	25
POM	7.5	2550	49
PE	6.8	1390	61
PC	2.4	1790	67
PET	4.2	1630	81

suggesting that individual molecules still have the random coil shape. If individual molecules are assumed to reside inside a sphere of radius slightly larger than r_g, then to achieve the melt density the spheres must overlap each other. For example, for polymethyl methacrylate (PMMA) $r_g = 25$ nm for a molecular weight $M = 10^6$. However, 40 such molecules would have to be packed into a sphere of radius 25 nm in order to achieve the density of the melt. Assuming that the centres of the spheres are evenly distributed in space, this means that the sphere for one molecule will overlap with the 320 others whose centres are within a radius of $2r_g$. Equation (2.5) shows that the volume of the sphere of radius r_g increases in proportion to $M^{1.5}$. As the polymer density is constant, the density of molecules per unit volume is proportional to M^{-1}. This means that the number of other molecules that are capable of interacting but a molecular coil increases in proportion to $M^{0.5}$.

The nature of the interaction between molecules in the melt is not exactly clear; entanglements occur where molecules are knotted around each other, and there will be a contribution from the van der Waals forces. One simple model of these effects uses the concept of an entanglement molecular weight M_e. For this length of polymer chain the intermolecular forces can be replaced by one temporary crosslink (see Section 2.3.5 for the effects of permanent chemical crosslinks in a rubber). For polyethylene $M_c \approx 1400$ (Table 2.1); this means that with commercial polyethylene where the number average molecular weight $M_N > 2M_e = 2800$, the average molecule will be entangled with at least two neighbours and will transmit forces between them. Fig. 2.9 shows some molecules in a flowing melt. If the forces transmitted by the entanglements become significant then the sections of the molecule between entanglements will become both elongated and oriented in the direction of flow.

Appendix B explains how a flow curve for a polymer melt can be derived, and defines apparent (shear) viscosity. It is difficult to correlate the apparent viscosity of a polymer with its molecular weight, because the width of the molecular weight distribution influences the result. In the limit of very low strain rates, when the entanglements between polymer chains produce

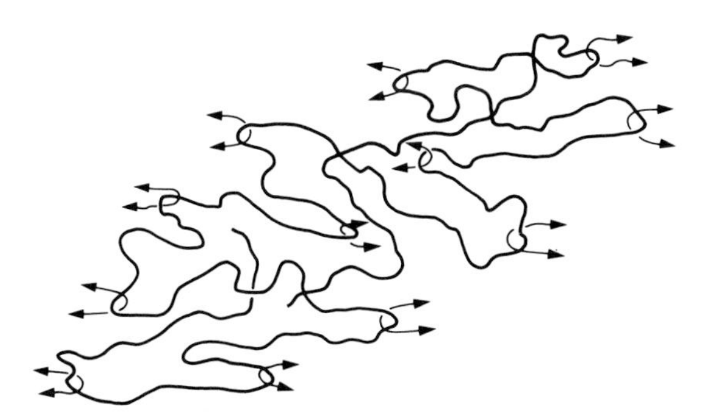

Fig. 2.9 Sketch of the molecular orientation of segments of a polymer molecule between entanglements that develops in melt flow

negligible extension of the molecules, the apparent viscosity approaches a limiting value

$$\eta_0 = \underset{\dot{\gamma}\to 0}{\text{Limit}} \left(\frac{\tau}{\dot{\gamma}}\right)$$

This limiting viscosity is found to depend on M_W (Fig. 2.10). When $M_W < 2M_e$ the limiting viscosity is proportional to M_W, but when $M_W > 2M_e$ the strong effect of entanglements between molecules makes

$$\eta_0 = AM_W^{3.5} \tag{2.7}$$

Some values of the entanglement molecular weights for melts are given in Table 2.1. They have been derived from transient melt modulus measurements using the equivalent of equation (2.20). There is some disagreement about the constant to be used in the equation, so the values in Table 2.1 should be taken as relative rather than absolute. The last column in Table 2.1 gives the entanglement density, calculated by dividing the amorphous density at 20 °C by the entanglement molecular weight. This assumes that the number of entanglements does not change on cooling from the melt. The entanglement densities vary considerably and these values will be used later to explain some of the mechanical properties of the polymers.

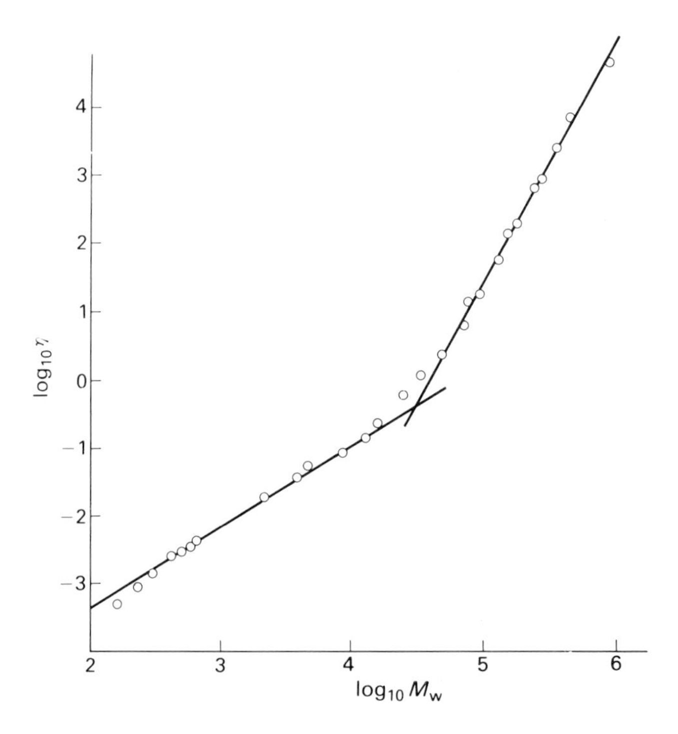

Fig. 2.10 Log melt viscosity of polydimethylsiloxane at 20 °C versus log weight average molecular weight (from N. J. Mills, *Eur. Polym. J.*, 1969, **5**, 675)

2.3.4 The effect of stress on the shape of a network chain

The case of tensile forces applied to the ends of a network chain is the easiest to analyse. Similar shape changes will occur to molecules in a flowing polymer melt, but the forces will be applied to various positions in the chain via entanglements with neighbouring molecules. We will use the *freely jointed chain model* as it is the simplest model that adequately explains rubber-like behaviour. Each rigid link has a free ball joint at each end. If the more realistic rotational isomer model were used, its internal energy would change slightly as the chain shape changed. The restricted flexibility of several C—C bonds could then be replaced by one longer rigid link with a flexible joint at each end.

When a single network chain is treated as a *thermodynamic system*, the internal energy change of the system $\mathrm{d}E$ is given by

$$\mathrm{d}E = \mathrm{d}q + \mathrm{d}w \tag{2.8}$$

As the chain is freely jointed, $\mathrm{d}E = 0$, and if the chain is stretched reversibly (slowly enough for equilibrium to be maintained) the heat input $\mathrm{d}q$ is equal to the product of the absolute temperature T and the entropy change $\mathrm{d}S$. As the work input $\mathrm{d}w$, when the forces f stretch the end to end vector by an amount $\mathrm{d}r$, is $f\mathrm{d}r$, equation (2.8) becomes

$$0 = T\mathrm{d}S + f\mathrm{d}r$$

Consequently the network chain acts as an *entropy spring* with the force being given by

$$f = -T\mathrm{d}S/\mathrm{d}r \tag{2.9}$$

This contrasts with glassy and semi-crystalline polymers where the elasticity is due to internal energy changes (equation 2.25). The entropy can be evaluated from the number of distinguishable shapes W that the chain has using

$$S = k \ln W \tag{2.10}$$

where k is Boltzmann's constant. The relationship between W and the length r will be calculated for a simple one dimensional chain (Fig. 2.11). Each of the steps is either in the positive or negative x direction. This can be considered as the projection of a three dimensional chain on to a single axis x, so long as the projection of each link is taken to have the same average length l.

For very short chains it is easy to see the possible chain shapes. If there are 5 links, then there is

1 way to achieve an r value of 5	$+++++$
5 ways to achieve an r value of 3	$++++-$, $+++-+$, $++-++$, $+-+++$, $-++++$
10 ways to achieve an r value of 1	$+++--$, $++-+-$, $+-++-$, $-+++-$, $++--+$ $+-+-+$, $-++-+$, $+--++$, $-+-++$, $--+++$

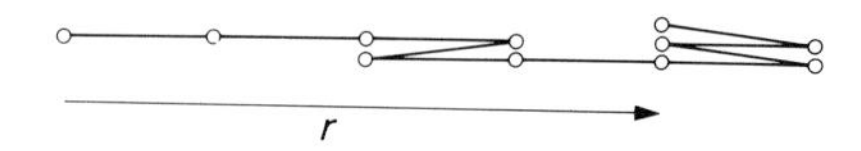

Fig. 2.11 One dimensional network chain

Symmetry shows that there are 10, 5 and 1 way to achieve r values of -1, -3 and -5 respectively. The shape of the histogram of chain shapes is already recognisable as approaching the Gaussian distribution. For a chain with the ends m steps apart, containing p positive and q negative steps, there are W distinguishable chain shapes where

$$W = \frac{n!}{p!q!} \tag{2.11}$$

where the ! sign means factorial. To allow analytical treatment the number of links n must be $\gg 10$ and the chain must not be fully extended, so $m \ll n$ and p and q are both large numbers. It is then possible to use Stirling's approximation

$$\ln n! = n \ln n - n + 1/2 \ln(2\pi n)$$

for n, p and q in equation (2.11). Since we only wish to evaluate dW/dr, we only need to consider the terms in the expansion that are functions of r. Thus we can write

$$\ln W = C_1 - (p + 0.5) \ln p - (q + 0.5) \ln q + p + q \tag{2.12}$$

where C_1 is a constant. The sum of the last two terms also is a constant, and we can expand

$$\ln p = \ln\left(\frac{n+m}{2}\right) = \ln\frac{n}{2} + \ln\left(1 + \frac{m}{n}\right)$$

which approximates to

$$\ln p \simeq \ln\frac{n}{2} + \frac{m}{n} - \frac{1}{2}\left(\frac{m}{n}\right)^2$$

When this, and the similar approximation for $\ln q$ which has a $(-m/n)$ term, are substituted into equation (2.12) we obtain

$$\ln W = C_2 - (p+q)\left(\ln\frac{n}{2} - \frac{m^2}{2n^2}\right) - (p-q)\frac{m}{n}$$

so

$$\ln W = C_3 - \frac{m^2}{2n}$$

where C_1, C_2 and C_3 are constants. W can be written in terms of the end to end length r as

$$W = A \exp\left(\frac{-r^2}{2nl^2}\right) \tag{2.13}$$

where A is a constant. This expression was compared with computer simulations in Fig. 2.7. The entropy of the chain, from equation (2.10), is

$$S = kC_3 - \frac{kr^2}{2nl^2} \tag{2.14}$$

The parabolic reduction in entropy as the chain extends in length means that the force, from equation (2.9), is

$$f = \frac{kTr}{nl^2} \tag{2.15}$$

The form of equation (2.15) reminds us that the thermal energy kT of each rotating C—C bond is necessary for rubberlike behaviour. The network chain acts as a linear spring which is stiffer for short chains than for long chains.

2.3.5 Rubbers

The result for a single network chain can be used to calculate the modulus of a block of rubber that is sheared. Fig. 2.12 shows what is assumed to happen when a shear strain γ is imposed on a rubber containing a network chain with an end to end vector **r**. It is assumed that the crosslink deformation is *affine* with the rubber deformation, which means that the components of **r** change in proportion to the macroscopic dimensions. The components r_y and r_z are unchanged, but r_x becomes

$$r'_x = r_x + \gamma r_y$$

From equation (2.14) the entropy change of the chain when it is sheared is

$$\Delta S = -\frac{k}{2nl^2}(r'^2_x - r^2_x) = -\frac{k}{2nl^2}(\gamma^2 r^2_y + 2\gamma r_x r_y) \tag{2.16}$$

When we sum this result for the N network chains in unit volume, as many of these will have negative r_y values as have positive r_y values, so the second

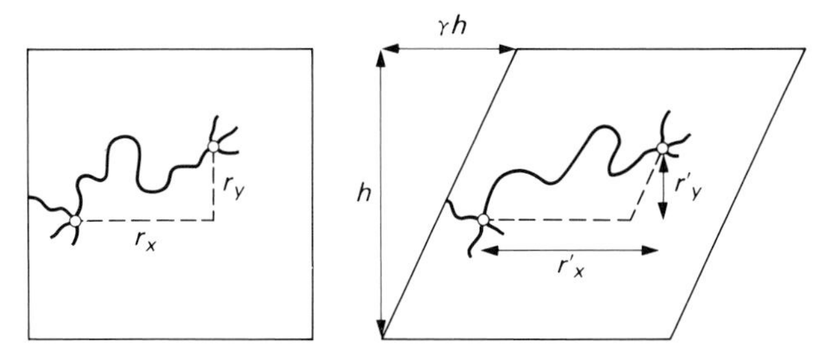

Fig. 2.12 The affine deformation of the ends of a network chain when a shear strain γ is imposed on a rubber

term in equation (2.16) will cancel, leaving

$$\Sigma\Delta S = -\frac{Nk}{2nl^2}\gamma^2\overline{r_y^2} \tag{2.17}$$

The average value of r_y in an unstressed chain can be calculated from the equivalent of equation (2.2) for a freely jointed chain as

$$\overline{r_x^2} = \overline{r_y^2} = \overline{r_z^2} = nl^2 \tag{2.18}$$

Consequently the shear stress can be calculated from the equivalent of equation (2.9) as

$$\tau = -T\frac{\mathrm{d}\Sigma S}{\mathrm{d}\gamma} = NkT\gamma$$

which means that the shear modulus of the rubber is

$$G = NkT \tag{2.19}$$

This remarkably simple result has been checked experimentally by preparing rubbers in which each crosslinking molecule produces exactly two network chains. Fig. 2.13 shows that the experimentally measured modulus is proportional to the number of network chains, but that there is a positive deviation from equation (2.19). The discrepancy has been attributed to physical entanglements which exist in the polymer before any crosslinks were introduced. Rubbers are one of the few solids for which the elastic modulus can be predicted, to within a small error, from the molecular structure.

The elastic moduli for rubbers can be predicted from the network chain molecular weight M_c, and vice versa. Equation (1.11) can be substituted into equation (2.19) to give

$$G = 1000\frac{N_A k\rho T}{M_c} = 8310\frac{\rho T}{M_c} \tag{2.20}$$

where the density ρ is in kg m^{-3} and the shear modulus G is in N m^{-2}. The minimum shear modulus is determined by the entanglement effect shown in Fig. 2.9, whereas the maximum shear modulus occurs when the network chains are so short that rubber like extensions are no longer possible. The approximation used to obtain equation (2.13) breaks down if the freely jointed chain has less than 10 links. For natural rubber each isoprene repeat unit, of molecular weight 68, is approximately equivalent to a freely jointed link, so the minimum M_c of 680 corresponds to a maximum G of 3.2 MN m^{-2}.

If the rubber is loaded in tension the strain measure used is the extension ratio λ defined by

$$\lambda \equiv \frac{\text{deformed length}}{\text{original length}} \tag{2.21}$$

The extension ratios for the two lateral directions will have the value $1/\sqrt{\lambda}$ as the rubber deforms without change of volume. When the calculations of the entropy change are repeated for the tensile deformation, the true stress

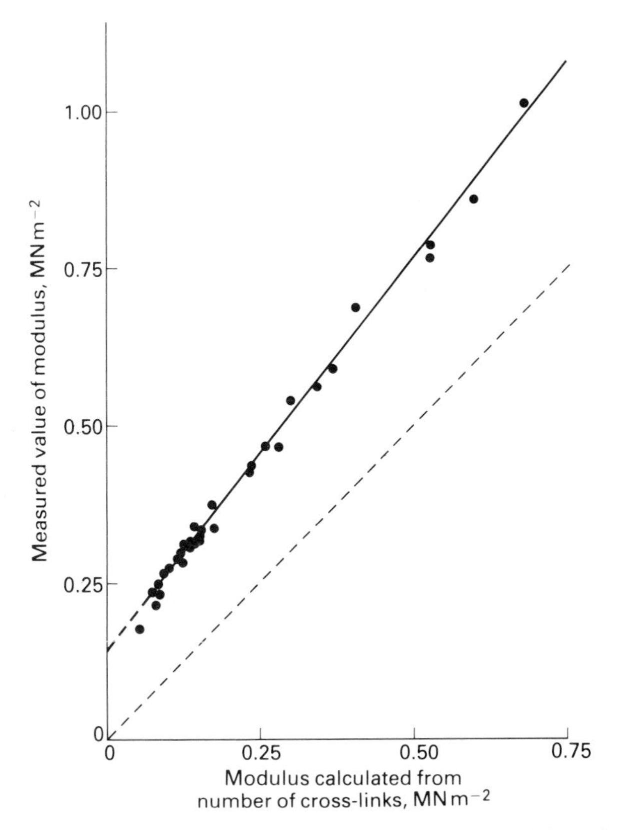

Fig. 2.13 Comparison of the measured shear modulus of a rubber with that calculated from the density of network chains (from L. G. R. Treloar, *Introduction to Polymer Science*, Wykeham, 1970)

(defined as the force divided by the deformed cross sectional area) is predicted to be

$$\sigma_t = G\left(\lambda^2 - \frac{1}{\lambda}\right) \tag{2.22}$$

The behaviour of most rubbers follows equation (2.22) but there are deviations if the rubber crystallises under stress. This happens with natural rubber and explains why it is impossible to stretch rubber bands to an extension ratio in excess of 7. The relationship will be used in Chapter 7 to predict the increase in stress when semi-crystalline polymers are deformed to high tensile strains.

2.3.6 The glass transition temperature

A detailed theory of the glass transition does not exist. It is possible to make generalisations, such as that aromatic in-chain groups lead to high T_g values, but it is not possible to calculate T_g from a knowledge of the polymer structure. It is useful to contrast the phenomena associated with the glass transition with those that occur at the melting point of a crystalline phase, to obtain some idea of the microstructural changes that occur at T_g. Fig. 2.14 shows that the specific volume versus temperature graph changes in slope at T_g. This contrasts with a melting point (Fig. 5.10a) where there is a step increase in specific volume, due to the decrease in molecular packing. When the glassy volume–temperature line in Fig. 2.14 is extrapolated above T_g, the difference between this and the melt volume is referred to as the *free volume*. In Chapter 6 a variety of viscoelastic effects are shown to change at T_g, in particular there is a maximum in the internal damping of mechanical vibrations. It is found that the value of T_g increases for higher frequency (or shorter time scale) observations, so the value of T_g cannot be defined in an absolute manner.

Some clues as to the nature of the glassy state can be obtained from the way in which the physical properties of a melt change as it is cooled. Fig. 2.15 shows how the melt viscosity increases with decreasing temperature to a level beyond which further measurement is impossible. The melt viscosity increases drastically as the free volume decreases, but there is no quantitative model of this effect. Properties such as specific volume or enthalpy do not change in a step-like way on cooling, as they would if crystallisation occurred. Instead the gradient of the graph against temperature changes at T_g, so that there is a step in the thermal expansion coefficient or the specific heat. One technique that does give a clear indication of a change at T_g is the line width in a nuclear magnetic resonance trace. The line width broadens when protons (H atoms) experience different environments, and narrows when rapid molecular motion gives every proton the same average environment. Experi-

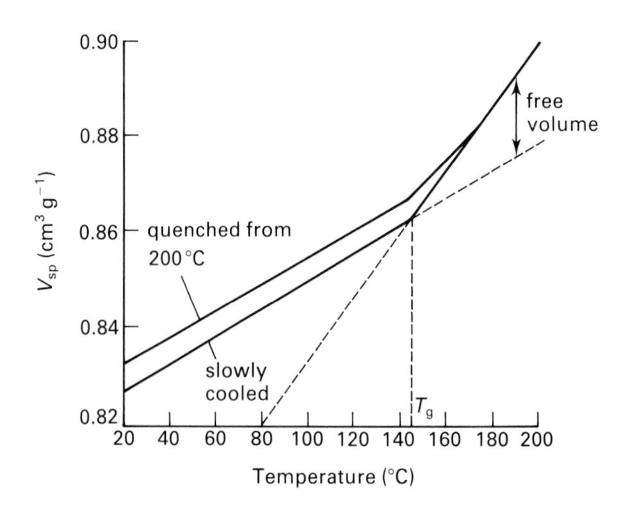

Fig. 2.14 Specific volume versus temperature on heating for polycarbonate having two thermal histories (data from Hachisuka *et al.*, *Polymer*, 1991, **32**, 2383), and the definition of free volume

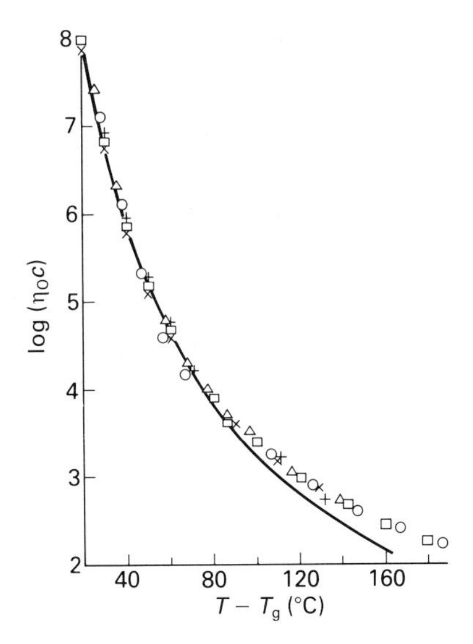

Fig. 2.15 Log melt viscosity of polysulphones of different molecular weights versus temperature. To make the curves superimpose they are shifted vertically. The curve is log $\eta = 2000/(T + 40 - T_g)$ (from Mills and Nevin, *J. Macromol. Sci. Phys.* 1970, **4**, 863)

ments on polystyrene, diluted with a small amount of carbon tetrachloride, show that main chain rotation ceases when the polymer is cooled below the T_g of 90 °C. It is still possible for side groups to rotate in certain polymers, but the overall molecular shape becomes frozen.

Since there is no step change in volume on heating through T_g the value of the transition temperature cannot be directly related to the strengths of the van der Waals bonds between the chains (E_0 in equation 1.1). The *thermal energy* of the polymer, a measure of the thermal vibrations of the atoms, will increase with the absolute temperature T. (For gas molecules the thermal energy is $\frac{3}{2}RT$, where R is the gas constant = 8.3 J K^{-1} mol^{-1}.) Therefore the linear increase in free volume with temperature seems to be a consequence of the thermal energy exceeding a critical value, but the molecular meaning of the critical value is unknown. It is likely that the free volume is non-uniformly distributed in the melt, and that a number of lower density regions move rapidly through the melt (rather as dislocations move through a crystal lattice to allow plastic deformation).

The glass transition phenomenon in polymers is the same as that in silicate and other inorganic glasses. Silicate glasses are often given heat treatments either to toughen them or to stress-relieve them after forming, and this has led to the use of the concept of the *fictive temperature*. This is the temperature at which the structure of the glass would be stable. If a glass is annealed at a temperature just below T_g the density slowly increases up to a limiting value which lies on the extension of the melt volume–temperature line (Fig. 2.14). If a heat-treated glass is heated up the extrapolation of the volume temperature line of the glass meets the melt line at the fictive temperature. Polymer glasses are invariably cooled rapidly after processing, and rarely given heat treatments, so the concept of fictive temperature has not been widely used. However it could be used for the aged polymer glasses discussed in Section 5.2.1.

2.3.7 Glass microstructure

There is no experimental method of imaging the shape of the molecules in a glass, so we rely on computer models to show the molecular arrangement. Limitation of computer power mean that only a small region of the glass is considered, and the chains are relatively short. The rotational isomeric model of section 2.2.1 is used to generate the initial molecular shapes. For polypropylene the probability of occurance of pairs of neighbouring bond rotational states is known from the interaction of the methyl side groups. The bond rotation angles are then varied somewhat from the exact 0° and ±120° positions to reduce the potential energy of the molecules in the glass. In Fig. 2.16 a cube of side 1.82 nm is shown, with the molecular arrangement in this cube repeated in all the neighbouring cubes. The experimental density of polypropylene of 890 kg m^{-3} at −40°C was reproduced with only a slight change to the distribution of rotational isomers. The radius of gyration of the molecules in the glass was almost identical to that in a dilute solution, in agreement with experimental data of the type shown in Fig. 2.8. There are no regions in the model where bundles of chains are parallel (often shown in sketches of microcrystalline regions in polymers). Neighbouring chains tend to be perpendicular to each other, but for neighbouring chains separated by more than 1 nm there is no correlation of the direction of main chain C–C bonds.

This modelling supports the indirect experimental evidence that glasses

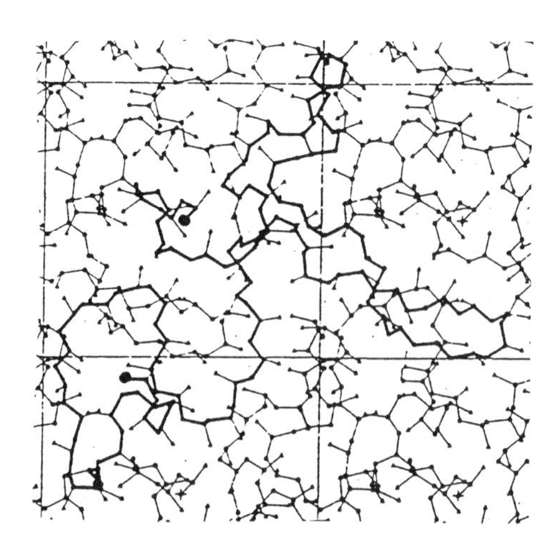

Fig. 2.16 Projection of a 76 monomer Polypropylene chain in the glassy state, with the parent chain shown in bold and the image chains shown as thin lines. The pattern in the cube of side 1.82 nm repeats in the neighbouring cubes. (from Theodorou and Suter, *Macromolecules*. **18**, 1985, 1467)

contain *frozen random coils*—the molecules have the shapes described in section 2.2 and are unable to change their shapes. The glassy state is a 'still photograph' of the structure of a polymer melt. There is an entanglement network between the molecules, but this will only come into effect if high strains are imposed.

For glassy polymers the angular variation of diffracted intensity in an X-ray diffraction experiment shows a single broad peak. This peak cannot be interpreted in terms of a crystal lattice; instead it is interpreted in terms of a radial distribution function (RDF). The RDF is a graph of the density $\rho(r)$ of other atoms at a radial distance r from any reference atom; it does not distinguish between the directions in which other atoms occur. In the RDF for polycarbonate (Fig. 2.17) the initial sharp peaks between 0.1 and 0.25 nm are in-chain distances, whereas the broad peak at 0.55 nm gives the range of nearest neighbour distances between chains. Atoms further away than 1.5 nm appear to occur at random distances.

One way of estimating the packing efficiency of polymer molecules in the glassy state is to divide the density of the glass by the density of the crystalline form of the same polymer. The results range from 0.80 to 1.02, and the average value is smaller at 0.88, for polymers with only H and F side groups, than the 0.96 value for polymers with bulky side groups. The packing efficiency in the glassy state is relatively high, but it cannot be inferred that there is local ordering in the glassy state.

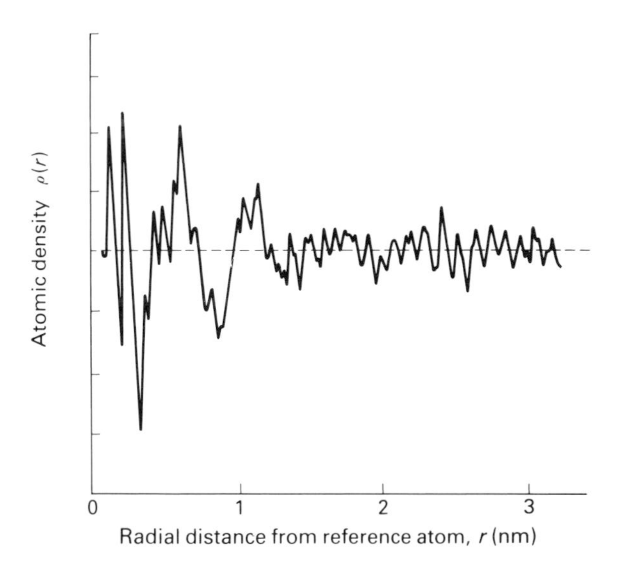

Fig. 2.17 The radial distribution function for glassy polycarbonate is a plot of the atomic density at a distance r from the reference atom (from Frishart, *The Physics of Non-crystalline Solids*, TransTech, Aedermansdorf, Switzerland, 1977)

3.2.8 The elastic moduli of glasses

The Young's moduli of glassy polymers at 20 °C are remarkably similar; they range from 2 to 3.5 GN m^{-2} for polymers that have glass transition temperatures from 80 °C to 225 °C. This suggests that the common microstructural feature, of the weak van der Waals forces between the chains, is responsible for the magnitude of the modulus. This postulate has been tested by analysing the bulk modulus of solid methane, in which only van der Waals bonding occurs. Equation (1.1) gave the potential energy of a pair of methane atoms a distance R apart. Above 20 K methane has a cubic close packed structure in which every molecule has 12 neighbours at a distance of 0.41 nm. Hence equation (1.1), when expressed in terms of the molar volume V, becomes

$$\frac{E}{E_{\mathrm{m}}} = \left(\frac{V_0}{V}\right)^4 - 2\left(\frac{V_0}{V}\right)^2 \tag{2.23}$$

where E_{m} is now the molar constant and V_0 is the equilibrium molar volume. When the solid is compressed by a pressure p the molecules move slightly closer together to produce a volume change $\mathrm{d}V$ but there is no change in the molecular disorder, and hence no entropy change. Equation (2.8) for the internal energy change therefore reduces to

$$\mathrm{d}E = p\,\mathrm{d}V$$

This can be substituted in the definition of the bulk modulus

$$K = -V\frac{\mathrm{d}p}{\mathrm{d}V} \tag{2.24}$$

to produce a relationship between the bulk modulus and the internal energy

$$K = -V\frac{\mathrm{d}^2E}{\mathrm{d}V^2} \tag{2.25}$$

When equation (2.23) is substituted this leads to the result

$$K = \frac{4E_{\mathrm{m}}}{V_0}\left[5\left(\frac{V_0}{V}\right)^5 - 3\left(\frac{V_0}{V}\right)^3\right] \tag{2.26}$$

for the variation of bulk modulus with applied pressure. The bulk modulus of solid methane at atmospheric pressure when $V = V_0$ is given by equation (2.26) as

$$K = \frac{8E_{\mathrm{m}}}{V_0} = \frac{N_{\mathrm{A}}E_0}{V_0} \tag{2.27}$$

From the density of solid methane of 547 kg m^{-3} and molecular weight of 16, the molar volume V_0 is 29×10^{-6} m^3 mol^{-1}, so equation (2.27) predicts a bulk modulus of 4.0 GN m^{-2}.

It is more difficult to predict the bulk modulus of a glassy polymer because there is a greater variety of van der Waals bonds present, and the exact intermolecular distances are not known. However, it can be assumed that the polymer chains are incompressible along their lengths, so that van der Waals forces only act in the two directions in the plane perpendicular to the chains. Therefore the bulk modulus of a polymer should be 50% higher than that of solid methane if the same strength van der Waals bonds exist. This is confirmed by the atmospheric pressure bulk modulus of $\sim 6\ \mathrm{GN\ m^{-2}}$ for PMMA in Fig. 2.18. Although the value of E_0 is not known independently for the van der Waals bonds in PMMA, the variation of the bulk modulus with the applied pressure agrees with the theory (Fig. 2.18). The Young's modulus E will be smaller than the bulk modulus because the relation between the two for isotropic materials is

$$E = 3K(1 - 2\nu) \tag{2.28}$$

and Poisson's ratio for glassy plastics ranges from 0.3 to 0.4.

2.4 SEMI-CRYSTALLINE POLYMERS

In spite of much research into the microstructure of crystalline polymers some details are still unknown. Modification of the microstructure, to improve important properties such as the toughness or the resistance to stress cracking, (Sections 9.5 and 13.2.2) has proceeded empirically with the microstructural knowledge acquired later. Nevertheless there is certain basic information about the microstructure of semi-crystalline polymers that will be needed later when the physical properties are discussed. The order of presentation of this material is: the bonding in the crystal unit cell, the size and shape of lamellar crystals, the microstructure of spherulites, measure-

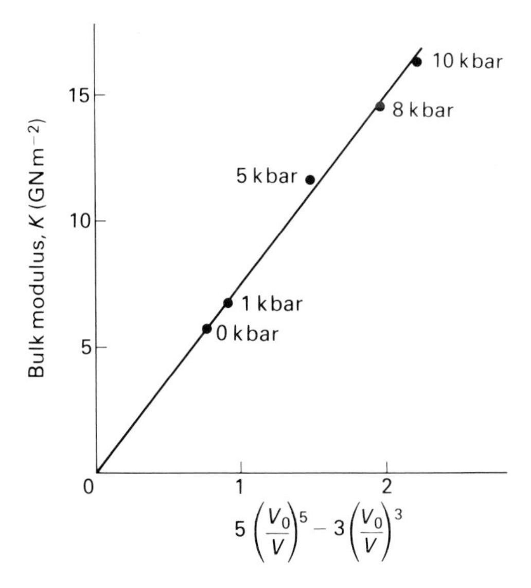

Fig. 2.18 The bulk modulus of PMMA at different pressures as a function of molar volume. The straight line is the prediction of equation (2.26) for a van der Waals solid with $E_m = 137\ \mathrm{kJ\ mol^{-1}}$

ment of the overall crystallinity and the processes of crystallisation. The detailed crystallography of polymers and descriptions of the varied microstructures can be found in the texts mentioned in *Further Reading.*

2.4.1 Chain shape and bonding in the unit cell

Although it is possible to describe in detail the symmetry of the unit cell, and to use Miller indices to specify particular sets of parallel planes in the lattice, these topics are of less importance than for metals. The reasons are firstly that polymers are always *semi-crystalline* so the amorphous phase plays an important part in the mechanical properties. Secondly the yield strength of plastic is not determined by the obstacles to dislocation movement as in metals. The methods of strengthening metals (solid solution strengthening, work hardening, second phase precipitation, . . .) do not exist in polymers. In contrast it is possible to fabricate highly anisotropic forms of semi-crystalline polymers, so the description of the crystal shape and orientation becomes more important.

The polymer chains form *helices* to minimise the internal energy of the molecule. Any large side groups are evenly spaced on the outside of the helix. In polyethylene all the rotational isomers are trans (Fig. 2.2), whereas in polypropylene t and g^+ rotational isomers alternate to form left handed helices or t and g^- altenrate for right handed helices. The isotactic polypropylene molecule forms a 2 * 3/1 helix, which means that the repeating motif of the helix contains 2 chain atoms, and that 3 motifs occur in 1 complete 360° turn in the length **c**. The carbon atoms in the polyethylene crystal can be described either as a 1 * 2/1 helix or as forming a *planar zigzag* which is the most extended form of an all-carbon-atom chain. More complex helices occur, such as 2 * 9/5 for polyoxymethylene in which O and C atoms alternate down the chain.

The chains pack together in the unit cell to maximise the density of the crystal. The helix axes are parallel to the **c** or chain axis of the unit cell. In the polyethylene unit cell there are two settings of the C—C planar zigzags, so that the hydrogen atom protuberances on one chain fit into indentations in the surrounding chains. In the isotactic polypropylene unit cell the helices are displaced by units of **c**/12 to maximise the packing density. If there is the possibility of hydrogen bonding between chains, then the number of hydrogen bonds will be maximised. This is done by displacing the chains in the **c** direction, and making all the hydrogen bonds occur in parallel sheets, as in nylon 6,6. This stronger intermolecular bonding stabilises the crystals so that the melting temperature is 260 °C, compared with 135 °C for polyethylene.

In contrast with metals and ceramics, there is high *anisotropy of bonding* in the unit cells of polymers. There is only continuity of covalent bonding along the **c** axis of the cell, and there is the much weaker van der Waals bonding in the **a** and **b** axes directions. Fig. 2.19 shows the unit cells of polyethylene and isotactic polypropylene. The crystal lattice is made up by translating the unit cell by an amount $h\mathbf{a} + k\mathbf{b} + l\mathbf{c}$ where h, k and l are integers and **a, b** and **c** are the primitive lattice vectors.

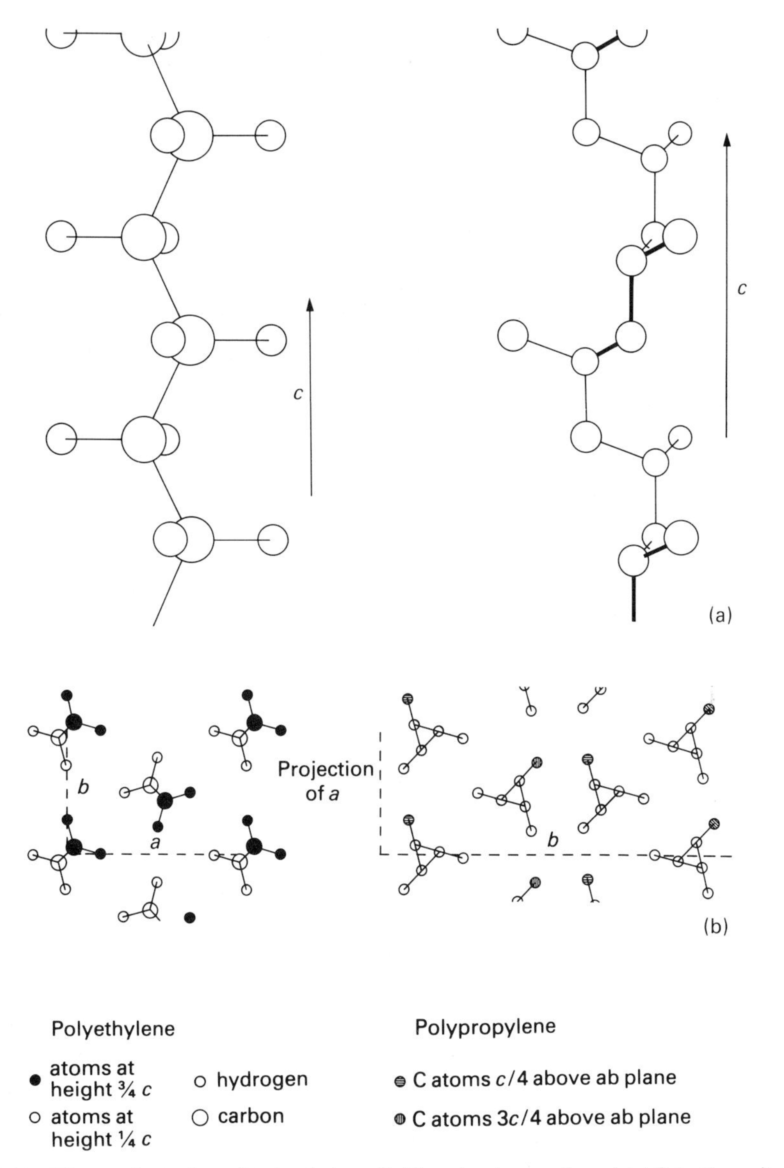

Fig. 2.19 The conformation of polyethylene (left) and polypropylene (right) in the unit cell. (a) Projection of a single chain on to a plane containing the **c** axis. (b) Projection of the unit cell on to a plane normal to the **c** axis

2.4.2 Shape of the crystals

In a two phase microstructure the shape and connectedness of the crystals is important. The usual form is lamellar (in the form of a thin plate or sheet), whereas a fibrous form is possible when the melt is under a high tensile stress when it crystallises. Lamellar single crystals were originally observed when dilute polymer solutions were crystallised slowly, but melt crystallised polymers also contain lamellar crystals. The lamellae are typically 10 to 20 nm thick, and of the order of 1 μm long and wide. They can be flat, or their surfaces may be curved (Fig. 2.20). The **c** axes of the unit cell lie within 40° of the normal to the lamella surface, which means that in single crystals grown from solution there must be chain folding at the upper and lower surfaces. The fully extended length of a polyethylene molecule of $M = 10\,000$ is 90 nm which is many times the lamella thickness. For melt crystallised lamella the situation at the lamella surfaces is not so simple. A number of nearly parallel lamellae grow together into the melt with layers of amorphous material between them. The geometry is similar to that of lamellar eutectics in metallic alloys, such as pearlite in steel. The time necessary for a randomly coiled molecule in the melt to change its shape into a regularly folded form so that it can join a single lamella is much longer than that available when the crystal grows at many μm per s into the melt. Consequently sections of the molecule are incorporated into neighbouring lamellar crystals, and the intervening lengths form part of the amorphous interlayers. These intervening lengths, often containing comonomers, behave like network chains in a rubber, by virtue of their ends being trapped in crystals. The crystallisation process induces molecular orientation near growth fronts, and some bundles of elongated molecules coalesce to form *intercrystalline links*. These are fibrous extended chain crystals in the order of 1 μm in length and less than 10 nm in diameter. They are clearly revealed when mixtures of polyethylene and $C_{32}H_{66}$ paraffin are crystallised, and when the paraffin is removed. Fig. 2.20 shows that the intercrystalline links connect the growing lamellae.

2.4.3 Spherulites in polarised light

A polymer melt crystallises from nuclei that are heterogeneous. This means that they are foreign particles, which either happen to be there (dust particles, catalyst residues, pigment), or which are deliberately added to reduce the spherulite size (sodium benzoate in polypropylene for example). The lamellae grow outwards from the nucleus, and through a process of branching develop through a sheaf like entity into a spherulite (little sphere) (Fig. 2.21).

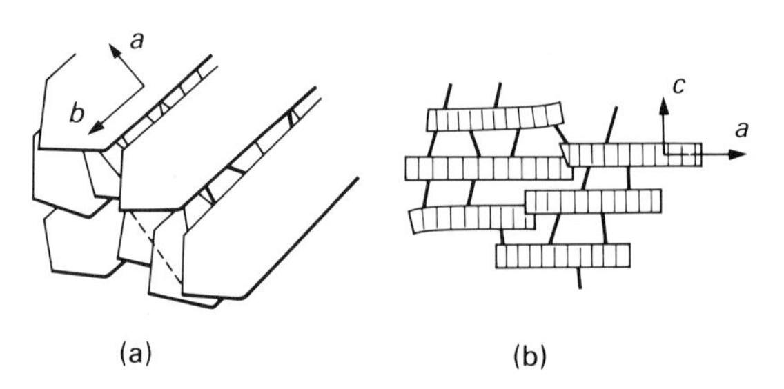

Fig. 2.20 Sketch of intercrystalline links near the surface of a growing spherulite. (a) Plan view of lamellar faces; (b) edge view of the lamellae

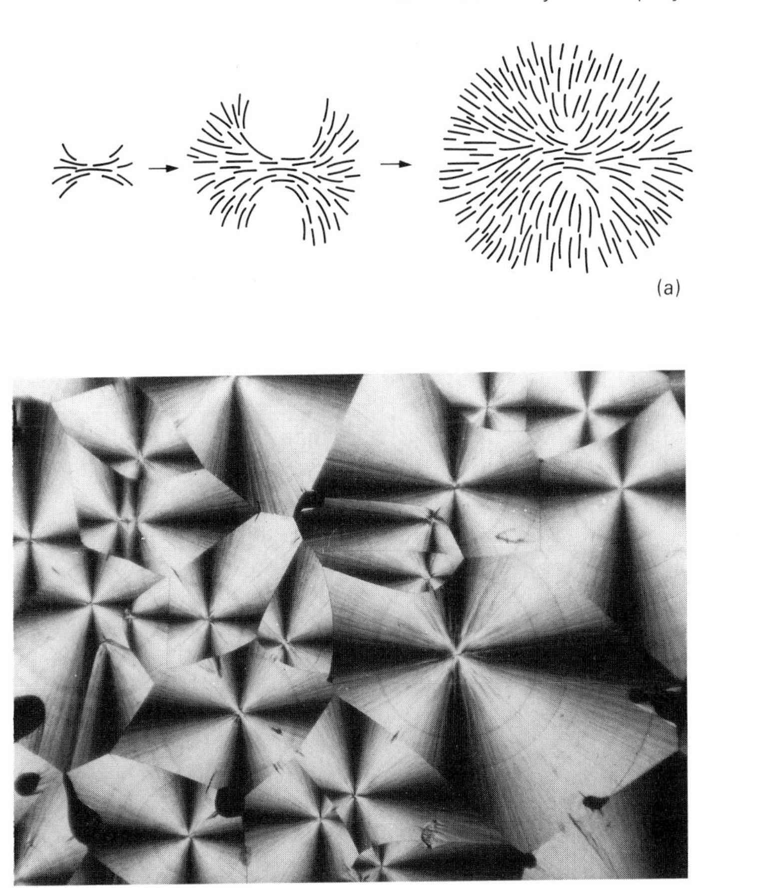

Fig. 2.21 (a) Stages in the formation of a spherulite from a stack of lamellae. (b) Polarised light micrograph of two-dimensional spherulites grown in a thin film of polyethylene oxide

The branching is not crystallographic as is dendritic growth in metals, nor does the term imply that a crystal splits into two separate ones; rather new lamellae can nucleate and grow in the widening gaps between the initial lamellae. The spherulites grow in the melt until they impinge, so their boundaries are polyhedral. The final diameter is proportional to $d^{1/3}$ where d is the total number of nuclei per m^3 that have become active. The range of d is considerable, so that some polymers (PE, POM) often have spherulites that are $<10\ \mu m$ in diameter, whereas low molecular weight polyethylene oxide can have spherulites 1 mm in diameter.

It is easy to observe the spherulites growing in a thin film of polyethylene oxide melt between a microscope slide and a cover slip. The spherulites in this case grow as discs once their diameter exceeds the melt film thickness. Two features are visible in polarised light; the radiating fibrous appearance is due to lamellae of different orientations, whereas the overall Maltese cross pattern, whose arms are parallel to the crossed polarising filters below and above the specimen, can be explained by the orientation of the crystal lattice axes.

Isotropic materials (such as a non-oriented glassy polymer) have a single refractive index n which determines the speed C of light in the material

$$C = C_0/n \tag{2.29}$$

where $C_0 = 3.00 \times 10^8$ m s^{-1} is the speed of light in a vacuum. Anisotropic materials, such as a polymer crystal, have 3 orthogonal optic axes for which the refractive indices are n_1, n_2 and n_3. These indices relate to the speed of propagation of plane polarised light (light is a transverse wave, and the direction of its transverse electric field can be constrained to one plane by passing it through a polarising filter, which absorbs any photons with the 'wrong' polarisation direction). In Fig. 2.22 a light ray passes along the **b** axis of a polyethylene crystal. As the polarisation direction x of the electric vector **E** is at an angle θ to the **a** axis, the light propagates in the crystal as component of magnitude $E \cos\theta$ and wavelength λ/n_a polarised along the **a** axis, and a component of magnitude $E \sin\theta$ and wavelength λ/n_c polarised along the **c** axis. After passing through a thickness t of crystal, these components emerge with a phase difference $(n_c - n_a)t/\lambda$ wavelengths. Unless the crystal thickness happens to make the phase difference an integer the emerging light will be elliptically rather than plane polarised. Consequently some of the light will pass through an 'analyser' filter, which is set with its transmission direction at 90° to the polarisation direction. Only if the angle θ is 0° or 90° will the light emerge from the crystal plane polarised. The contrast pattern on the spherulite in Fig. 2.21b can now be explained. In the dark cross region one of the optic axes of the crystals is parallel to the polariser filter. The cross remains stationary relative to the filters if the polymer is rotated on the microscope stage. This proves that the optic axes are tangential and radial in the spherulite. Electron diffraction shows that it is the crystal **c** axes that are tangential in spherulites.

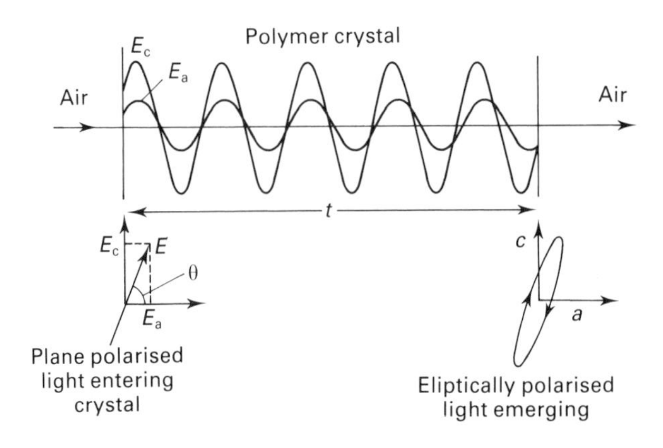

Fig. 2.22 The effect on plane polarised light of passage through a polymer crystal. The plane of polarisation is initially at θ to the crystal **a** axis. The emerging light is elliptically polarised. The number of wavelengths in the crystal has been reduced for clarity

2.4.4 Crystallisation rate

The rate of crystallisation can be measured at different temperatures using bulk specimens in dilatometers. Alternatively individual spherulites can be measured during their growth between glass slides on a hot stage microscope. It is found for all crystallisable polymers that the rate of crystallisation increases with supercooling, until a maximum occurs, then decreases to zero at T_g (Fig. 2.23). For polyethylene the peak rate ($\sim 100\ \mu m\ s^{-1}$) and the nucleation density are so high that the left hand side of the curve is never observed, and polyethylene can never be quenched into the amorphous state. For some polymers the maximum growth rate is so low ($0.6\ \mu m\ h^{-1}$ for polycarbonate) that they are glassy at room temperature unless special heat treatments are used. For nylon 6, PET and polypropylene the intermediate spherulite growth rates mean that both the spherulite sizes and the crystallinity are affected by the cooling rates experienced.

When spherulites grow into a polymer melt, different molecular weight chains may end up in different parts of the spherulite. This happens because the rate of crystallisation is a function of molecular weight (Fig. 2.24). The data for polyethylene stop at 125 °C because the growth rate becomes too fast for measurement. As the lamellae grow into the melt there is time for short chains, that crystallise slowly, to diffuse a distance of half the lamellar thickness. This low molecular weight fraction crystallises at a lower temperature, in the spaces between the primary lamellae. Irregular chains will also tend to be rejected into the amorphous interlamellar material. If the rate of cooling is slow there may be time for low molecular weight material to diffuse away ahead of the growing spherulites, and end up as weak regions as the interspherulitic boundaries.

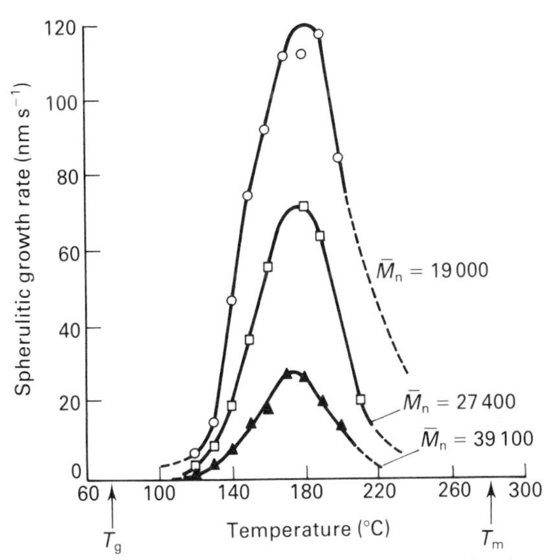

Fig. 2.23 Growth rate of spherulites in PET of different molecular weights versus the crystallisation temperature (from Polymer Engineering: Unit and Advanced Processing. Open University Press, Milton Keynes, 1985)

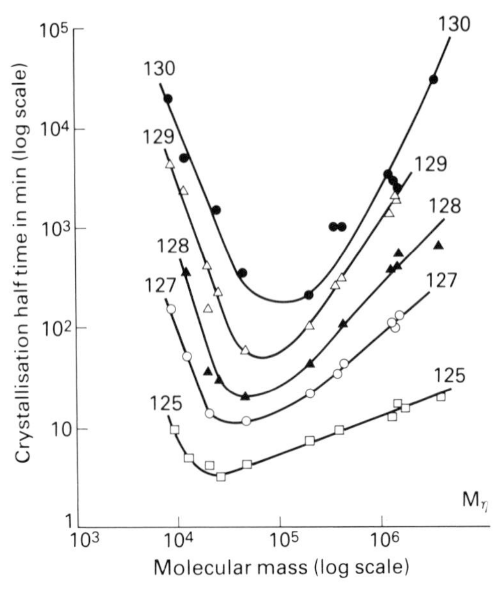

Fig. 2.24 Logarithm of crystallisation half time versus logarithm of molecular weight for polyethylene crystallised isothermally at the temperatures indicated (from Mandelkern, *J. Mater. Sci.*, 1968, **6**, 615, Chapman and Hall)

2.4.5 Percentage crystallinity

Although spherulites usually fill the whole of space in the microstructure, the overall crystallinity can be in the range 40 to 90%. Measurement of percentage crystallinity implies the existence of separate crystalline and amorphous phases, whereas a small amount of the amorphous content may actually be crystal defects or folds on the surfaces of lamellar crystals. The most convenient methods of measuring crystallinity are measurements of density or of the enthalpy of fusion, although these give slightly different results than X-ray diffraction methods. For polyethylene the density of the crystal unit cell $\rho_c = 997 \text{ kg m}^{-3}$ is considerably greater than the (extrapolated) density of the amorphous phase $\rho_a = 854 \text{ kg m}^{-3}$. The polymer density ρ_p can easily be measured by hydrostatic weighing in a liquid of known density, and in air, or by using density gradient columns, then equation (1.16) can be used to determine the volume fraction crystallinity. Alternatively a differential scanning calorimeter (DSC) can be used to record the variation of specific heat with temperature (Fig. 2.25). Integration of the area above the baseline gives the enthalpy of fusion H_f, which can then be divided by the value for the crystal to give the fraction crystallinity. ($H_f = 295 \text{ kJ kg}^{-1}$ for polyethylene and values for most common polymers are available.) The DSC trace also gives valuable information on the perfection of the lamellar crystals present, since the thinnest, and lowest molecular weight lamellae may melt some 30 °C below the final melting point. If a rapidly cooled polymer, e.g. polyethylene, is subsequently annealed in this temperature range, the lamel-

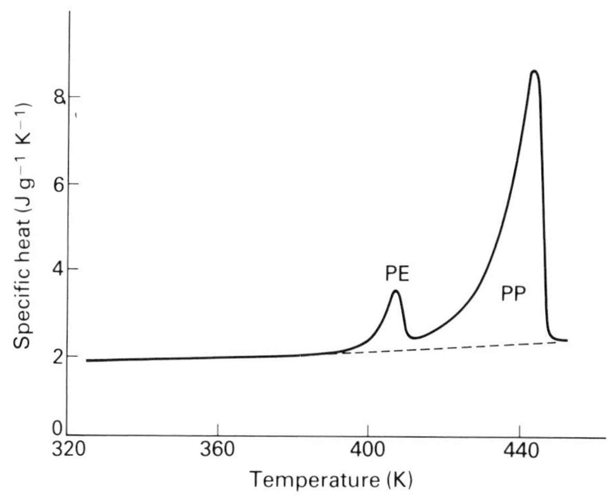

Fig. 2.25 Specific heat versus temperature trace for heating a polypropylene copolymer at 20 K min^{-1} in a differential scanning calorimeter. The separate melting peaks indicate a polyethylene component, and a range of crystalline perfection

lae will thicken by a process of partial melting and recrystallisation, and the shape of the DSC trace will change.

2.4.6 X-ray diffraction of semi-crystalline polymers

X-ray diffraction gives no information about the location of crystals, but the information that it gives about the range of orientations in the irradiated volume can be used with optical microscopy to build up a picture of the microstructure. The crystalline phase in semi-crystalline polymers diffracts X-rays in the same way as other crystalline solids. In the crystal lattice model used to interpret the diffraction patterns there are many sets of parallel planes. Polymer crystals have lower lattice symmetry than metals, so the relationship between the interplanar spacing d and the Miller indices (hkl) of the plane are complex (Kelly and Groves, 1970). The Bragg condition

$$n\lambda = 2d \sin \theta \tag{2.30}$$

gives the diffraction angle θ for the nth order diffraction peak from the (hkl) planes, where λ is the wavelength of the monochromatic X-ray beam. Fig. 2.26 shows that the vectors **i** for the incident X-ray beam, **j** for the diffracted X-ray beam and **n** for the plane normal or *pole* are coplanar. The pole must have this specific orientation for diffraction to occur, otherwise the X-ray beam is not diffracted. The diffraction pattern for a spherulitic semi-crystalline polymer consists of several complete concentric rings, corresponding to different sets of planes. Fig. 2.27 shows the angular intensity scan across such a pattern for polypropylene. There are four main peaks but one of these is a composite peak. The inset diagram shows the position of the (040)

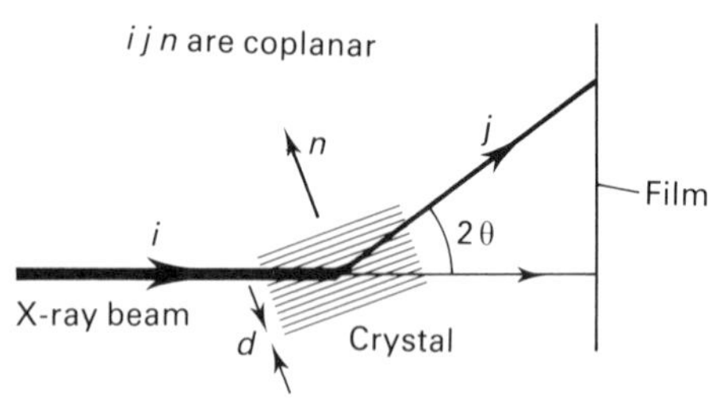

Fig. 2.26 The Bragg diffraction condition for a set of planes in a crystal

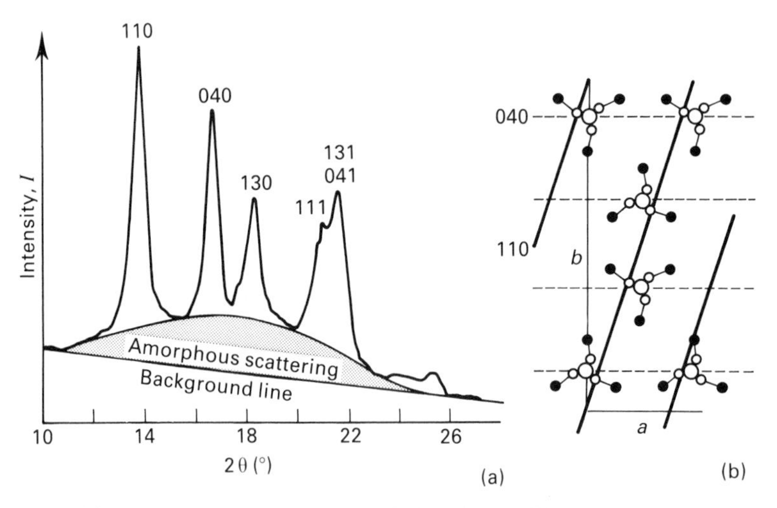

Fig. 2.27 (a) An angular intensity scan of the X-ray diffraction pattern for spherulitic polypropylene. (b) The 110 and 040 planes in the unit cell projected on to the **ab** plane

and (110) planes in the unit cell of polypropylene (shown in more detail in Fig. 2.19). The most useful information would be from (001) planes, but the (001) diffraction is of negligible intensity. Consequently information on the **c** axis orientation must be inferred from diffraction from other planes. For the spherulitic sample the (110), (040), (130) and (111) poles are randomly distributed in space, so it is reasonable to assume that the (001) poles are also randomly distributed. This fits with our model of the spherulite having radial symmetry.

When the X-ray diffraction pattern from a highly oriented polypropylene tape, used for parcel strapping, is examined (Fig. 2.28), there are seen to be the remnants of the four diffraction rings; the orientation of the crystal planes relative to the X-ray beam has changed. The length axis L of the tape is vertical, so the (110), (040) and (130) diffraction peaks lie on the 'equator' of the figure. When semi-crystalline polymers are stretched the crystal **c** axes tend to align with the tensile direction, so the assumed *orientation distribution*

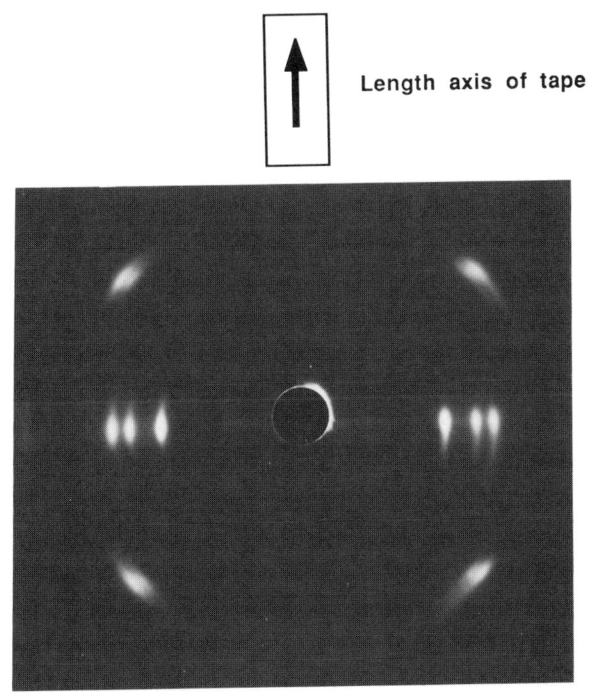

Fig. 2.28 The X-ray diffraction pattern for polypropylene strapping that is uniaxially oriented; the strapping length axis is vertical relative to the pattern

is that the **c** axes are perfectly aligned with the L axis, with the **a** and **b** axes randomly distributed in the plane at 90° to L. To confirm this assumption would need further diffraction patterns taken as the sample is rotated around its L axis. The diffraction peaks in Fig. 2.28 are consistent with the assumption; the $(hk0)$ poles are at 90° to the **c** axis, so are at 90° to the L axis of the tape. Fig. 2.26 can be used to illustrate this—assuming the L axis is normal to the paper, the diffracting planes shown will be of the $(hk0)$ type. As the $(hk0)$ poles are randomly oriented in the plane of the diagram, there will always be crystals positioned to produce diffraction spots on either side of the 'equator' of the pattern.

2.4.7 The average orientation of the crystalline phase

The two most common types of oriented polymer product are:

(a) A uniaxially stretched fibre, tape or film where the measure of orientation is the angle θ between the crystal **c** axis and the product length axis L. The average orientation measure used is the second Legendre polynomial

$$P_2 = 0.5(3\,\overline{\cos^2\theta} - 1) \tag{2.31}$$

Since the angle θ between the crystal **c** axis and the tape axis can only vary between 0° and 90° the range of values of P_2 is -0.5 to 1.0. There are several techniques that measure P_2 including measurement of the *birefringence* $n_L - n_T$, where n_L and n_T are the refractive indices for light polarised in the length and transverse directions of the product. If light with L polarisation passes through a thickness t of product and emerges with a phase difference of f wavelengths from the light with T polarisation, the birefringence is calculated as

$$n_L - n_T = \frac{f\lambda}{t} \tag{2.32}$$

The crystalline phase birefringence can be divided by the refractive index difference $n_c - n_a$ for the crystal, to give P_2. However there is also a contribution to the birefringence from the molecular orientation of the amorphous phase. Fig. 2.29 shows the contributions to the overall birefringence of polypropylene films that were hot stretched at 110 °C by different amounts. The increase in the orientation with strain is non-linear and it differs between the phases. There are various theories for the variation of P_2 with the deformation of a semi-crystalline polymer. In the *pseudo-affine model* it is assumed that the distribution of the crystal **c** axes is the same as the distribution of the network chain end-to-end vectors **r**, in a rubber that has undergone the same macroscopic strain. Fig. 2.12 showed the affine deformation of an **r** vector with the shear of a rubber block. When there is a tensile extension ratio λ, with an extension ratio $1/\sqrt{\lambda}$ in the lateral direction, the angle θ in the deformed material is related to the value θ' in the undeformed material by

$$\tan\theta = \frac{\lambda^{-0.5}}{\lambda}\tan\theta'$$

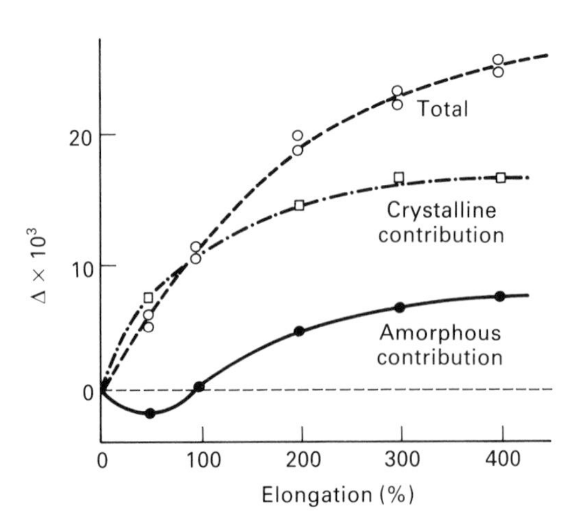

Fig. 2.29 The variation with the elongation of polypropylene film of the amorphous and the crystalline contributions to the birefringence (from Samuels, *Structured Polymer Properties*, Wiley, 1974)

If the initial distribution of θ' values is random it can be shown that in the deformed material

$$\overline{\cos^2\theta} = \lambda^3\left(\frac{1}{a^2} - \frac{1}{a^3}\tan^{-1}a\right) \tag{2.33}$$

where $a^2 = \lambda^3 - 1$. This relationship is substituted into equation (2.31) and the theoretical curve plotted in Fig. 2.30. Given the simplicity of the model, the agreement is very good.

For a rubber under tension the molecules elongate, as well as the **r** vectors moving towards the tensile axis. Hence the variation of P_2 (for the rigid links of the freely jointed chain model) with extension ratio will differ from the pseudo-affine model. For moderate strains the increase of P_2 with extension ratio is linear, but at high extensions the approximation used in equation (2.12), that both p and q are large, breaks down. Treloar (1975) gives details of the refined models which take into account the number of links in the network chains. Fig. 2.30 shows that the orientation function abruptly

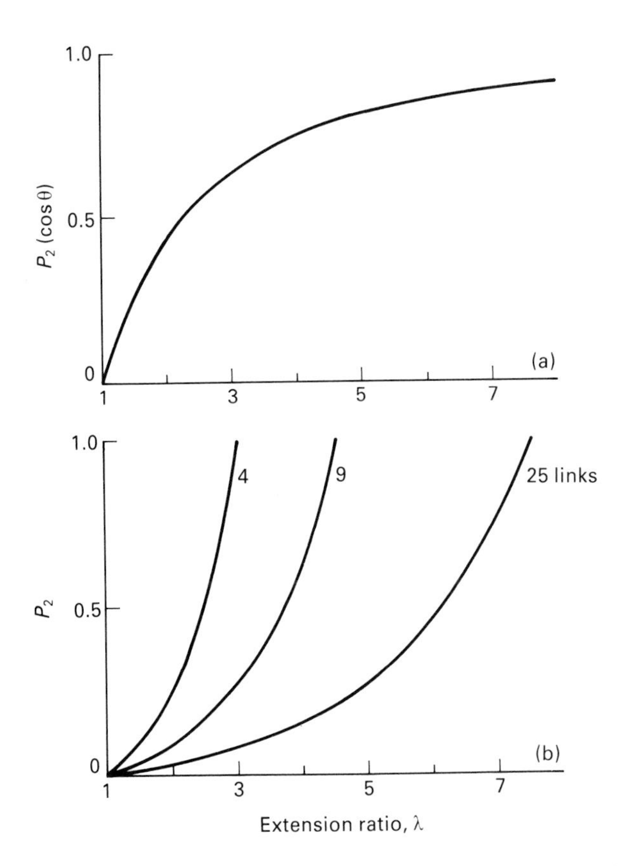

Fig. 2.30 The predicted orientation function P_2 versus tensile extension ratio for (a) crystals according to the pseudo-affine model and (b) rubber networks of 4, 9 and 25 link chains according to Treloar

approaches 1 as the extension ratio of the rubber exceeds $\sqrt{\lambda}$. Although the model is successful for rubbers it fails for the amorphous phase in polypropylene (Fig. 2.29), presumably because the crystals deform and reduce the strain in the amorphous phase.

(b) For biaxially stretched products the aim is to have the **c** axes in the plane of the product, with a greater number of **c** axes in the direction of the highest stress. For cylindrical pressure vessels the hoop stress is twice the longitudinal stress (Chapter 7) so there should be more crystal **c** axes in the hoop direction. In the stretch blow moulding of PET bottles for carbonated drinks (Section 4.5.2) the extension ratio in the hoop direction is greater than that in the longitudinal direction. X-ray diffraction *pole figures* are used to quantify the orientation distribution. Fig. 2.31 shows the projection of the PET unit cell along the **c** axis. Although the unit cell is triclinic, with none of the cell angles equal to 90°, it is convenient to interpret the X-ray diffraction as if there was an **a′** axis at 90° to **b**. The **a′** axis is perpendicular to the planes of the in-chain benzene rings. The pole figures for a bottle include that for the (100) poles which is the direction of the **a′** axis PET. The intensity of the (100)

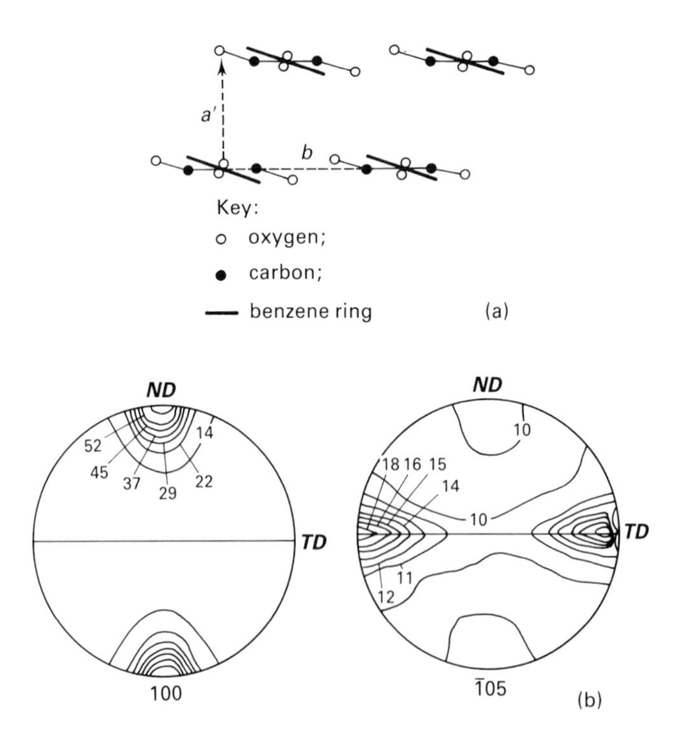

Fig. 2.31 (a) The projection of the PET unit cell along the **c** axis, with the direction of the **a′** and **b** axes shown. (b) The 100 and 105 pole figures for the wall of a PET bottle, shown as a stereographic projection, with ND being the wall normal, and TD the hoop direction. Intensity level 10 is the same as for a random texture (from Cakmak *et al.*, *Polym. Eng. Sci.*, 1984, **24**, 1390.

poles rises to 5 times random in the direction of the normal to the bottle wall, which means that the benzene rings in the crystals tend to lie in the plane of the bottle wall. The (105) pole figure is for an axis that is close to the **c** axis. The intensity of (105) poles rises from 1 times average in the bottle vertical direction to about 1.8 times average in the hoop direction. The complete interpretation of the pole figures is difficult, as the amount of **a**′ axis orientation is greater than that for any other axis.

3

Polymeric composites

3.1 INTRODUCTION

The mechanical properties of any of the solid forms of polymer can be modified by adding another material, be it a glass, metal, air or another polymer. These composite materials can be classified into macroscopic composites (Sections 2 and 3), where the constituent materials or phases can be distinguished with the naked eye, and microscopic composites (Sections 4 to 8) where optical or electron microscopes are needed to see the constituent phases. The examples described have been chosen to illustrate the methods of increasing stiffness, toughness and energy absorption, and to illustrate the varied phase geometries found. Short fibre composites are treated briefly in Section 8; for further information on fibre composites, see the appropriate texts.

The concept of upper and lower limits on the modulus of a composite will be used to analyse the modulus of spherulitic polymers in Section 7. The limits become exact for specific geometries and load types, which will be analysed in Section 2. Thus the shear modulus of a stack of parallel plates is a lower limit for the modulus of all composite geometries, and the Young's modulus parallel to continuous fibres in a composite is the upper limit.

3.2 ELASTIC MODULI AND STIFFNESS OF COMPOSITES

3.2.1 Shear modulus of rubber/steel laminated springs

We have seen in Chapter 2 that rubber is an isotropic material with a very low elastic modulus. It can be subjected to much higher elastic strains than can a metal. Consequently a simple solid rubber block can replace a metal spring of complex shape; the latter uses the bending or twisting of a long slender structure to compensate for the high modulus of the metal. There are some applications where springs or bearings with anisotropic stiffness are required. The original design for the application may have been a metal spring (multiple steel leaf springs in heavy vehicle suspensions) or may not (steel

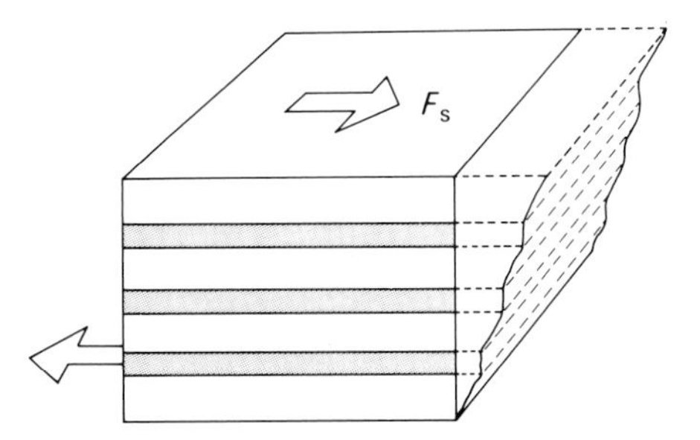

Fig. 3.1 Laminated steel and rubber spring, loaded in shear

rollers between two steel plates as an expansion bearing at one end of a bridge deck). We shall examine laminated rubber designs which can replace these products, and which require no maintenance (Fig. 3.1).

The stack of steel and rubber plates is loaded in series, whether there is a compressive or a shear force applied to the end plates. Thus the same force F is transmitted to each layer. As the plates have equal areas A, apart from edge effects the compressive or shear stress will be the same everywhere. This *uniform stress condition* is the essence of the analysis.

In shear the volume of the materials remains constant; for rubber with a Poisson's ratio of 0.499 this is important, as it is much more difficult to change the volume than it is to change the shape at constant volume. If the stack has total height H, then the *volume fraction* V_R of rubber is

$$V_R = \frac{h_R}{H} \tag{3.1}$$

where h_R is the total height of the rubber layers. The total shear deflection x_R in the rubber layers is

$$x_R = h_R \gamma_R \tag{3.2}$$

where γ_R is the shear strain in the rubber. Similar expressions to equations (3.1) and (3.2) can be derived for the steel layers, and the total shear deflection of the stack

$$\begin{aligned} x &= x_R + x_S \\ &= \gamma_R V_R H + \gamma_S V_S H \end{aligned}$$

when this equation is divided through by H we obtain the average shear strain in the composite

$$\gamma_C = \gamma_R V_R + \gamma_S V_S$$

and further division by the constant shear stress τ gives the relationship between the shear moduli G_C of the composite, G_R of the rubber and G_S of the steel

$$\frac{1}{G_C} = \frac{V_R}{G_R} + \frac{V_S}{G_S} \tag{3.3}$$

As the shear moduli are so disparate in magnitude ($G_R \cong 1$ MN m^{-2}, $G_S = 81$ GN m^{-2}) the second term in equation (3.3) can be neglected, and substitution of equation (3.1) leads to

$$\frac{1}{G_C} = \frac{h_R}{HG_R} \tag{3.4}$$

For design purposes the shear spring constant k_S is needed; by definition

$$k_S = \frac{F}{x} = \frac{AG_C}{H}$$

so substituting from equation (3.4) gives

$$k_S = \frac{AG_R}{h_R} \tag{3.5}$$

Hence the shear stiffness only depends on the rubber properties, and is the same regardless of the number of layers into which the total thickness h_R is split.

3.2.2 Compressive Young's modulus of rubber/steel laminates

The compressive behaviour of a rubber layer is more complex because the bond to the steel plates prevents the rubber expanding sideways at its top and bottom surfaces. The effect of this restraint on the compressive behaviour depends on the *shape factor* S of the layer (Fig. 3.2) defined as

$$S = \frac{\text{top loaded area}}{\text{bulge area}} \tag{3.6}$$

It is assumed that the width and length of the rubber block are comparable in magnitude. The analysis of the compressive stiffness of the stack of rubber and metal plates can be made in exactly the same way as for the shear deformation. The equivalent to equation (3.5) for the compressive stiffness k_C is

$$k_C = \frac{AE(S)}{h_R} \tag{3.7}$$

where the compressive modulus $E(S)$ of the rubber is a function of the shape factor S of the individual layers. When $S < 0.25$ (a cube) the end surfaces have very little influence on the bulging of the sides of the rubber layers so the conventional relationship with the shear modulus holds

$$E = 2G(1 + \nu) \tag{3.8}$$

As Poisson's ratio ν for a rubber $= 0.499$ this simplifies to $E = 3G$. If the shape factor is less than 0.1 then the block of rubber is too tall and thin to deform uniformly. It will buckle elastically (see Section 7.3) and this mode of deformation is used in rubber fenders for low speed ship impacts.

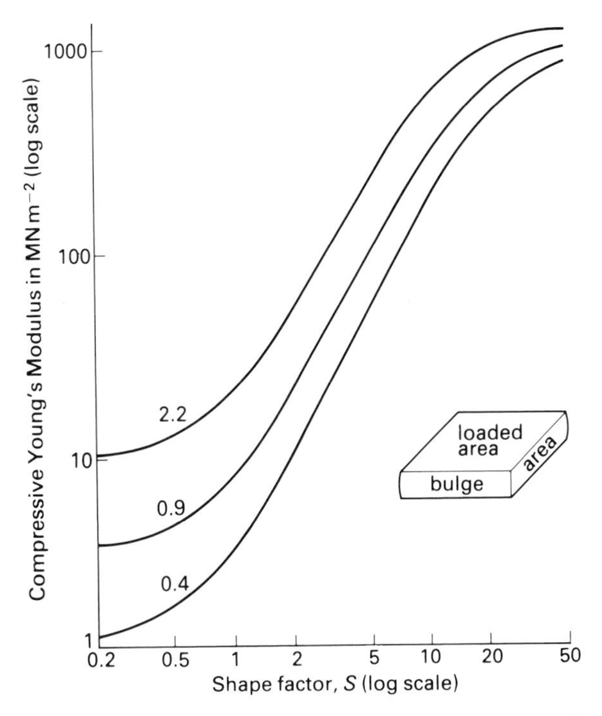

Fig. 3.2 Compressive Young's modulus of a rubber spring versus the shape factor. The curves are labelled with the shear modulus of the rubber in MN m^{-2} (from P. B. Lindley, *Engineering Design with Natural Rubber*, 4th ed., Malayan Natural Rubber Producers Association, 1974)

When $S \gg 1$ bulging can only occur near the sides of the block, and near the centre the rubber is compressed in volume with zero sideways strain. Therefore the volume strain dV/V is approximately equal to the vertical compressive strain, and uniform stress conditions exist everywhere except at the edges of the plates. Using the definition of the bulk modulus K in equation (2.30) for the case of a linear pressure-volume relationship, we find that

$$E(S \to \infty) = \frac{\sigma_x}{e_x} = -p\frac{V}{\mathrm{d}V} = K \tag{3.9}$$

The bulk modulus of rubber is 2 GN m^{-2} since it depends on the strength of the van der Waals bonds between the molecules. Therefore the compressive modulus of a rubber layer can vary by a factor of a thousand as the shape factor increases (Fig. 3.2). When a laminated rubber spring is designed, equations (3.5) and (3.7) allow the independent manipulation of the shear and compressive stiffnesses. The physical size of the bearing will be determined by other constraints, possibly the load bearing ability of the abutting material (if it is concrete), or a limit on the allowable shear strain to $\gamma < 0.5$ and the compressive strain $e < -0.1$.

3.2.3. The modulus parallel to fibres

The highest composite occurs when continuous fibres (or ribbons or other constant cross-section estrudates) are aligned parallel to a tensile stress (Fig. 3.3). In this case the matrix and the fibres experience *uniform strain e*, so long as there is interfacial bonding to prevent the fibres sliding in their holes. In this case the average tensile stress σ_C in the composite can be calculated by summing the forces on the ends of the fibres and matrix. In 1 m^2 of end face there is V_f m^2 of fibres (where V_f is the volume fraction of fibres), which are under a stress $E_f e$, hence the total force is $V_f E_f e$. There is a similar expression $V_m E_m e$ for the matrix, so summing the forces on the unit area gives

$$\sigma_c = V_f E_f e + V_m E_m e$$

Dividing through by the strain gives the Young's modulus $E_{\|}$ when the stress is parallel to the fibres

$$E_{\|} = V_f E_f + V_m E_m \tag{3.10}$$

3.2.4 Bounds for the moduli of other composites

Equation (3.3), for the composite modulus under uniform stress conditions, is used as a *lower bound for the modulus* of any composite; the quantities needed are the appropriate moduli E_1 and E_2 of the two materials, and their volume fractions V_1 and V_2. The phase geometry of the materials does not need to be known, and it is not necessary for either to be a continuous matrix. If there is specific knowledge of the geometry and the direction of the stresses then it may be possible to compute a better estimate of the modulus.

The proof that this is a lower limit is based on minimisation of the stored elastic energy of the composite and is beyond this text. The minimum modulus E_{min} (or G_{min} or K_{min} for the shear or bulk moduli) occurs under uniform stress conditions and is given by

$$\frac{1}{E_{min}} = \frac{V_1}{E_1} + \frac{V_2}{E_2} \tag{3.11}$$

Similarly the uniform strain condition leads to an *upper bound for the modulus* of a composite material.

$$E_{max} = V_1 E_1 + V_2 E_2 \tag{3.12}$$

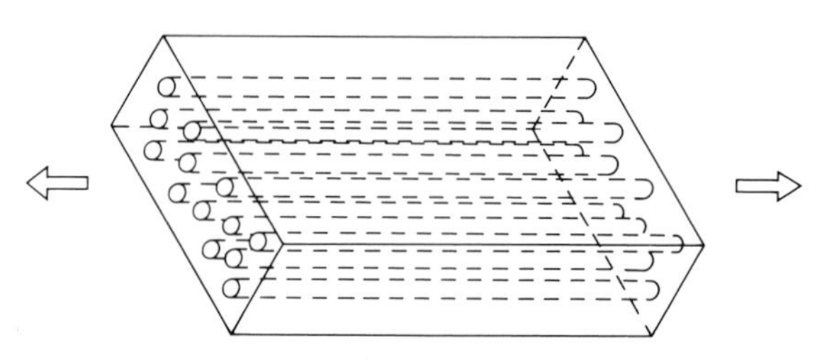

Fig. 3.3 Model for E_1, the Young's modulus parallel to continuous fibres

The modulus of any composite must then fall between the bounds $E_{\min}$ and $E_{\max}$.

3.3 LAYERED STRUCTURES IN BENDING

3.3.1 Bending stiffness theory

Some layered structures are easy to recognise, such as the use of glass reinforced skins bonded to a lightweight honeycomb core (Fig. 3.4a), to produce stiff panels for transport applications. Others are less obvious; the space between the steel outer container and the toughened polystyrene liner of a refrigerator is filled with rigid polyurethane foam. Still others may need sectioning to reveal the structure; Fig. 3.4b shows a section through a 'structural foam' injection moulding, revealing that the surface layers are denser than the core. Nevertheless, in all these applications there are two common factors; the layers are bonded together, and the effect is to increase the bending stiffness of the product.

The major difference between the bending of a beam or a plate, and compression and shear loadings examined in the last section, is that the in-plate tensile strain e_x varies linearly through the thickness. It is assumed that we are dealing with beams or plates that are much longer than they are thick, so that the effects of shear strains can be neglected. Consider a beam that is bent into a circular arc of radius R (in the general case R is the local radius of curvature of the beam). There is a layer in the beam that neither contracts nor extends in length; there is zero tensile strain in this *neutral surface*. It can be shown by comparing similar triangles that a parallel layer at distance y above the neutral surface has a tensile strain e_x in it given by

$$e_x = -y/R \tag{3.13}$$

The sign of the right hand side of equation (3.13) will depend on the direction of bending. The tensile stresses in the beam are usually calculated on the assumption that Poisson's ratio is zero, so that there is no lateral strain in the beam, and the tensile stress is given by

$$\sigma_x = E e_x = -Ey/R \tag{3.14}$$

Fig. 3.5 shows the resulting stress variation in a symmetrical sandwich beam, using equation (3.14) separately for the skins with high Young's modulus E_S, and the core with low modulus E_C. The applied bending moment M, about the z axis in the plane shown, must be in equilibrium with the moments of the internal stresses, so

$$M - \int yw\sigma_x \mathrm{d}y = 0$$

where w is the width of the beam (which can vary). When equations (3.13) and (3.14) are substituted this becomes

$$M - \frac{1}{R}\int Ewy^2 \,\mathrm{d}y = 0 \tag{3.15}$$

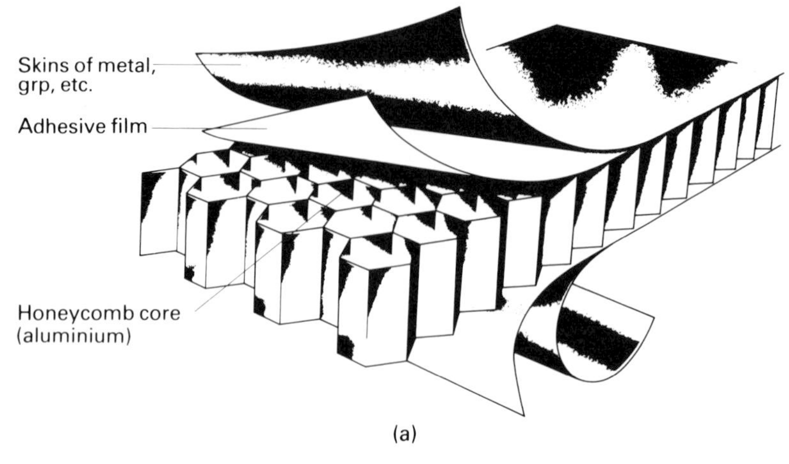

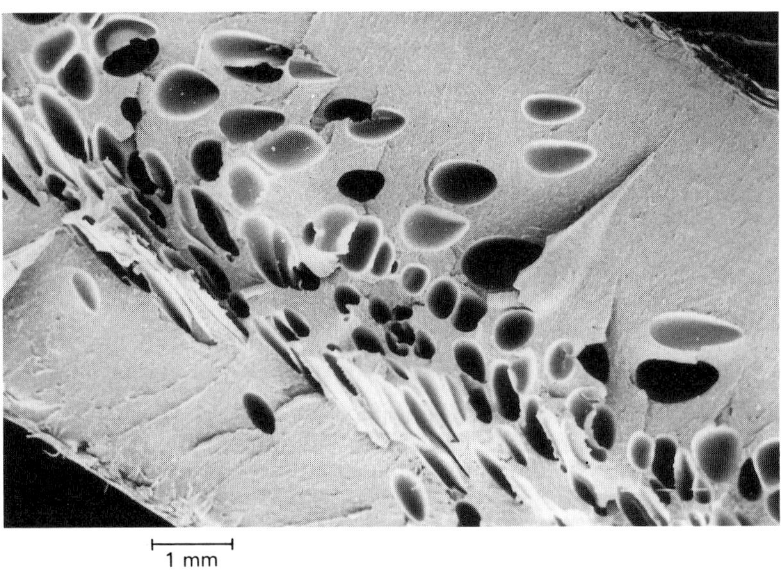

Fig. 3.4 Layered structures with improved bending stiffness. (a) Glass fibre reinforced skins on an aluminium honeycomb core. (b) A structural foam injection moulding with the maximum density at the skins

The definition of the *second moment of area I* of a beam cross section is

$$I \equiv \int wy^2\, dy \tag{3.16}$$

Because there are different Young's moduli E_S and E_C in the skin and core, the integral in equation (3.15) has to be separated into contributions from the skin area and the core area

$$MR = E_S \int w_S y_S^2\, dy_S + E_C \int w_C y_C^2\, dy_C$$

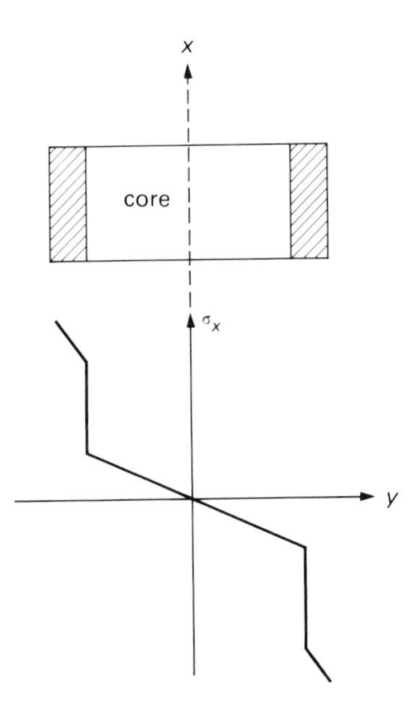

Fig. 3.5 Variation of the in-plane tensile stress through the thickness of a composite panel like Fig. 3.4a, when loaded in bending

or

$$MR = E_S I_S + E_C I_C \tag{3.17}$$

Equation (3.17) can be used to show that a lightweight sandwich structure with a high bending stiffness can be constructed by using thin high modulus skins bonded to a low density core of moderate modulus. The relatively thick core means that I_S will be large because the skins are well away from the neutral surface. The core then need only play the secondary roles of supporting the skins when surface compressive forces are applied, and transmitting shear forces from skin to skin. It is possible to reduce the mass of the beam, while maintaining the bending stiffness by increasing the core thickness and decreasing the skin thickness. This process must not be taken too far because the surface stresses, given by combining equations (3.17) and (3.14), increase in proportion to the total beam thickness. Therefore the danger of a tensile failure of a skin, or the puncturing of the skin by sharp objects, prevents the skins becoming too thin.

3.3.2 Structural foam injection mouldings

Another type of layered structure can be made by the 'structural' foam injection moulding process, in which the average density ρ_f of the foamed product is in the range 60 to 90% of the density of the solid polymer ρ_0. A chemical blowing agent produces enough gas to form a foam when the melt is at atmospheric pressure, but the gas will completely dissolve in the melt at a relatively low pressure of about 20 bar. The leading edge of the melt as it enters the mould is at a low pressure so it foams. These foam bubbles are sheared against the mould surface and solidify instantly leaving a corrugated surface (visible in Fig. 3.4b). Enough melt is injected so that the cavity pressure rises to about 40 bar; consequently the layers that solidify at a 40 bar pressure contain no gas bubbles. The volume shrinkage of the cooling polymer (see Section 5.2) reduces the pressure in the mould, and when the pressure drops below 20 bar the core of the moulding foams. In the final product there is a continuous increase in density from the core to close to the surface (Fig. 3.4b). It is easy to make mouldings where there is a variation of the moulding thickness up to 8 mm without there being sink marks on the surface—the foamed core is thicker in those regions. (Morton-Jones' Chapter 13 further reading) Chapter 13 will give more details of a washing machine tank moulding made from structural foam.

Section 3.5.2 describes how the Young's modulus of closed cell foams varies with the square of the density. The bending stiffness of the structural foam panel can be calculated using a generalisation of equation (3.15) in which the Young's modulus is a known function of the distance y from the neutral surface. The bending stiffness can typically be increased by 75% compared with a solid moulding of the same mass. Partly this is a result of the lower average density of the structural foam (typically 80% of the solid density) but it is also a result of having the highest modulus material in the solid surface layers. Structural foam mouldings are however less tough than solid ones; partly this is a result of the rough outer surface which contains the equivalent of short cracks, and partly it is the foamed core which allows the easier propagation of these cracks.

3.4 RUBBER TOUGHENING

3.4.1 Microstructure

We shall see in Chapter 6 that the toughness of most plastics is not high. Therefore in applications where the product is subjected to impacts rubber toughened grades may well be used. The chemistry of specific polymers determines whether it is possible to incorporate rubber in the form of dispersed spheres that are *well bonded* to the matrix. One of the difficulties that has had to be overcome is how to make this rubber phase stable at high melt processing temperatures (for example for polycarbonate). This aside, the principles of rubber toughening are the same for all polymers. The brittle glasses polystyrene and styrene–acrylonitrile copolymer were among the first to be toughened and consequently these systems have been studied in the greatest detail. The crosslinked rubber used in them is polybutadiene or a butadiene–styrene copolymer. Some of the combinations have developed their own names; thus high-impact polystyrene (HIPS) is polystyrene reinforced with polybutadiene and ABS is styrene–acrylonitrile copolymer reinforced with styrene–butadiene copolymer.

The rubber particles in these materials are in the form of spheres, unless the moulding orientation is high (see Section 5.2) when the spheres are distorted to ellipsoids. Fig. 3.6 shows that in the glassy matrix of SAN copolymer there are rubber spheres of a range of sixes up to 0.5 μm. Within the larger spheres are smaller spheres of white SAN glass. The reason for this *phase separation* is that the two polymers are immiscible. The complex spheres-within-spheres microstructure is the product of the polymerisation sequence. First an emulsion of butadiene droplets in water is polymerised; the resulting rubber latex contains crosslinked rubber spheres less than 1 μm in diameter. The sizes of the latex spheres determine the main sphere sizes in the ABS. Next the monomers styrene and acrylonitrile are added to the emulsion, and the polymerisation is re-initiated. Some of the droplets contain styrene and acrylonitrile dissolved in polybutadiene (PBD), others just styrene and and acrylonitrile. The latter will form SAN copolymer, but in the former some grafting of SAN to the double bonds in the PBD will occur. Insoluble SAN copolymer separates inside the rubber spheres, and also forms a shell around them. Finally the emulsion is coagulated, washed, dried and extruded into pellets.

The volume fraction of rubber particles is often around 30% for ABS, and the particle sizes in the range 0.1 to 1 μm. These two attributes have been chosen to give the optimum toughness for the particular polymer. There is linear reduction of modulus with rubber volume fraction, and a V_f of 20% lowers the Young's modulus of the composite by about 35%. The high modulus glassy matrix is the continuous phase, so the composite modulus falls close to the upper limit given by equation (3.12). The reduction is much greater at high V_f values when the glass is no longer the continuous phase.

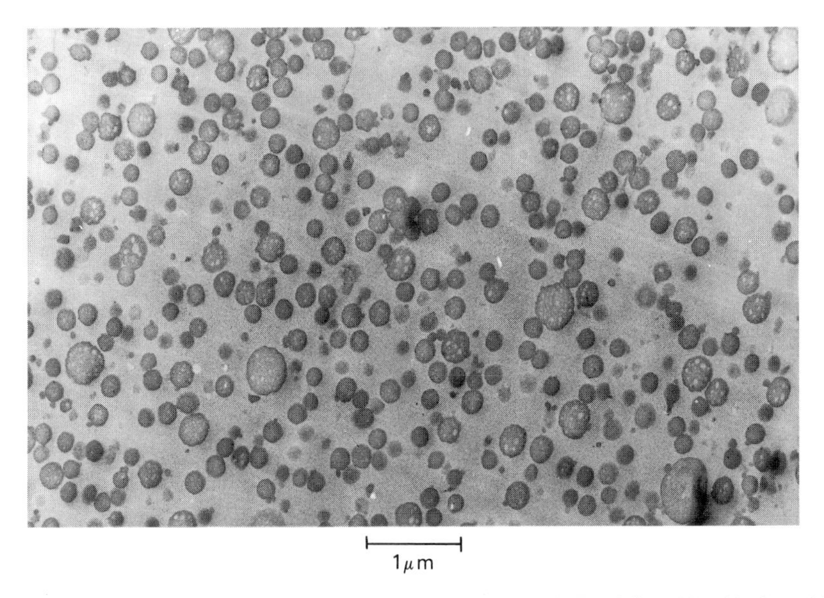

Fig. 3.6 Transmission electron micrograph of ABS, moulded 'Ronfalin TX' with the rubber phase stained dark (courtesy of Dutch State Mines)

The interparticle distance for the rubber phase is very small. An over-simplified microstructural model of uniform spheres of radius r μm, arranged into a cubic lattice with 2 μm between the sphere centres, has a rubber volume fraction of $\pi r^3/6$. It has been found that the brittle to ductile transition temperature of toughened nylon 6 decreases with the interparticle spacing (Fig. 3.7) so it is important to produce uniform small diameter spheres of rubber.

3.4.2 Stress concentrations and toughening mechanism

The mechanism of toughening can be understood by considering the stress concentrations (Section 8.3.1) that occur around the 'equator' of each rubber sphere when a tensile stress is applied to the matrix (Fig. 3.8a). As the modulus of the rubber is less than 0.5% of the Young's modulus of the glassy

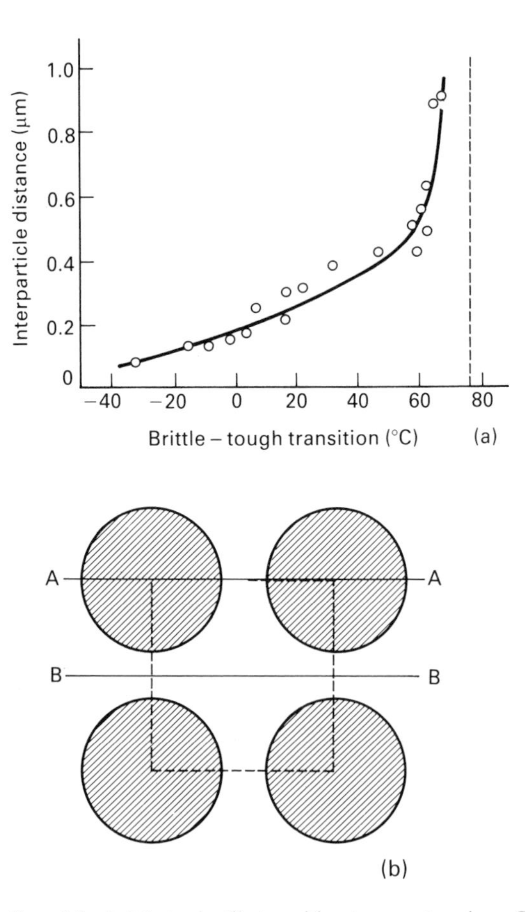

Fig. 3.7 (a) Variation of the brittle to ductile transition temperature in an Izod impact test with the interparticle spacing in toughened nylon 6 (from Borggreve *et al.*, *Polymer*, 1987, **28**, 1489). (b) A cubic array of rubber spheres when the rubber volume fraction is 0.25. On the cross section AA the area fraction of rubber is a maximum whereas on section BB it is zero

matrix (2 to 3 GN m^{-2}) the stress concentration factor at the equator of an isolated sphere of 1.9 is nearly as great as that of 2 for a spherical hole. These stress concentration values will be higher when the rubber spheres are in close proximity, and the exact value will depend on the spacing and arrangement of the spheres. The presence of the graft copolymer across the glass–rubber interface prevents any cracks occurring at the interface. The high local stresses around the rubber particles means that the yielding or crazing mechanisms (see Chapter 7) will start these first, and spread outwards. Fig. 3.8b shows that in HIPS crazing spreads out as an equatorial ring at 90° to the tensile stress, whereas Fig. 3.8c shows that shear yielding in bands at 45° to the tensile stress occurs in ABS.

Although the elastic stress concentration of the spheres explains the *onset of crazing* or shear banding, the whole microstructure cannot yield until the crazes spread across the cross-section. The stress for this *through-section yielding* can be estimated if the area that must craze is known. Fig. 3.7b shows a simple cubic array of equal sized spheres that can be used to obtain an estimate of the section yield stress. A rubber volume fraction of 0.25 ($= \pi r^3/6$) requires a sphere radius $r = 0.781$ μm if the spheres centres are 2 μm apart. In the cubic array there will be cross-sections such as BB that avoid the spheres altogether. For others, like AA in Fig. 3.7b, there is a minimum area fraction of matrix of $1 - 0.25\,\pi r^2$, which for $V_r = 0.25$ is 0.52. Hence for this rubber content the section will yield at a stress that is 52% of the yield stress of the non-toughened matrix.

Table 3.1 compares experimental yield stresses for the matrix materials alone with those for the rubber toughened versions. There is data both for tensile tests, where crazing can occur, and for compressive tests where crazing cannot occur. The rubber particles roughly halve the yield or fracture stress in the three cases where the deformation mechanisms do not change, indicating that the net section yielding calculation is approximately correct. The odd case is for SAN in tension, which fails by crazing. Adding rubber to produce ABS promotes shear yielding as the initial failure mechanism, and the tensile strength only falls by 1/3.

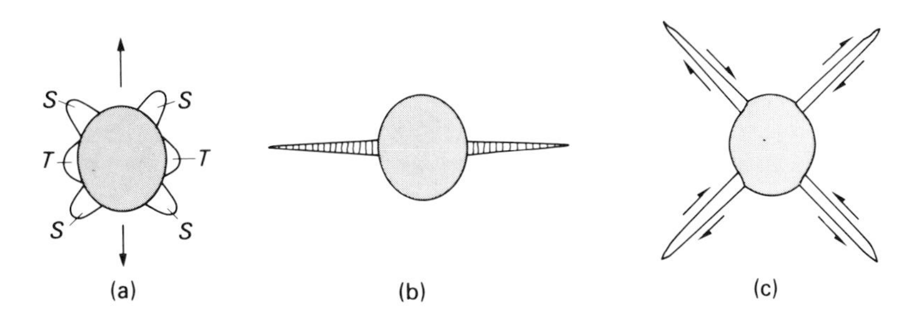

Fig. 3.8 (a) Positions around a rubber sphere where the tensile (T) and shear (S) stresses are greatest, when the glassy matrix is pulled in the vertical directon. (b) Craze that has initiated at a rubber sphere in high impact polystyrene. (c) Shear bands that have initiated at a rubber sphere in ABS

Table 3.1 Yield (or fracture) stresses ($MN\ m^{-2}$) at 20 °C

Polymer	Tension	Elongation	Compression
PS	40	*	100
HIPS	22	>60%	55
SAN (30% AN)	65	*	118
ABS	44	20%	52

* Brittle with zero permanent elongation.

When a high tensile stress is applied crazing or shear bands will emerge from every sphere. This contrasts with the untoughened polymer where crazes will be separated by distances ~1 mm, and consequently only a small fraction of the total material will craze. In this case the average permanent elongation at fracture will be <1%. In the rubber toughened plastic the growing crazes interact strongly. This may be due to their running into another rubber particle or merely due to their close proximity (Section 7.6). The overall effect is to provide a high elongation at break in tensile loading at the cost of a loss of <50% of the tensile strength. Consequently the energy absorbed before failure, proportional to the area under the stress–strain curve (Fig. 3.9), is much larger than for the untoughened polymer.

3.5 FOAMS

3.5.1 Foam geometry

When gas bubbles grow in a liquid, the foam goes through a number of stages. Fig. 3.10 shows a two-dimensional view in which the regularly spaced bubbles are all the same size. As the density of the foam decreases, the bubbles come into contact, forming a *closed cell foam.* Finally the cell faces burst and liquid drains to the cell edges leaving an *open cell foam.* Beyond this the foam collapses back to a liquid. When plastic foams are examined by scanning electron microscopy (SEM) two main types are found. If the gas bubbles have grown in highly viscous thermoplastic melt, the thinning of the cell walls is a slow process, and solidification can stabilise the closed cell foam. Fig. 3.11a shows that polystyrene foam of density 40 $kg\ m^{-3}$ has closed cells with 4, 5 and 6 sided faces.

A section through the cell walls of a similar PVC foam shows that the cell faces are uniformly thick. The faces, rather than the edges, contain most of the polymer. It is easier to form such foams using glassy polymers, but recently the melt rheology of polypropylene has been modified to allow stable foams to be formed in the narrow process temperature window above the melting point of the crystalline phase. It is also possible to crosslink polyolefins before foaming, which makes the cells more stable.

When a low viscosity thermoset prepolymer is foamed the polymer can eaily drain from the cell walls before the crosslinking reaction stabilises the foam cell edges. Fig. 3.11b shows a polyurethane open cell foam in which only

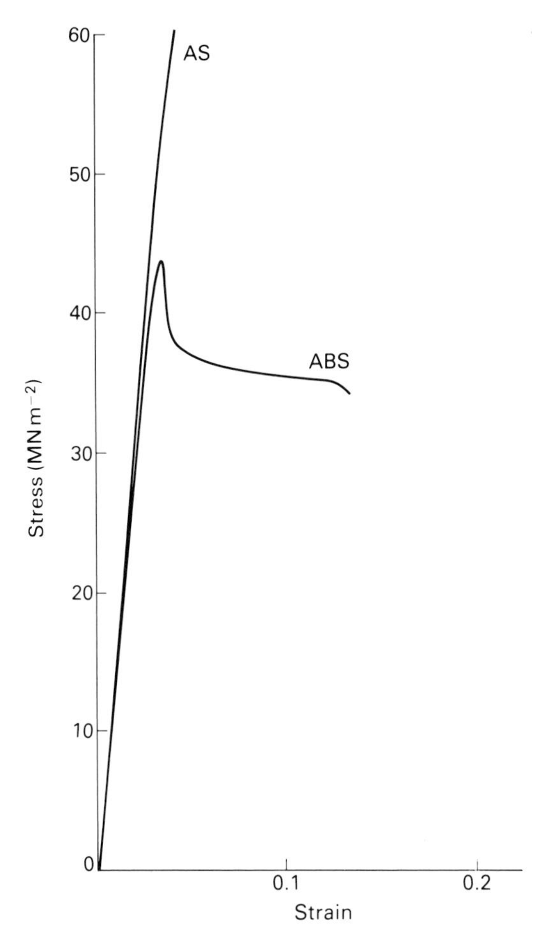

Fig. 3.9 Tensile stress–strain curve of acrylonitrile–styrene copolymer and of the rubber toughened modification ABS

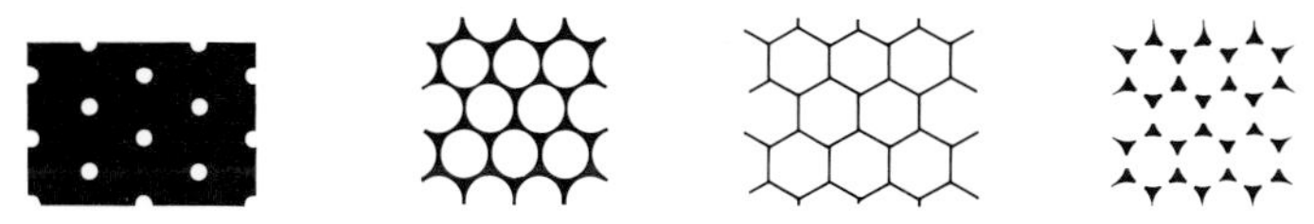

Fig. 3.10 Change in the structure of a foam containing a regular array of bubbles as the volume fraction of gas increases

the cell edges remain. These have three concave sides, the shape being fixed by the surface tension of the liquid prepolymer.

The microstructure of a foam can be specified by using a number of parameters. One is the cell size, or the distribution of cell sizes. In the case where the foam rises while it is supported on a table, the cells can be anisotropic in shape, with greater height than diameter. The volume fraction of gas in the foam can be related to the average density of the foam. In closed

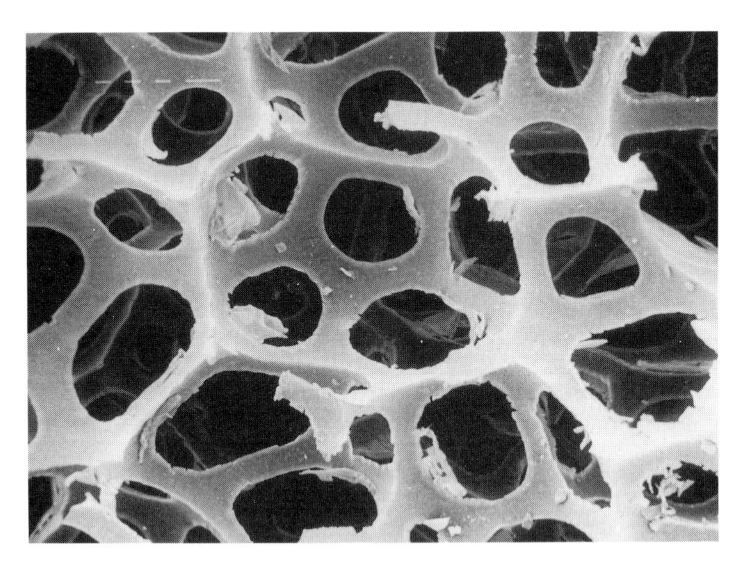

Fig. 3.11 Scanning electron micrographs of (a) a closed cell polystyrene foam and (b) an open cell polyurethane foam

cell foams it is useful to know the volume fractions of polymer in the cell faces, and in the edges. The thickness of the faces and the cross-section of the edges affect the mechanical properties of these elements. The cell shapes are irregular polyhedra, with the faces having on average ~5 edges. The regular dodecahedron has been used as an approximation for the cell shape. It has 12 faces and 30 edges meeting in 20 vertices. As each edge is shared by 3 cells there are on average 10 edges per cell. It is not possible to fill space with regular dodecahedra, so other less regular polyhedra must occur between the dodecahedra.

The geometrical parameters are interrelated for a family of polymer foams made by the same process. For polystyrene bead foam, where the density is determined by the initial mass of beads placed in the mould, the cells become larger and the cell walls thinner as the density of the foam decreases.

3.5.2 Model for the elastic modulus of an open cell foam

In open celled foams the loads are taken by the cell edges. In order to obtain a simple derivation of the relationship between the foam density, the polymer density and the foam elastic modulus it is necessary to use a gross simplification of the cell microstructure (Fig. 3.12a). Even if prediction of the model agrees with experimental data, this does not mean that it is the best model, as

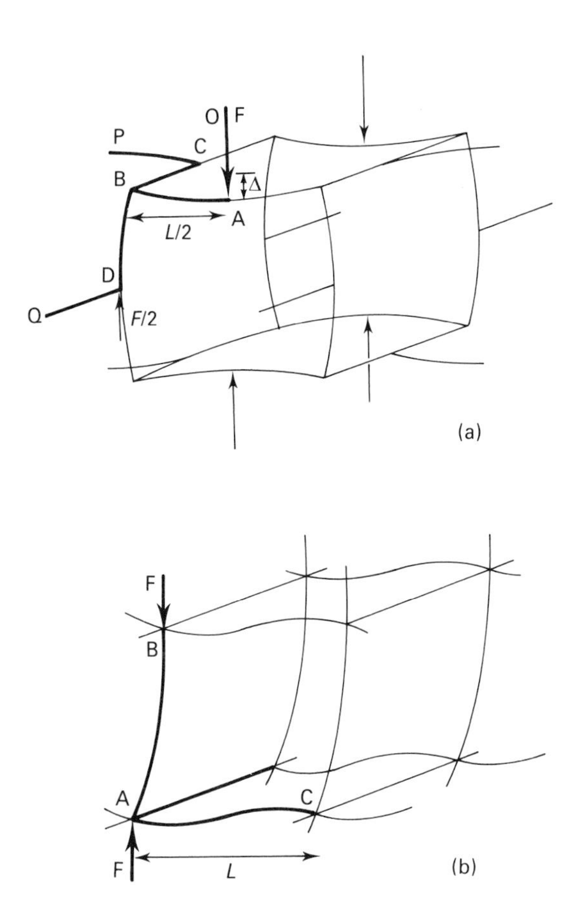

Fig. 3.12 Simplified open cell foam microstructures, with the repeat structure drawn in bold, used to predict (a) the elastic modulus—edges AB, BD and CP bend while BC twists, and (b) the collapse stress in compression—edges AB and AC buckle

other models may give the same result. The models are chosen to emphasise particular deformation mechanisms. The most important microstructural parameter is the *reduced density* of the foam R defined by

$$R \equiv \frac{\rho_F}{\rho_0} \tag{3.18}$$

where the densities are of the foam and the polymer respectively.

Figure 3.12a shows a highly simplified microstructure that is easy to analyse. There are cubic cells linked loosely by pairs of edges in the x, y and z directions. The set of 6 edges forms a motif that is repeated through the structure by mirror planes. The deformation is analysed when there is a compressive stress in the vertical z axis. The edges OA and DQ do not deform, while edges AB, BD and CP bend while BC twists. The edges are joined rigidly which means that the structure is *statically indeterminate*, with the moment at any point depending on the elastic stiffness of the edges. To keep the analysis simple the edge AB is treated as a cantilever of length $L/2$, built in at A and loaded at its free end B by a force $F/2$.

From beam theory the deflection Δ is given by

$$\Delta = \frac{FL^3}{48EI} \tag{3.19}$$

where the Young's modulus E is that of the polymer. The additional deflection due to the bending of BD will effectively double this. The second moment of area I for a strut of equiliateral triangular cross-section of side D is

$$I = \frac{D^4}{32\sqrt{3}} \tag{3.20}$$

As the cell size is uniform, the number of struts crossing unit cross sectional is L^{-2}, hence the compressive stress σ is FL^{-2}. The foam strain in the vertical direction is Δ/L, hence the foam Youngs modulus E_F is

$$E_F = \frac{F}{\Delta L} = \frac{\sqrt{3}}{4} E \left(\frac{D}{L}\right)^4 \tag{3.21}$$

The relative density of the foam R is equal to the volume fraction of edges (the average number of edges per cell is 3, and the volume of each edge is $0.25\sqrt{3}D^2L$), so

$$R = \frac{3\sqrt{3}}{4} \left(\frac{D}{L}\right)^2 \tag{3.22}$$

Substituting for the cell size parameters in equation (3.21) leads to the result

$$E_F = \frac{4}{9\sqrt{3}} ER^2 \tag{3.23}$$

Experimentally this relationship is obeyed for a number of open cell foams, but the numerical constant is ~1 rather than the 0.26 predicted. In a real open cell foam the cell edges are connected more than in Fig. 3.12a so the cells are more rigid. Hence better models of the cell geometry are required to predict realistic elastic moduli.

3.5.3 Model for the collapse stress of an open cell foam

The compressive stress strain curves of open celled polyurethane foams used for seating has a plateau region between 10% and 50% strain. The deformation mechanism is the *elastic buckling of struts*. In the simplified microstructural model there are a cubic array of struts, and the ones parallel to the stress buckle (Fig. 3.12b). This oversimplified model would predict an infinite Young's modulus for stresses acting along the cell edges. The critical compressive force F_c for elastic buckling of a strut of length L with one end free to move out of line and rotate (see Section 7.2.3) is

$$F_c = \left(\frac{\pi}{2L}\right)^2 EI \qquad (3.24)$$

The second moment of area of the strut cross-section is given by equation (3.20), and equation (3.22) for the relative density holds, so the following relation for the collapse stress σ_c is obtained

$$\sigma_c = \frac{\pi^2}{216\sqrt{3}} ER^2 \qquad (3.25)$$

The collapse stress of the foam should increase in proportion to the square of its density, and a high modulus foam will give a higher collapse stress at a given relative density. Experimentally the data for various flexible open celled foams follows equation (3.25) but the value of the constant is 0.05 rather than the 0.026 value predicted. Part of the difference can be accounted for by the model in Fig. 3.12b predicting an anisotropic strength. If a stress is applied at an angle of 45° to the axes the collapse stress will be less than if the stress is along the axes. A more realistic model would consider the co-ordinated buckling of a set of polyhedral cell edges.

3.5.4 Yielding in closed cell foams

The behaviour of closed cell foams at high compressive strains depends on the polymer used. It can vary from elastic if flexible polyurethanes are used, to strongly viscoelastic with high density polyethylene foams, through yielding with polystyrene foams, to brittle with methacrylate foams. The treatment of yielding in polystyrene foams will be left to the end of Chapter 7, when the analysis of yielding has been completed.

3.6 BLOCK COPOLYMERS

The polyurethane and polyurea systems are chosen to typify block copolymers because of their commercial importance in the reaction injection moulding process (see Section 4.7). In these materials the polymer chains contain both 'hard' and 'soft' segments. The hard segments are crystalline with melting points in excess of 150 °C, and the soft segments are crosslinked rubbers. Phase separation occurs rapidly as the materials are polymerised in the mould, and the systems are chosen to fit the stringent reaction rate requirements of the process. The variety of materials that can be made by varying the percentage of hard segments is similar to the variety of polyethylenes that can be made (see the next section). A typical polyurea hard block is a condensation polymer with the repeat structure

$$\left[\mathrm{C(=O)-NH-C_6H_4-CH_2-C_6H_4-NH-C(=O)-O-(CH_2)_4-O} \right]$$

It is prepared by reacting a diisocyanate (here 4,4′-diphenylmethane diisocyanate; MDI) with an aromatic diamine extender (here a mixture of the 2,4 and 2,6 diamine isomers of 3,5 diethyl toluene-DETDA). Typical polyurethanes use a diol extender such as 1,4-butanediol. The NH groups form hydrogen bonds with CO groups on neighbouring chains, and build up a crystalline structure (Fig. 3.13). The soft blocks are crosslinked polyethers or polyesters. A typical polyether structure is poly(propylene oxide)

$$\left[\mathrm{O-CH_2CH_2} \right]_n$$

which has been prepolymerised to a molecular mass M_N of about 6000, a degree of polymerisation $n = 140$. PPO has a glass transition temperature of about −60 °C, so the crosslinked PPO is an effective rubber. The medium molecular weight of the soft segment makes phase separation more marked, and thereby increases the modulus of the material at room temperature.

When the shear modulus versus temperature graph of RIM polyurethanes is examined (Fig. 3.14) there are seen to be two major transitions in modulus,

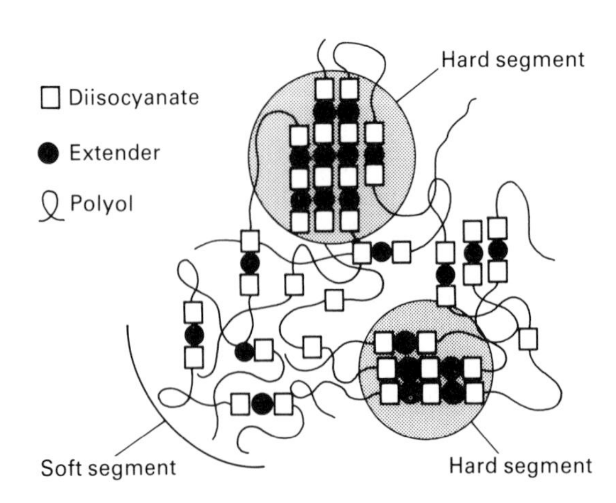

Fig. 3.13 Schematic drawing of the soft and hard segments in a polyurethane or polyurea block copolymer (from C. W. Macosko, *Fundamentals of Reaction Injection Moulding*, Hanser, 1989)

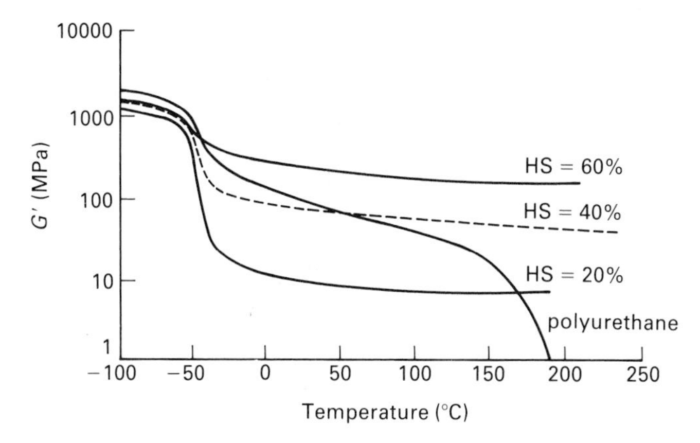

Fig. 3.14 Temperature variation of the shear modulus of a polyurethane, and of RIM polyureas (MDI–DETDA–PPO) with varying hard segment content (from C. W. Macosko, *Fundamentals of Reaction Injection Moulding*, Hanser, 1989)

which occur at the T_g of the soft segment and the T_m of the hard segment. This is the evidence for phase separation, with the polymer chain at room temperature threading alternately through soft rubbery regions and hard crystalline regions. In the graph for the polyureas the T_g of the PPO rubber is visible but the T_m of the hard block occurs above the 250 °C limit of the data.

The soft segments are typical of rubbers. For a network chain of $M_c = 3000$, the rubber elasticity theory of equation (2.20) predicts a shear modulus of about $0.8\ \text{MN m}^{-2}$. The hard blocks will have the typical $3\ \text{GN m}^{-2}$ Young's modulus of glassy polymers, and as Fig. 3.14 shows, the volume % of hard segment is capable of being varied. For automobile panel applications it is usual to have a high percentage of hard blocks so that the room temperature flexural modulus is $\sim 500\ \text{MN m}^{-2}$.

The block thicknesses are too small (<10 nm) for the microstructure to have been resolved—it is likely, however, that it bears a resemblance to the microstructure of other block copolymers such as styrene–butadiene–styrene where the block sizes are larger. Fig. 3.15 shows that in these the butadiene can be present as cylinders of diameter ~20 nm when the butadiene content is 40%, or as lamellae when it is 60% of the total. Neither the uniform strain model nor the uniform stress model seems appropriate for this type of microstructure. Consequently in polyurethanes the elastic moduli lie between the limits set by equations (3.12) and (3.11). Variation in the hard block content permits the Young's modulus to be varied in the range 30 to $500\ \text{MN m}^{-2}$ (Fig. 3.14).

A microstructural change occurs when polyurethanes are stretched ~150%. This elongation is sufficient to nearly straighten out the short polytetramethylene adipate soft segments, and they crystallise. This behaviour has a parallel with that of natural rubber which crystallises at high elongations. Both have crystal melting points in the region 25 °C to 60 °C, and in both the influence of crosslinking prevents crystals forming in the un-

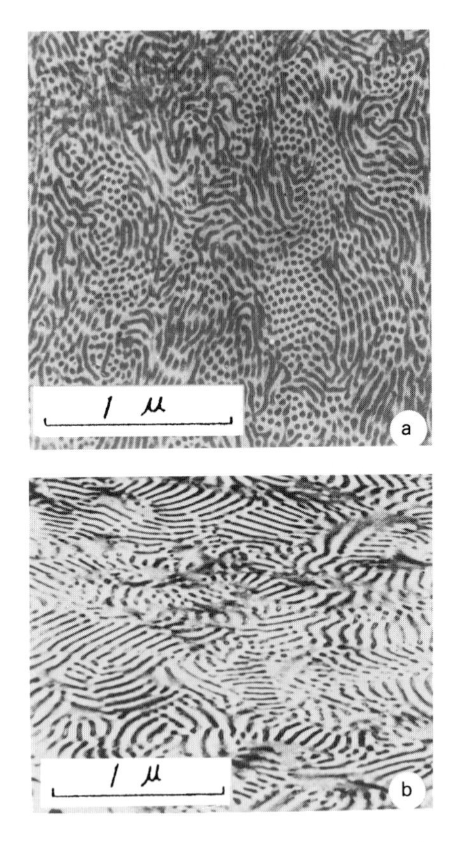

Fig. 3.15 Microstructure of styrene–butadiene–styrene block copolymers at (a) 40% butadiene content with spherical inclusions and (b) 60% butadiene content with lamellar inclusions (from J. A. Manson and L. H. Sperling, *Polymer Blends and Composites*, Plenum Press, 1976)

stretched state. This crystallisation on stretching plays a major part in the strength and abrasion resistance of polyurethanes.

3.7 THE MODULUS OF POLYETHYLENE SPHERULITES

The major phase in most spherulitic polymers is the crystalline one, and the crystals have highly anisotropic elastic moduli because the continuous covalent bonding is only along the **c** axis. The Young's modulus E_c in the **c** direction is proportional to the number of polymer chains per unit area. For the orthorhombic polyethylene unit cell, with $a = 0.49$ nm and $b = 0.74$ nm and two chains per unit cell, each chain has a cross sectional area of 0.18 nm^2. As the size of side groups increases the chain density decreases; the polypropylene chain has a cross sectional area of 0.35 nm^2. The stiffness of each chain depends on whether it has a planar or a helical conformation. Taking polyethylene as an example of the former, about 40% of the compliance is due to the stretching of the C—C bonds, and 60% from the opening of the C bond angle. If the chain is helical the high compliance of C—C bond rotation dominates the other deformation modes. Therefore polymer crystals can be divided into two categories: those without large side groups have planar chains and high E_c values in the range 250 to 350 GN m^{-2} (PE, PET), and those with large side groups have helical chains and low E_c values (50 GN m^{-2} for PP, 10 GN m^{-2} for isotactic PS). All these values are measured at 20 °C. The elastic moduli for stretching in the *a* or *b* directions, and for all the shear modes, are expected to be dominated by the high compliance of the van der Waals bonds, as discussed for glassy polymers. There will be anisotropy because the interchain distances are different in the two directions; thus $E_a = 8$ and $E_b = 5$ GN m^{-2} for polyethylene. If there is hydrogen bonding, as occurs in one direction in polyamides, this direction will have the higher Young's modulus.

The concepts of the micro-mechanics of composites have been used to provide a partial explanation of the elastic moduli of spherulitic semi-crystalline polymers. Fig. 2.20 shows a stack of parallel lamellar crystals with the interleaved amorphous layers. There is a close geometric similarity with Fig. 3.1 of a rubber/metal laminated spring. The crystals have different Young's moduli E_a, E_b and E_c as explained above, and different shear moduli when the shear stresses are in the ab, bc or ac planes. The amorphous layer is isotropic, with a shear modulus higher than that of a rubber because of the intercrystalline links.

If the shear stresses are applied in the ac or bc planes, the lamellae move parallel to each other in the manner of Fig. 3.1. There are uniform stress conditions and equation (3.3) can be used, so the shear compliances of the phases can be added in proportion to their volume fractions. If a tensile stress is applied in the 3 direction normal to the lamellar surfaces, the uniform stress conditions mean that the tensile compliances in the 3 direction can be added giving

$$\frac{1}{E_3} = \frac{V_c}{E_c} + \frac{V_a}{E_a} \tag{3.26}$$

where the Vs are the volume fractions of crystalline and amorphous material. The high shape factor of the amorphous layers means that the amorphous Young's modulus E_{am} will be close to the amorphous bulk modulus of 2 GN m^{-2}. The tensile compliance in the 3 direction will be dominated by the amorphous contribution because E_c of the polyethylene crystal is 250 GN m^{-2}. Hence

$$E_3 \cong \frac{E_{am}}{V_{am}}$$

Although E_3 is relatively high the shear moduli G_{31} and G_{32} are very small and this means that the interlamellar layers will shear if at all possible.

For stress components that act in the ab plane the amorphous and crystalline layers act as a composite laminate, with uniform strain conditions prevailing. Consequently the tensile moduli E_1 and E_2 or the shear moduli

G_{12} can be added in proportion to their volume fractions, using

$$E_1 = V_c E_a + V_{am} E_{am} \tag{3.27}$$

The crystal moduli are so high compared with the amorphous Young's modulus that the second term can be neglected, hence

$$E_1 \cong E_2 \cong V_c E_a$$

The elastic moduli of the stack of lamellae need to be averaged for all the orientations of the stack that occur in a spherulite. Neighbouring parts of the spherulite deform in different ways, and the spherulite boundaries are constrained to remain in contact with the neighbours. Fig. 3.16 shows that where 3 spherulites meet, the lamellar stacks have different orientations. For an applied tensile stress in the vertical direction the stack in the top spherulite is relatively stiff, whereas those in the lower spherulites shear easily and are soft.

Current models for the modulus of polyethylene ignore the interactions between spherulites, and between various parts of a spherulite. They provide upper and lower bounds for the modulus by averaging respectively the stiffnesses and the compliances of the lamellar stacks. Fig. 3.17 shows the bounds for the shear modulus as a function of crystallinity, at a temperature of −80 °C. The main weakness of such a model is that there are no experimental values of the amorphous moduli; they are chosen in Fig. 3.17 to suit the experimental data. Secondly polyethylene does not behave as an elastic material, so it is impossible to define a single modulus value. Chapter 6 shows that the creep modulus can change by a factor of three or more according to the duration of loading. Nevertheless the model provides closer bounds for the modulus than is possible when the microstructure is unknown, represented by the dashed curves in Fig. 3.17.

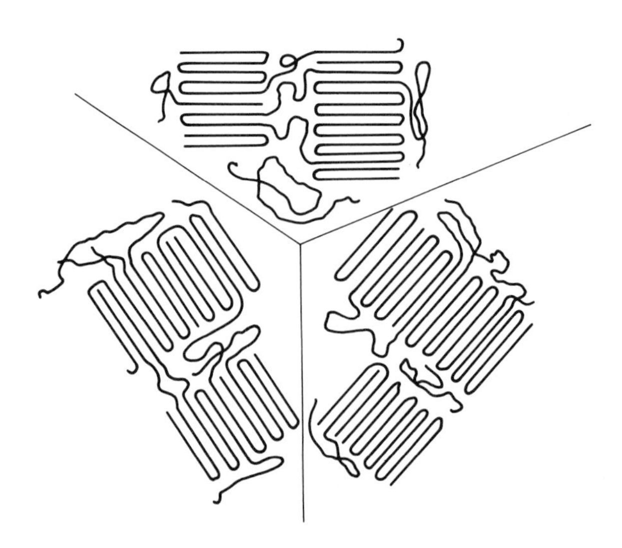

Fig. 3.16 Lamellar crystal stacks near the boundary between 3 spherulites

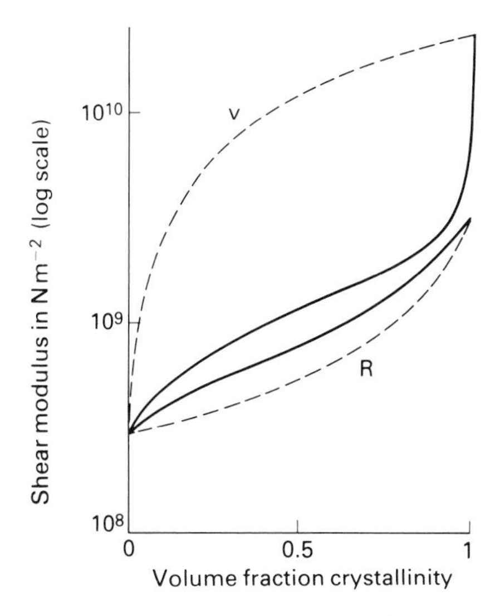

Fig. 3.17 Predicted upper and lower bounds for the Young's modulus of spherulitic polyethylene at −80 °C, versus the volume fraction of crystallinity. The dashed curves are the bounds calculated from equations (3.11) and (3.12) (from R. Boyd, *J. Polym. Sci. Polym. Phys.*, 1983, **21**, 500, Wiley)

3.8 GLASS FIBRE REINFORCEMENT

Glass fibres, with Young's moduli of 72 GN m^{-2}, are much stiffer than polymers, and with tensile strengths of 1 to 2 GN m^{-2} if undamaged are much stronger. The temperature resistance, with a T_g exceeding 500 °C, is much higher. However glass is a brittle elastic solid, and the fibres are easily damaged in plastics processing. The total volume fraction of glass has to be restricted to 0.2 or less to prevent the melt viscosity becoming excessive. Manufacturers quote the percentage *by weight* of glass fibres; as the glass has a density of about 2700 kg m^{-3} compared with a range of 900 to 1400 kg m^{-3} for the plastic matrix, 30% by weight of glass reduces to a volume fraction of about 0.15. The length of the glass fibre may start at 3 mm before it is incorporated into the plastic, but the effects of the high stresses in the shear flow in an extruder barrel rapidly comminutes the fibre. A typical fibre length distribution in a moulded part is shown in Fig. 3.18a. We will now investigate the effects of this relatively small content of short glass fibres on the mechanical properties of the polymer.

The microstructural characterisation of a glass reinforced plastic requires knowledge of the fibre volume fraction V_f, the length distribution and the orientation distribution, all of which can vary with position. The fibres are only about 10 μm in diameter, thus they can have aspect ratios of length to diameter of up to 100:1. One method of finding the fibre orientation is by using contact micro-radiography. Fig. 3.18b shows the X-ray 'shadows' of the fibres in an injection moulded bar. Two features are apparent: the orientation

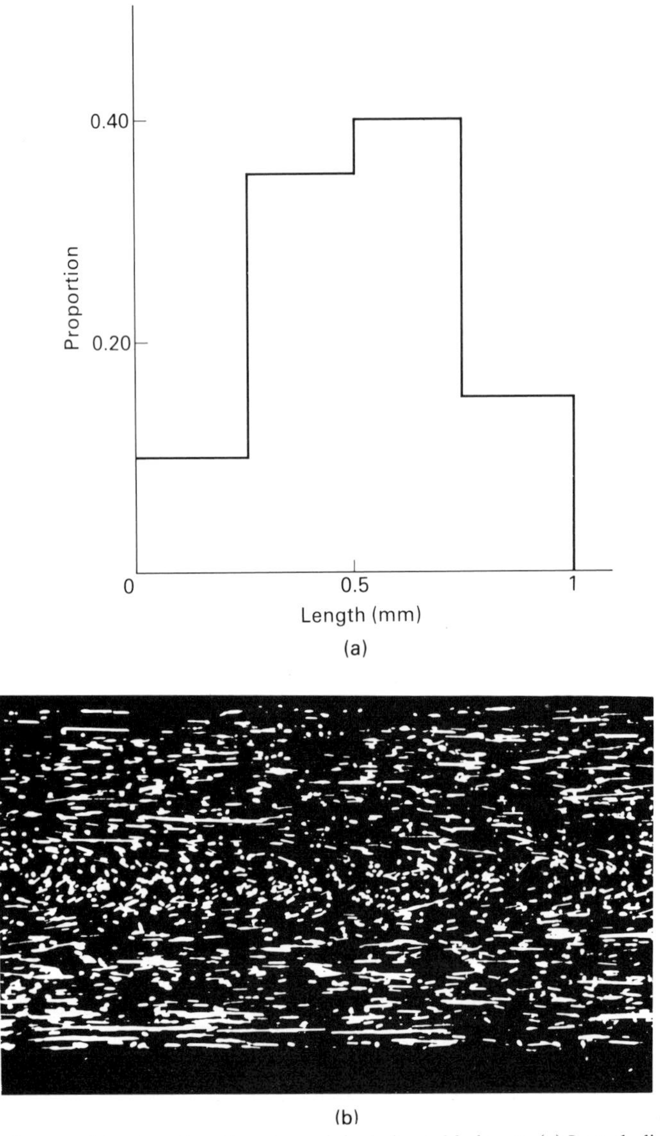

Fig. 3.18 Characterising the glass fibres in an injected moulded part. (a) Length distribution of extracted fibres. (b) Longitudinal section through a 20 weight % glass fibre reinforced polypropylene injection moulding (courtesy of M. J. Folkes, Brunel University)

varies from the skin to the core of the moulding, and the orientation is only partially towards the flow direction of the melt.

There is significant mechanical property anisotropy caused by the fibre orientation. Fig. 3.19 shows how the bending stiffness of an injection moulded plate varies with the direction of the axis of bending. The stiffness is highest when the tensile and compressive bending stresses are along the flow direction, and this value is about twice that across the flow direction. The

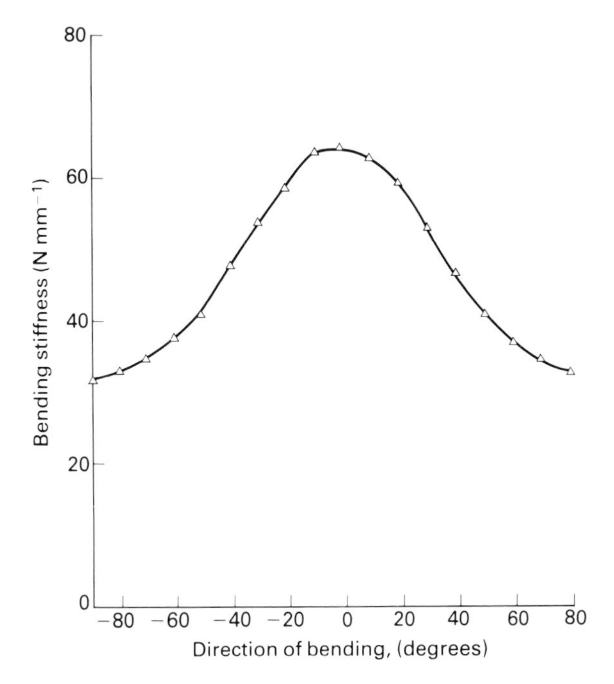

Fig. 3.19 Variation of the bending stiffness of a disc cut from an edge gated plate moulding. The polypropylene contains 25% by weight glass fibres, and in the 0° direction the bending stresses are along the melt flow direction(from Stephenson, *Plastics and Rubber: Materials and Applications*, 1980, **7**, Elsevier Applied Science)

anistropy makes the prediction of the stiffness of fibre reinforced mouldings difficult. There is a corresponding anistropy in the strength of the mouldings. We do not expect the stiffness at 90° to the length of the fibres to be as low as that predicted by equation (3.11) because there are stress concentrations around each fibre of the same kind as shown in Fig. 3.8a for rubber spheres. It is instructive to compare the predictions of equations (3.11) and (3.10) with the actual moduli of a short fibre reinforced thermoplastic. Glass fibre has $E_f = 70$ GN m^{-2} whereas polypropylene has $E_m = 1.52$ GN m^{-2} (100 s creep modulus). For $V_f = 0.111$ the equations predict $E_1 = 9.12$ GN m^{-2} and $E_{min} = 1.71$ GN m^{-2}. The actual modulus parallel to the flow direction was measured as 5.4 GN m^{-2} showing that the situation is neither uniform strain nor uniform stress.

The strength of short fibre composites is smaller than that of continuous fibre composites, because it is difficult to transfer a high stress to a short fibre. Fig. 3.20 shows a single fibre parallel to a tensile stress. The force transfer through the ends of the fibre is negligible compared with that via the shear stress τ_i at the cylindrical interface. A force balance calculation on a length dx of fibre gives

$$\sigma_f \pi r^2 = (\sigma_f + d\sigma_f)\pi r^2 + \tau_i 2\pi r \, dx$$

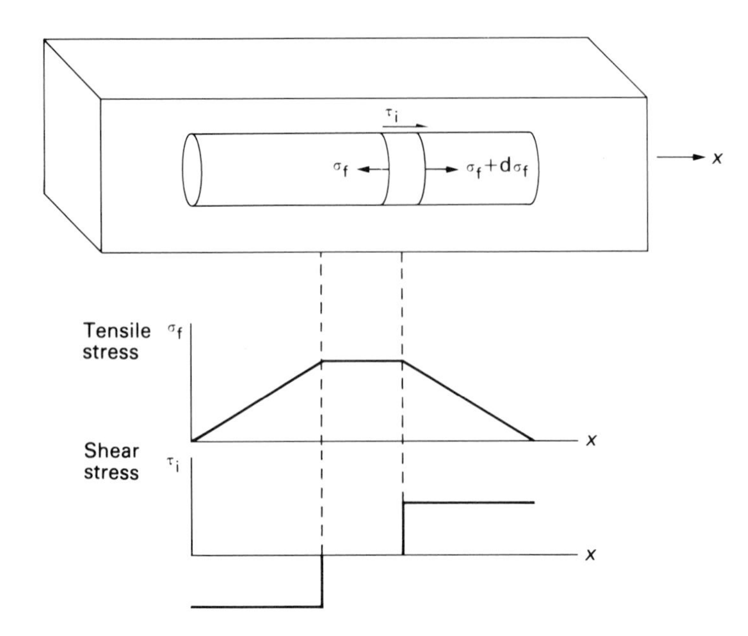

Fig. 3.20 Stress transfer to a single fibre in a thermoset matrix that is under tension in the x direction. The graphs below show the variation of the fibre tensile stress and the interfacial shear stress, when the interface yields at both ends of the fibre

hence

$$\frac{d\sigma_x}{dx} = -\frac{2\tau_i}{r} \tag{3.28}$$

Equation (3.28) can only be integrated when some assumption has been made about the behaviour of the matrix and interface. If the matrix remains elastic the shear stress rises to a maximum value at the ends of the fibre where the tensile strains in the fibre e_f and the matrix e_m differ most. However, it is more realistic for ductile matrices, such as polypropylene with a low shear yield stress $\tau_y < 20\ \text{MN m}^{-2}$, to assume that the matrix yields in shear near the ends

$$\begin{aligned} \tau_i &= \pm\tau_y \quad \text{when} \quad e_f < e_m \\ \tau_i &= 0 \quad \text{when} \quad e_f = e_m \end{aligned}$$

This leads to the simple stress variations shown in Fig. 3.20, with

$$\sigma_f = \frac{2\tau_y x}{r} \tag{3.29}$$

at the left hand end of the fibre. For the fibre to contribute its full tensile strength σ_f^* to the composite there are two conditions that have to be met. Firstly the average matrix tensile strain must exceed the tensile fracture strain of the fibre e_f^*; this usually is possible since thermoplastics have much higher failure strains than most fibres. Secondly the tensile stress in the fibre, given by equation (3.29), must build up to σ_f^*. As the maximum value of the build up length x is half the fibre length L, the condition is

$$\frac{2\tau_y L}{2r} > \sigma_f^*$$

or

$$\frac{L}{D} > \frac{\sigma_f^*}{2\tau_y} \qquad (3.30)$$

For glass fibres in polypropylene this condition means that for $\sigma_f^* = 1000\ \mathrm{MN\,m^{-2}}$ and $\tau_y = 20\ \mathrm{MN\,m^{-2}}$, $L > 25D$ or $L > 250\ \mu\mathrm{m}$ if $D = 10\ \mu\mathrm{m}$. Fibres of this minimum length will only ensure that the average tensile stress in the fibre is half of σ_f^*; in order to obtain a higher average stress in the fibres, and thereby the best stiffness and strength for a given glass content, the fibre length must be several times longer. Re-examination of Fig. 3.18a now reveals that the fibre lengths in a typical glass reinforced thermoplastic are not sufficient for the optimum reinforcement. This is why process development has aimed at increasing the length of the fibres to closer to 5 mm. It should not be forgotten that the fibre surfaces need special chemical treatment to achieve an adequate interfacial strength between the polymer and the fibre. This is especially important for non-polar polymers of low surface energies such as polyethylene and polypropylene. For these the glass fibres need to be treated with silane or other coupling agents to achieve optimum strength and toughness in the composite.

4

Processing

4.1 INTRODUCTION

The initial aim of describing polymer processes is to explain the shapes that can be made with a particular process, and to indicate the order of capital cost that is necessary. If a deeper understanding of the limits to the rate of production is required, then the heat flow processes must be analysed to see which limit the rate of melting and solidification, and the melt flow processes analysed to see if the flow rates are feasible. The knowledge gained by this analysis will be essential in understanding the effects of processing, which are dealt with in Chapter 5.

Comprehensive descriptions of all the methods of processing plastics are available (see *Further Reading*). Here the processes are selected and analysed, to bring out the main concepts of processing. Excepting the rare cases of solid state processing, the processing of thermoplastics can be split into the stages shown in Fig. 4.1. First they must be heated into the melt state, so the methods and rates of heat transfer should be studied. The characteristics of highly viscous yet strongly non-Newtonian melts are completely different from those of other liquids, so the analysis of the flow is different than for other materials. The low power consumption and the low temperatures involved should be compared with the processing of other materials. Finally the products can only solidify when the heat has been conducted away. This has repercussions in both the product design and the productivity of the process.

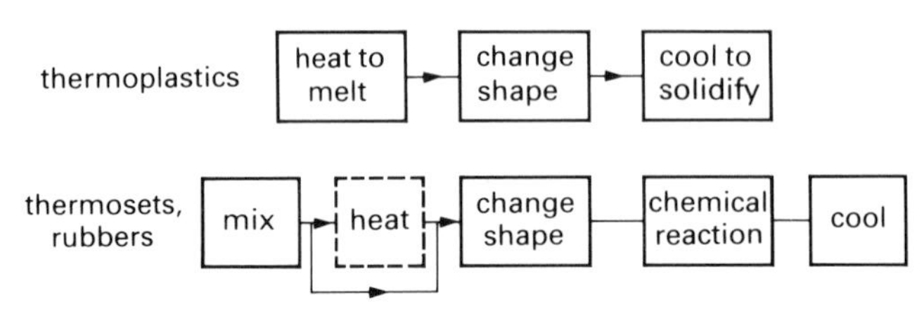

Fig. 4.1 Block diagrams of the stages in processing thermoplastics and thermosets of rubbers

The processing of plastics which undergo chemical crosslinking reaction during processing is rather different. The starting materials may be of low viscosity and therefore easy to mix and pump into the mould. Reaction injection moulding can offer a lower cost route to certain plastics products, and it will be examined at the end of the chapter.

4.2 HEAT TRANSFER MECHANISMS

The main heat transfer processes met in plastics processing are conduction, convection and viscous heating, with radiation playing a minor role. Most products are much thinner than they are wide so to simplify matters we shall only consider one-dimensional heat flow. The x axis is taken to be perpendicular to the surface of the product, so there will be planar isotherms perpendicular to the x axis.

4.2.1 Conduction

The thermal conductivity k is defined by *steady state conduction* in which the heat flux Q across an area A down a temperature gradient $\mathrm{d}T/\mathrm{d}x$ is

$$Q = -kA\frac{\mathrm{d}T}{\mathrm{d}x} \tag{4.1}$$

The thermal conductivity is of the order of 0.2 W m^{-1} K^{-1} for polymers. This low value compared with steel (50 W m^{-1} K^{-1}) is due to the lack of free conduction electrons, and the weak intermolecular bonding. Equation (4.1) implies that there is a linear temperature gradient through the plastic, but there are very few cases where this occurs in polymer processing.

Appendix A derives the differential equation for *transient conduction*

$$\frac{\mathrm{d}T}{\mathrm{d}t} = \alpha\frac{\mathrm{d}^2T}{\mathrm{d}x^2} \tag{4.2}$$

and shows how it can be solved in the one-dimensional case; α is the *thermal diffusivity*, which is the combination $k/\rho c_p$, where ρ is the density and c_p the specific heat. For most polymer melts α is approximately equal to 0.1 mm^2 s^{-1} (Fig. 4.2). Appendix A gives some solutions, when the surface of a sheet is held at a constant temperature, and the thermal diffusivity α is assumed to be independent of temperature.

One important case is when a thick sheet of polymer, initially at a uniform temperature T_0, is placed at time $t = 0$ in contact with a metal surface, which is kept at a constant temperature T_b. Equation (A.17) for impurity diffusion can be re-written for heat diffusion to give

$$T - T_b = (T_0 - T_b)\,\mathrm{erfc}\,(x/2\sqrt{\alpha t}) \tag{4.3}$$

where erfc is an error function. If we take some typical temperatures for the melting of low density polyethylene: a barrel temperature $T_b = 220$ °C, an initial polymer temperature $T_0 = 20$ °C and melting complete at $T = 120$ °C,

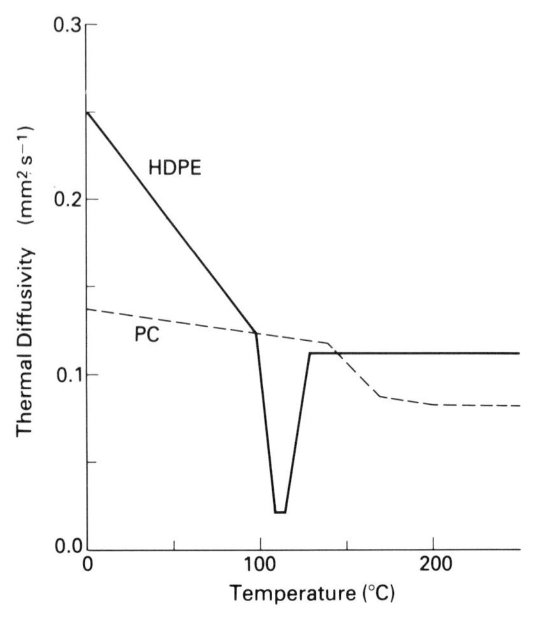

Fig. 4.2 Variation of the thermal diffusivity with temperature for amorphous polycarbonate (PC) and semi-crystalline polyethylene (HDPE)

then the melt front is at a position where $\mathrm{erfc}(x/2\sqrt{\alpha t}) = 0.5$, i.e. where

$$x = 0.94\sqrt{\alpha t} \tag{4.4}$$

Substituting the value $\alpha = 0.1$ shows that the thickness of melt x_m in mm is related to the contact time t in s by

$$x_m \cong 0.3\sqrt{t} \tag{4.5}$$

This result will only apply while x_m is much less than the thickness of the polymer product, but it gives a guide to the rate of melting both in an extruder and in hot-plate welding. It shows that conduction alone cannot be used to melt thick layers of plastic; it would take 100 s to melt a 3 mm surface layer.

For the cooling stages of processing the time for the product to completely solidify usually determines when the mould can be opened. The analysis is detailed in Appendix A where Fig. A4 shows the dimensionless temperature profiles that exist at different times since the start of cooling. Processes where the product is cooled from both sides (injection moulding) can be differentiated from those (blow moulding, and thermoforming) where one side of the product is effectively thermally insulated. When the dimensionless time, referred to as the Fourier number

$$F_0 \equiv \frac{\alpha t}{L^2} \tag{A.25}$$

is used to evaluate the cooling time t, the dimension L is the half thickness of an injection moulding, but the full thickness of a blow moulding. It is a reasonable approximation to say that the product is completely solidified when $F_0 = 0.3$, so the solidification time in the mould t_{solid} in s is given by

$$t_{solid} \cong 0.3 \frac{L^2}{\alpha} \cong 3L^2 \tag{4.6}$$

where L is measured in mm, using the approximation that $\alpha = 0.1$. Equation (4.6) shows the great necessity to keep the thickness $2L$ of injection mouldings to a minimum to a minimum if the productivity of the mould is to be high.

In reality α is not a constant quantity; Fig. 4.2 shows that for amorphous polycarbonate α increases on passing from the melt to the glassy state. For semi-crystalline polyethylene on cooling there is a very low value through the crystallisation temperature range, because the specific heat has a high value. Computer methods are needed to obtain solutions to transient conduction problems using realistic thermal data.

4.2.2 Convection

Convection cooling does not occur inside polymer melts, because the melts are far too viscous for convection currents to be established. The viscosity of water is at least a factor of 10^6 smaller, hence the small density decrease in warmer water creates a significant convection current. Convection cooling does occur at polymer surfaces surrounded by liquid or gaseous environments. The heat flow across the polymer/liquid interface is

$$Q = hA(T_s - T_0) \tag{4.7}$$

where h is the heat transfer coefficient, T_s the polymer surface temperature and T_0 the environmental temperature. The heat transfer coefficient depends to some extent on the size, shape and orientation of the object, but approximate values for different cooling media are given in Table 4.1.

Table 4.1 Heat transfer coefficients

Medium	Heat transfer coefficient (W m^{-2} K^{-1})
Still air	10
Air at velocity 5 ms^{-1}	50
Water at 5 °C	1000
Water spray	1500

We are familiar with the cooling effect of wind and rain on our own exposed skin: the 'wind chill factor' and the consequent risk of hypothermia. The data in Table 4.1 show how effective water sprays are in removing heat from the surface of extruded plastic pipes. In reality the heat transfer is limited by the conduction process within the plastic—see the next section.

4.2.3 Conduction and convection—Biot's modulus

When two heat transfer mechanisms operate at the same time, a *dimensionless group* can be used to indicate which mechanism is dominant. For instance for a slab of thickness L which is being cooled on one side by convection, and which has conduction in the interior, the Biot's modulus

$$B \equiv \frac{hL}{k} \tag{4.8}$$

indicates whether there is a significant temperature gradient inside the material during cooling. The group of variables used to define B has no dimensions, so it is a pure number, similar to the Fourier number in Section 4.2.1. If $B \gg 1$ then there is marked internal temperature gradient, if $B \ll 1$ then there is not. If any sheet of polymer of thickness greater than 1 mm is cooled by water or by contact with a steel mould then $B \gg 1$, and the total heat flow is limited by the internal conduction. When $B > 10$ it is a good approximation to say that the polymer surface temperature is immediately reduced to that of the cooling medium.

4.2.4 Radiation

The radiation heat flux from a black body of area A and absolute temperature T is

$$Q = A\sigma T^4 \tag{4.9}$$

where the constant $\sigma = 5.72 \times 10^{-8}\ \mathrm{W\,m^{-2}\,K^{-4}}$. In the thermoforming process the temperature of the radiant heaters is ~400 °C or 673 K. Use of equation (4.9) shows that the heat flux from such a heater is 12 kW $\mathrm{m^{-2}}$. The power spectrum of black body radiation shifts to shorter wavelengths as the temperature of the body increases. For a heater at 673 K the maximum of the spectrum is in the infrared region at wavelengths of 2 to 5 μm. All polymers strongly absorb in the infrared region so even those which are transparent in the visible region of the spectrum will absorb the radiant power in their surface layers. There will be small thermal losses as the plastic surface re-radiates, but the surface temperature rarely exceed 200 °C and the T^4 term in equation (4.9) means that the losses will be small.

4.2.5 Viscous heating

The flow of a viscous fluid generates heat throughout the fluid. This should not be confused with frictional heating, which occurs at the interface between two solids in relative motion. The power dissipated, in a small cube of melt in a shear flow, is the product of the shear force on the top and bottom surfaces and the velocity difference between these surfaces. When this quantity is divided by the volume of the cube, the power dissipated per unit volume W is found to be

$$W = \tau\dot{\gamma} \tag{4.10}$$

The greatest power dissipated in a pressure flow in a channel is at the channel walls because the shear strain rate is highest there. Hence the viscous heating will lead to temperature differences between the core and the surface of the melt. If we just want to estimate the average temperature rise when a volume V of melt falls through a pressure drop Δp in passing down a channel, we can assume adiabatic conditions, so that no heat is transferred to the channel walls and use

$$\Delta p = \rho c_p \Delta T \tag{4.11}$$

For a pressure drop $\Delta p = 500$ bar $= 50\,\mathrm{MN\,m^{-2}}$ passing into an injection mould, the temperature rise $\Delta T = 40\,°\mathrm{C}$.

4.3 MELT FLOW OF THERMOPLASTICS

Flows can be classified into *streamline*, which means that the pattern of the velocities in the fluid remains constant with time, and *turbulent*, which means that vortices cause unpredictable changes in the flow with time. The changeover occurs at a critical value of the Reynolds number which is defined as the melt velocity, divided by the viscosity times the channel diameter. The high viscosity of thermoplastic melts causes the velocities to be low. Hence the Reynolds number is very low and the flows are streamline. In contrast in the RIM process (Section 4.6) the flow of the low viscosity constituents in the mixing head uses turbulent flow to achieve intimate mixing.

4.3.1 Shear flows

In a steady *shear flow* the velocity along each streamline (the path of a particle) remains constant, and there is a velocity gradient at right angles to the streamline. If the x axis lies along the streamline and the y axis lies in the direction of the greatest velocity gradient, the shear strain rate is defined by

$$\dot{\gamma} \equiv \frac{\partial V_x}{\partial y} \tag{4.12}$$

where V_x is the velocity component along the streamline. The simplest constant shear rate flow can be produced by *drag flow* where one metal boundary surface moves parallel to the other at a velocity V (Fig. 4.3a). Since polymer melts adhere well to metals there is no slip at the metal/polymer interface, so at the interface the polymer and metal velocities are equal. In such a flow the velocity field is

$$V_x = \dot{\gamma} y \tag{4.13}$$

where the shear strain rate $\dot{\gamma}$ is constant. This flow field occurs whatever the flow law of the melt. Drag flow occurs in an extruder barrel, as a result of the rotation of the screw.

The other cause of shear flow is *pressure flow*, as a result of a pressure gradient in the melt. Fig. 4.3b shows that the pressure p falls down the

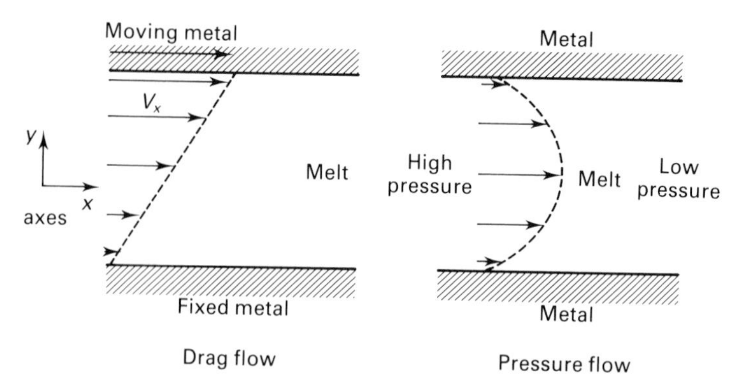

Fig. 4.3 The velocity field in shear flows caused by (a) drag flow and (b) pressure flow

streamline, and that the metal boundaries of the flow are fixed. Appendix B derives the relationship between the pressure gradient, the dimensions of the channel and the flow law of the fluid. For channels of rectangular, circular, or annular cross-section the shear stress τ varies linearly across the channel. We met in Chapter 2 the *Newtonian flow law*

$$\tau = \eta\dot{\gamma} \tag{4.14}$$

where the constant η is the viscosity. Polymer melts do not behave in this way except at very low shear rates, and a more realistic approximation to the shear flow law is the *power law fluid* for which

$$\tau = k\dot{\gamma}_a^n \tag{4.15}$$

where k and n are constants and γ_a is the apparent shear strain rate. Because of the wide use of the term viscosity it has been customary to define an apparent (shear) viscosity for polymer melts by

$$\eta_a \equiv \frac{\tau}{\dot{\gamma}_a} \tag{4.16}$$

even though the apparent viscosity will be a function of the shear strain rate. Fig. 4.3b shows that the velocity profile in a rectangular channel has its maximum at the mid-channel.

There is no detailed molecular explanation for the variation of the apparent viscosity with the shear strain rate. The qualitative explanation is that the high apparent viscosity of polymer melts is due to entanglements between the molecules. An increase in the polymer molecular weight increases the apparent viscosity, because each molecule is entangled with more of its neighbours. In commercial processes the high melt stresses cause the sections of molecules between entanglements become extended. Hence the average value **r** of the end to end vector becomes larger than the equilibrium value. In a shear flow this elongation of the molecules causes the apparent viscosity to fall; this may be due to a decrease in the number of entanglements, or a reduction in the time for which they act, or both.

4.3.2 Extensional flows

In a steady extensional flow the velocity increases along a streamline, but there is no velocity gradient in the direction at 90° to the streamline. Fig. 4.4 shows the simple example of fibre melt spinning where the velocity V_x increases as the result of a tensile stress σ_x along the fibre. The tensile strain rate $\dot{e}_x$ is defined by

$$\dot{e}_x \equiv \frac{\partial V_x}{\partial x} \tag{4.17}$$

and the tensile viscosity is defined by

$$\eta_T \equiv \frac{\sigma_x}{\dot{e}_x} \tag{4.18}$$

Because of the decrease of the fibre cross-section with the distance x from the die exit the tensile stress increases. This increase is exponential if the melt does not cool, and the tensile velocity η_T is independent of the strain rate. The velocity under these circumstances increases according to

$$\frac{V_x}{V_0} = \exp\left(\frac{\sigma_{x0} x}{\eta_T V_0}\right) \tag{4.19}$$

where σ_{x0} is the tensile stress and V_0 the average velocity, at the die exit. In the melt spinning of polymer fibres the temperature decrease of the melt causes the tensile viscosity to rise which limiting the increase in the fibre velocity. Finally the crystallisation of the polymer prevents any further flow.

4.3.3 Molecular weight influences on flow

Much effort has been expended in trying to explain melt rheology in terms of molecular processes. The physical chemists attempted this via the modes of vibration of short chain molecules (to explain the low molecular weight shear viscosity data in Fig. 2.10), applied mathematicians attempted to explain the non-Newtonian and elastic properties of melts in terms of the lifetime functions of temporary entanglements between molecules, and physicists

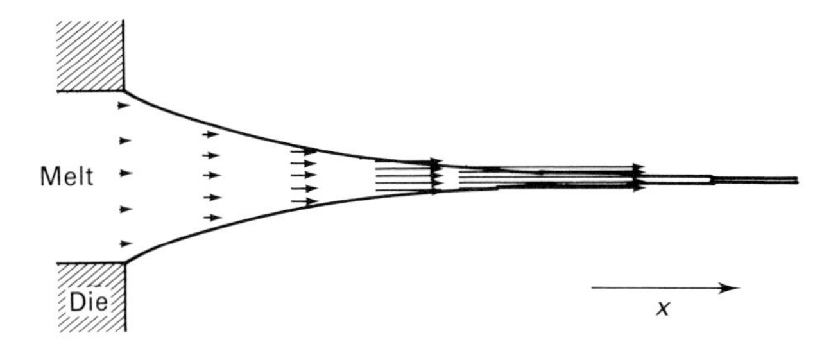

Fig. 4.4 The velocity field in a tensile extensional flow for fibre spinning

have recently used the worm-like motions of sections of polymer chains (reptation). None of these approaches has been completely successful. However there is a dramatic effect of the MWD on the melt flow properties.

Consider first the behaviour in shear flow of a monodisperse polyethylene of molecular weight M. Fig. 4.5 shows on logarithmic scales how the shear stress τ (or σ_{xy}) varies with the shear strain rate $\dot{\gamma}$. At low $\dot{\gamma}$ values the value of τ is proportional to $\dot{\gamma}$ and the elastic stresses are insignificant. In this Newtonian region the shape of polymer molecules is still the equilibrium random coil of Fig. 2.4.

The *elastic effects* in polymer melts are associated with the deformation of the shape of the molecular coils shown in Fig. 2.9. Such effects include the die swell which occurs when polymer melt exits from a die and increases in diameter, and various flow instabilities at high flow stresses, such as melt fracture. One measure of the elastic effects is the tensile stress difference $\sigma_{xx} - \sigma_{yy}$ that occurs in shear flow in the xy axes. It may appear as a tensile stress σ_{xx} in the direction of flow, or it may appear as a compressive stress σ_{yy} on the channel walls, or as a combination of the two. Fig. 4.5 shows that, as the shear rate increases, the value of $\sigma_{xx} - \sigma_{yy}$ increases with $\dot{\gamma}^2$ until it is of the same magnitude as τ. The sections of the molecules between entanglements have now become elongated by the elastic stresses, and the increase in shear stress is no longer proportional to $\dot{\gamma}$. Elastic deformation of the molecules is always associated with this non-Newtonian viscous behaviour.

No commercial plastics melts are monodisperse, and the effect of the MWD on the shear flow curve is important. Fig. 4.5 shows that the flow curve of the broad MWD polyethylene is much more non-Newtonian. It has the same M_W as the monodisperse polyethylene, so it has the same zero shear rate viscosity. The elastic stresses at very low shear rates are more influenced

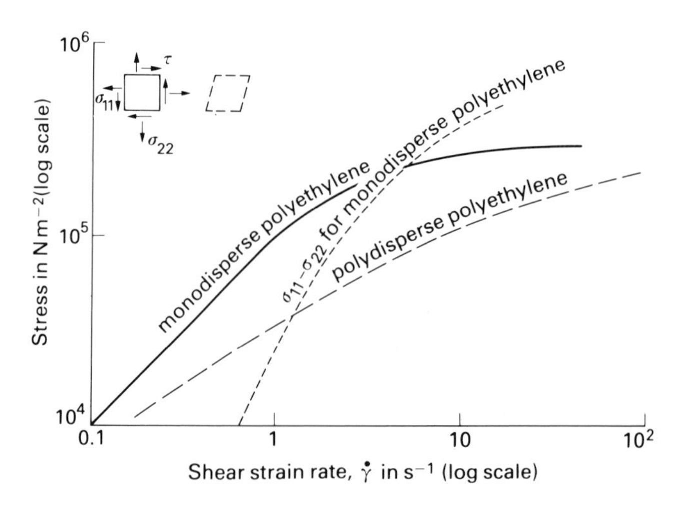

Fig. 4.5 Shear stress versus shear strain rate for a monodisperse polyethylene, and for a polydisperse polyethylene of the same M_W value. For the monodisperse polyethylene the shear stress curve becomes non-linear when the tensile stress difference is of the same order of magnitude

by the high molecular weight tail of the MWD than is the viscosity, so it is found that

$$\sigma_{xx} - \sigma_{yy} = B\dot{\gamma}^2(M_W M_Z)^{3.5} \tag{4.20}$$

where B is a constant. Therefore increasing the breadth of the MWD, in particular the parameter M_Z/M_W, increases the elasticity of the melt and thereby decreases the exponent n in the power law relationship between shear stress and shear strain rate. This applies particularly to the addition polymers such as PS, PE and PP, where the melt flow properties can be tailored for particular processes. This does not mean that all polymers can be processed by every technique; processes involving the shaping of molten polymer by air pressures require a combination of high tensile viscosity and thermal stability that certain polymers do not possess.

4.3.4 Interactions between heat flow and melt flow

The mixing flows in an extruder influence the heat transfer process, and conversely the flow of a viscous fluid generates heat according to equation (4.10). The resulting rise in the melt temperature reduces the apparent melt viscosity according to

$$\eta_a = A \exp(-B/T) \tag{4.21}$$

where A and B are constants and T is the absolute temperature. These interactive effects mean that any realistic calculations of melt flows and pressures must be computer based, with the temperature and viscosities of melt elements being updated at the end of every calculation step.

4.4 EXTRUSION

4.4.1 Melting and plasticisation

The extrusion process produces products that have a constant cross section, whether it is a pipe, a sheet of material, or a complex profile for a window frame. Fig. 4.6 shows part of an extrusion line for making pipe. We shall concentrate on the analysis of the output of the extruder screw and die, and the size of the cooling section. The continuous output from the extrusion line must either be coiled, if it is sufficiently flexible, or cut into lengths and stacked for distribution. All parts of the line must be carefully controlled to keep the product dimensions within the acceptable limits.

The extruder screw (Fig. 4.7) has 3 main sections. Solid granules fall under gravity into the *feed section.* In the newer machines the feed section of the barrel wall has longitudinal grooves in it to aid the positive forwards conveying of solid granules. The flight of the screw is usually at an angle $\theta = 18°$ to the direction of rotation, so that the pitch of the screw is equal to its diameter D.

The main function of the feed section is to melt the granules. This occurs by two mechanisms, conduction from the electrically heated barrel, and by

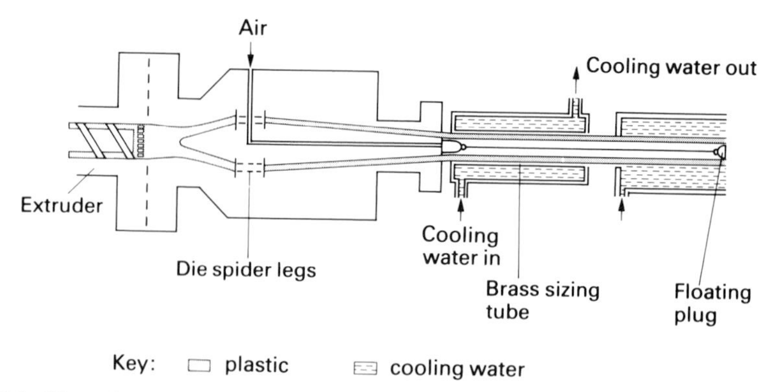

Fig. 4.6 Die and cooling sections in an extrusion line for the manufacture of plastic pipe

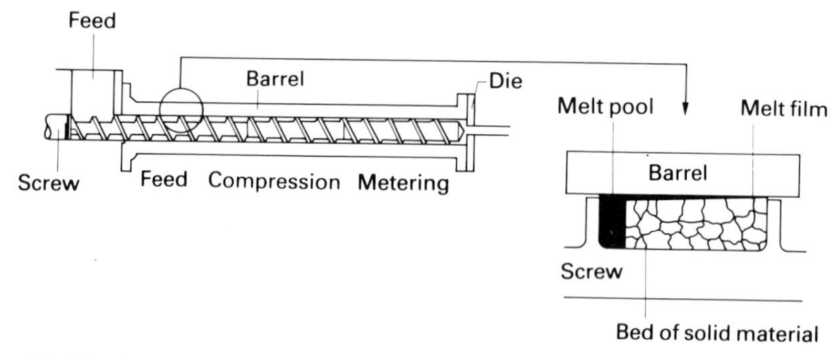

Fig. 4.7 The three sections of an extruder screw, with a detail showing the melting mechanism in the feed section

viscous heating from the mechanical work input of the screw rotation. The inset in Fig. 4.7 shows that a thin film of melt develops in contact with the barrel. Since the screw rotates at about 60 r.p.m. this layer of melt is scraped from the barrel wall every second by the relative motion of the screw flight, to form a melt pool on the forward face of the land. As the polymer progresses down the screw channel the width of the melt pool increases, and the average temperature of the polymer increases. The thickness of the melt layer can be calculated by using equation (4.5). A layer 0.3 mm thick of melt will develop in the 1 s between passes of the screw land. This will happen for each revolution of the screw so after 10 revolutions the equivalent of a 3 mm layer of melt has accumulated in the melt pool. This contrasts with the rate of melting should the screw be stationary, when it would take 100 s for a similar amount of melt to be plasticised. Once the polymer is molten the viscous dissipation mechanism, described by equation (4.10), comes into play. Reference to typical flow curves for polyethylenes shows that the shear stress τ is of the order of 10^5 N m^{-2} when the shear strain rate $\dot{\gamma}$ is 100 s^{-1}, so the power input is of the order of 10^7 W m^{-3}. This power is only dissipated in the molten layer so its thickness will increase rather more rapidly than the rate calculated for conduction heating alone.

The simplest form of central section of an extruder is a *compression section* where the channel depth decreases. Screws are designed with different numbers of turns and different compression ratios (the ratio of the channel depth in the feed section to that in the metering section) to suit the rheology of polymer being extruded. The pressure generated will squeeze out any gas bubbles, or cause the gas to dissolve in the melt. Unless a foamed extrusion is being manufactured, bubbles must not be allowed to reform in the melt when it returns to atmospheric pressure after the die. Consequently there may be a vent to the atmosphere or to a vacuum line just before the compression section to aid degassing.

The process of melting is not necessarily stable, as breaks can occur in the continuity of the solid bed. This will cause pressure fluctuations at the die, and hence fluctuations in the volume output rate which will cause the pipe wall thickness to vary. These fluctuations become more probable as the screw speed is increased, and problems will arise if the screw speed is so fast that the residence time of the polymer in the extruder is insufficient for complete melting. There are various ways of overcoming these problems. One is to use a screw design in which a barrier is placed in the compression section; this barrier leaves a gap through which melt, but not granules, can pass and the final metering section smooths out any pressure variations. Another is the use of a pack of wire-mesh screens, supported on a perforated steel 'breaker plate' between the screw and the die. As well as filtering out any large foreign particles, the screens control the pressure build-up in the extruder.

4.4.2 Output of the extruder

The *metering section* of the screw should contain nearly 100% melt at a nearly constant temperature. It controls the output from the whole extruder. Only the velocity components parallel to the flight (Fig. 4.8a) contribute to the output; the other two velocity components are part of a circulatory flow that mixes the polymer. The ouput is a combination of a drag flow and a pressure flow. The drag flow is due to the motion of the screw surface with circumferential velocity.

$$V = \pi D N \tag{4.22}$$

where N is the rotation speed (rev s^{-1}). If we ignore the effects of the curvature of the channel, and of the edges of the channel (as the breadth b is much greater than the height h) the drag flow velocity components, relative to the screw surface, are shown in Fig. 4.8b. V_y increases linearly from zero at the screw surface to $V \cos \theta$ at the barrel surface. Consequently the average value of V_y is $\frac{1}{2} V \cos \theta$, and the drag flow output is

$$Q_{drag} = \tfrac{1}{2} V \cos \theta \, bh \tag{4.23}$$

The shear strain rate in the metering section is relatively low; for an extruder of diameter $D = 50$ mm and a channel depth of $h = 2$ mm rotating at 1.5 rev s^{-1} it is

$$\dot{\gamma} = \frac{\pi d N}{h} = \frac{\pi \times 50 \times 1.5}{2} = 118\ \mathrm{s}^{-1}$$

This is an order of magnitude smaller than the shear strain rates in injection moulding; consequently higher molecular weight polymers can be processed by extrusion.

The pressure flow output depends on the pressure p at the end of the metering section. The pressure is assumed to be zero at the start of the metering section, which has a channel length L. The analysis of pressure flow in a rectangular slot, in Appendix B, shows that the shear stress varies linearly across the channel, with a maximum value at the walls (equation B.5) of

$$\tau_w = \pm \frac{h}{2} \frac{p}{L} \tag{4.24}$$

It would appear that we should use the non-Newtonian shear flow curve to calculate the strain rates. However, the pressure flow is much smaller than the drag flow, and its shear rates are superimposed on the higher shear rate $V \cos \theta / h$ in the drag flow. Therefore for the pressure flow we can assume that the melt is approximately Newtonian, with a viscosity equal to the apparent viscosity η_d in the drag flow, and put

$$\tau_w = \eta_d \dot{\gamma}_w \tag{4.25}$$

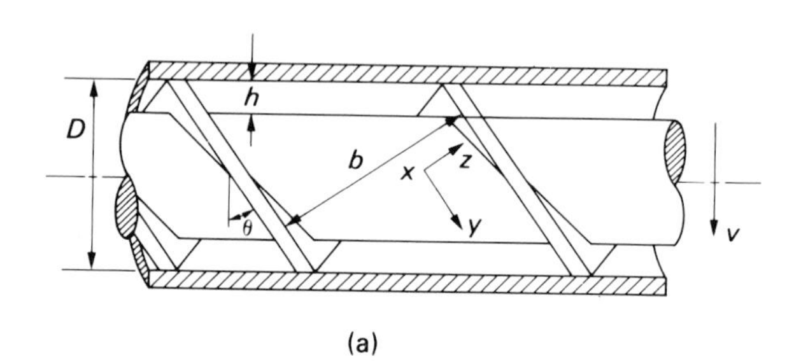

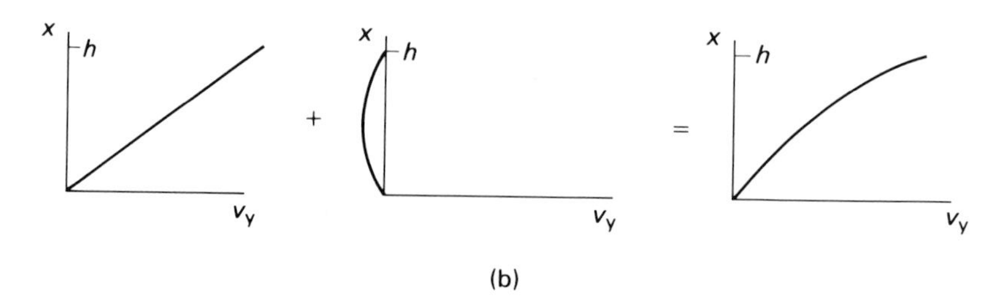

Fig. 4.8 (a) Geometry of the flow in the metering section of an extruder screw. The axes are taken relative to the moving screw surface. (b) The velocity components V_y due to drag flow, pressure flow and to a combination of the two

where $\dot{\gamma}_w$ is the shear rate at the wall. We then use the Newtonian version of equation (B.9) to calculate the pressure flow output rate

$$Q_{\text{press}} = -\frac{b\dot{\gamma}_w h^2}{6} = -\frac{bh^3 p}{12\eta L} \quad (4.26)$$

When the drag and pressure flow are added the total output is given by

$$Q_{\text{total}} = \tfrac{1}{2} bhV \cos\theta - \frac{bh^3 p}{12\eta_a L} \quad (4.27)$$

This equation cannot be solved until the flow equation for the die is known. If this is a pipe die (Fig. 4.6) consisting of a channel of circumference b, height h and length L (or a set of such channels in series) its flow equation from equations (B.5) and (B.9) is

$$Q = \frac{bh^{2+1/n}}{2(2+1/n)} \left(\frac{p}{2kL}\right)^{1/n} \quad (4.28)$$

Equation (4.27) for the metering section and (4.28) for the die can be solved graphically as shown in Fig. 4.9; the solution lies at the intersection of the two lines and is known as the *operating point.* The performance of a real extruder at different screw speeds (varying V) and with different dies (varying the pressure flow component) can be used to construct the screw and the die characteristics, and confirm the above analysis.

Extruder/die combinations are designed so that the pressure flow is less than 10% of the drag flow, so the latter is a reasonable estimate of the total output. They are run as fast as possible consistent with the melting process being complete, and the output being stable.

4.4.3 Solidification of the extrudate

When the melt emerges from the extruder die its shape must be fixed within a

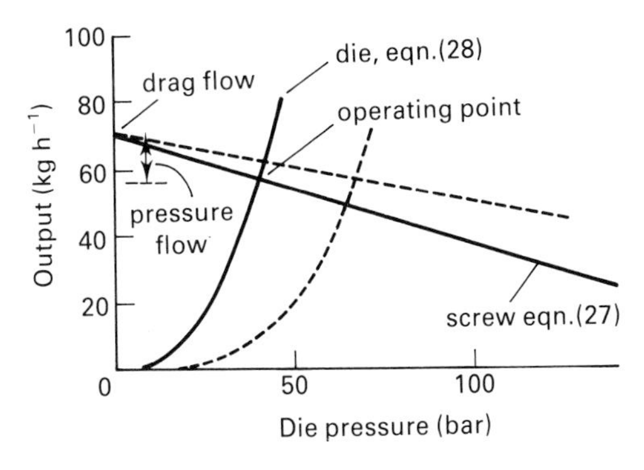

Fig. 4.9 Extruder operating diagram of pressure at the end of the screw versus output rate. The solid lines are for a lower viscosity melt than for the dashed lines

short distance. It is pulled forward by the haul-off mechanism acting on the solid extrudate, and either air pressure or vacuum is used to bring the melt surface into contact with a cooled metal calibrating section. Once the outer skin of the extrudate has solidified the process can be completed by cold water baths or sprays. The cooling section is relatively long because conduction is the only mechanism of removing heat from the extrudate; calculation of the length of this section is an important part of the design of the plant.

Fig. 4.10 shows another type of cooling for extruded sheet; one sheet surface is in contact with a cooled metal roll for a distance $\pi D/2$, then the other surface is cooled. It is possible to use an analytical solution (see Fig. A4 in Appendix A) for the cooling of a sheet of thickness L on the first roll; the roll surface temperature T_0 is constant, and the sheet surface in contact with the air is effectively insulated.

The temperature distribution after the cooling of the second surface on the lower roll can only be calculated numerically. Cooling of one then the other surface for a certain total time is more efficient than cooling on one side only for the same time.

4.5 PROCESSES INVOLVING MELT INFLATION

There are a number of processes in which a film or tube of polymer melt is changed in shape by the application of a gas pressure difference across the

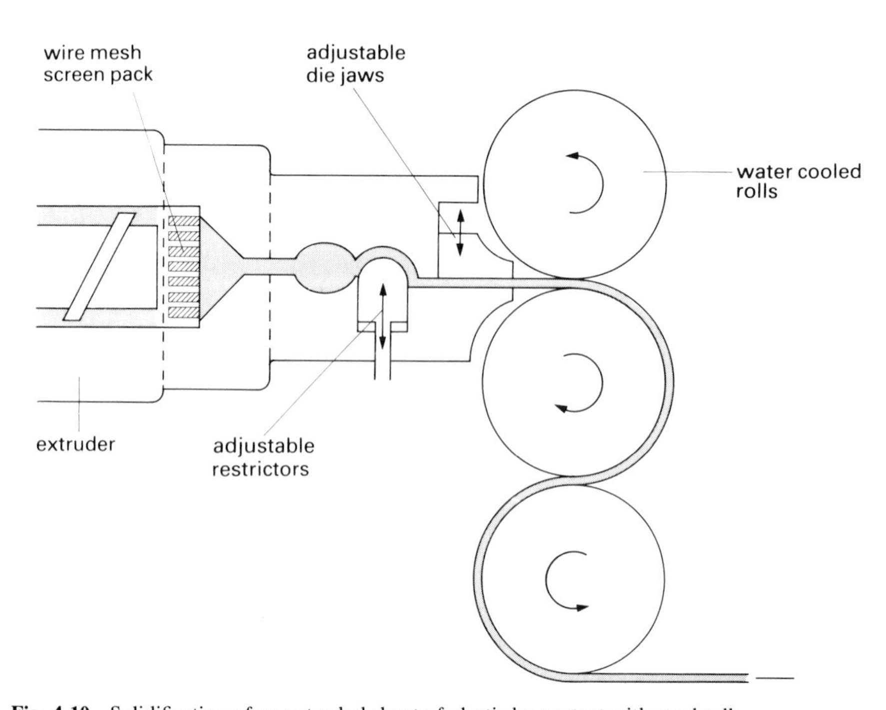

Fig. 4.10 Solidification of an extruded sheet of plastic by contact with steel rolls

melt. We shall examine one continuous process—blown film production—and two cyclic processes—blow moulding and thermoforming. They all involve some extensional flow of the melt with the resulting thinning of polymer, and at least one side of the polymer solidifies without the constraint of contact with a mould. In the cyclic processes the melt bubble expands with time, so the tensile strain rate can no longer be calculated by equation (4.17). Instead it is calculated from the instantaneous extension rate of a gauge length in the melt, as for a tensile test on a solid.

4.5.1 The blown film process

In blown film production an annulus of melt rises vertically from a die attached to the end of an extruder (Fig. 4.11). It is stretched vertically and circumferentially by a factor of 2 or more, so that an initial melt thickness of about 1 mm is reduced to between 250 and 100 μm. The majority of the output is polyethylene film. Fig. 4.12 shows the temperature profile of a LDPE film which has a blow-up ratio (= final bubble diameter/die diameter) of 3.5. The melt temperature falls nearly linearly with height until crystallisation occurs at about 120 °C. The cooling is provided by an annular air jet which blows upwards on the outside of the bubble with an initial velocity of

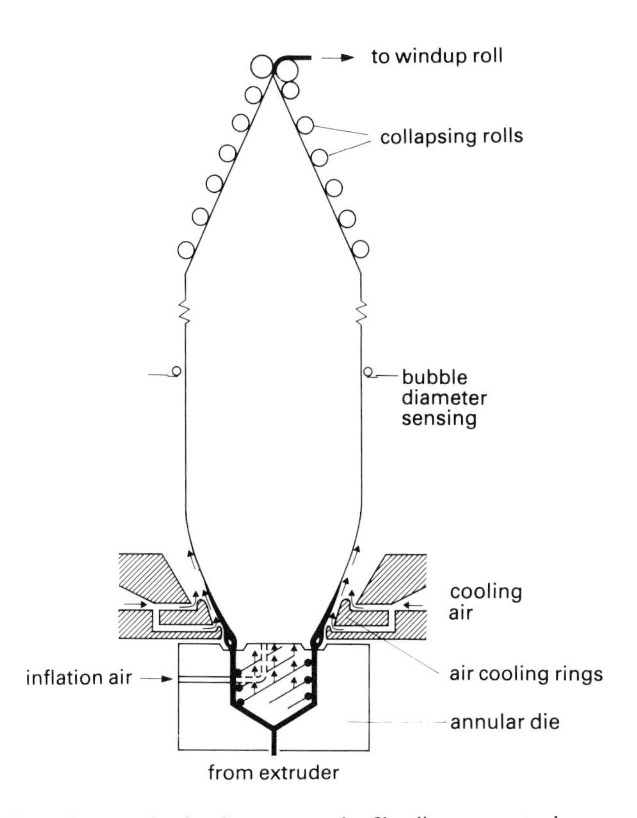

Fig. 4.11 Blown sheet production from an annular film die on an extruder

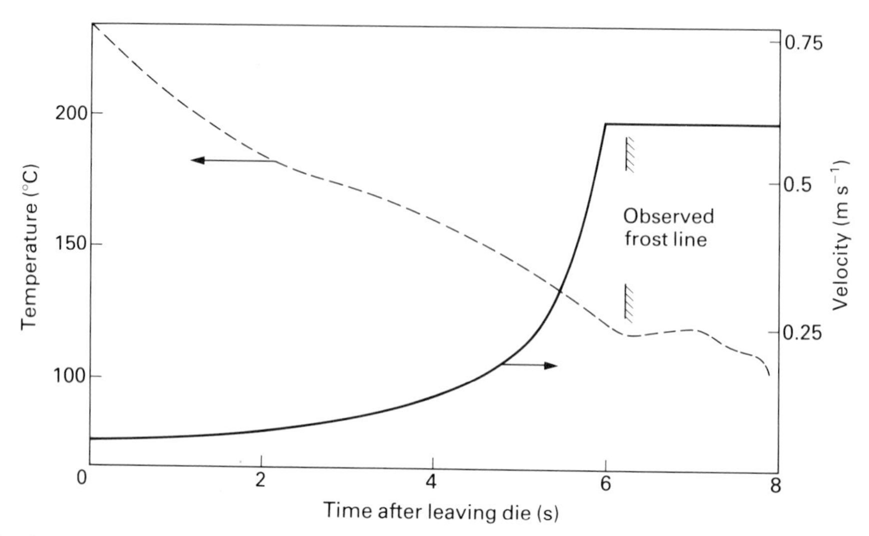

Fig. 4.12 The temperature and the velocity of a LDPE film versus the time after leaving the die (from Dowd, *Soc. Plast. Eng. J.*, 1972, **28**, 22)

about 1 m s^{-1}. The heat transfer coefficient becomes smaller as the air velocity falls but the reduction in thickness means that the surface area to volume ratio increases. There is a constant height 'frost line' where crystallisation occurs, and the collapsing rolls which flatten the bubble must be above the frost line. Output has been increased by increasing the efficiency of cooling, either by using a second ring of external air cooling above the first (Fig. 4.11) and/or by using internal cooling. This latter involves blowing cold air up through the die, and removing hot air, without changing the volume of air inside the bubble. The process has the advantage that the film thickness and width (when laid flat) can be varied through a considerable range by adjusting the bubble diameter and the wind up speed.

Blown film production is an example of biaxial tensile flow. The melt stresses in the longitudinal (L) and hoop (H) directions can be calculated from the pressure p inside the bubble, the current bubble radius r and thickness t, and the tensile force F from the wind up as

$$\sigma_L = \frac{pr}{2t} + \frac{F}{2\pi rt} \qquad \sigma_H = \frac{pr}{t} \tag{4.29}$$

The gas pressure is very low, less than 0.05 bar, but the very high r/t value at the top of the bubble means that the melt crystallises under a significant tensile stress. Fig. 4.12 shows that the melt accelerates in the longitudinal direction until just before it crystallises. It is almost impossible to find data for this type of unsteady biaxial tensile flow except by instrumenting a blown film machine. The tensile viscosity, defined by equation (4.18), hardly changes with the tensile strain rate. Fig. 4.13 shows some data for the uniaxial stretching of an LDPE and an HDPE. The apparent viscosity even increases

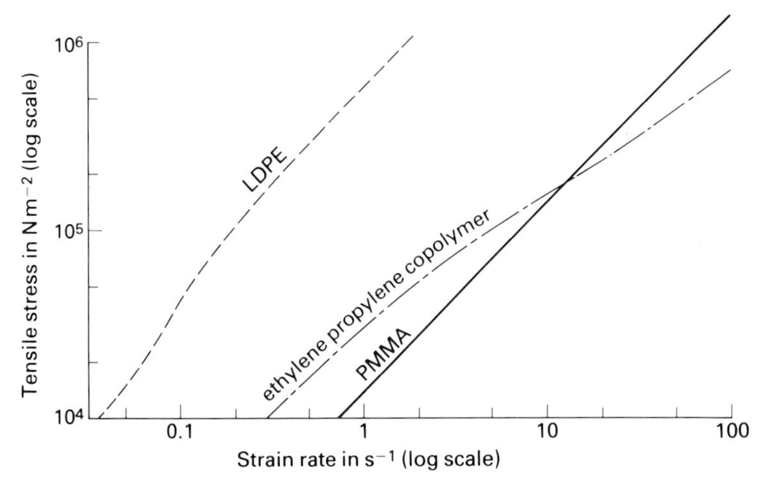

Fig. 4.13 Tensile stress versus tensile strain rate (log scales) for the uniaxial extensional flow of LDPE, ethylene–propylene copolymer and PMMA

with strain rate for the more elastic LDPE. This is in complete contrast with the marked non-Newtonian behaviour in shear flows.

The tube of melt, during its passage through the die, must pass over two or more spider legs that support the core of the die against high melt pressures. The melt streams weld together above each spider leg, and these weld lines could be regions of weakness in the melt bubble. To overcome this dies incorporate one or more spiral mandrels, in which the main melt supply spirals upwards, gradually leaking the flow into a vertical motion (Fig. 4.11).

4.5.2 Blow moulding

In *extrusion blow moulding* machines the heavy extrudate hangs under gravity from the die (Fig. 4.14). The products are hollow containers for liquids, and air ducting, or any other product that approximates in shape to a long hollow tube, e.g. a wind surfing board or canoe. For the larger products the melt generated by an extruder is held temporarily in an accumulator chamber (preferably annular in shape) before being extruded rapidly by a piston. The parison of melt must emerge from the die without any surface roughness due to over-rapid flow and yet be able to support its own weight without sagging for the few seconds before the two halves of the mould close.

When the melt is emerging from a die gap of width h at an extrudate velocity V the shear strain rate at the die wall is (see appendix B)

$$\dot{\gamma}_{aw} = \frac{6V}{h} \tag{4.30}$$

This result for a Newtonian fluid provides an approximation for the non-Newtonian polymer melts; typical shear strain rates are $300\,s^{-1}$. To avoid surface roughness with high molecular weight polyethylenes (see Fig. 5.14a)

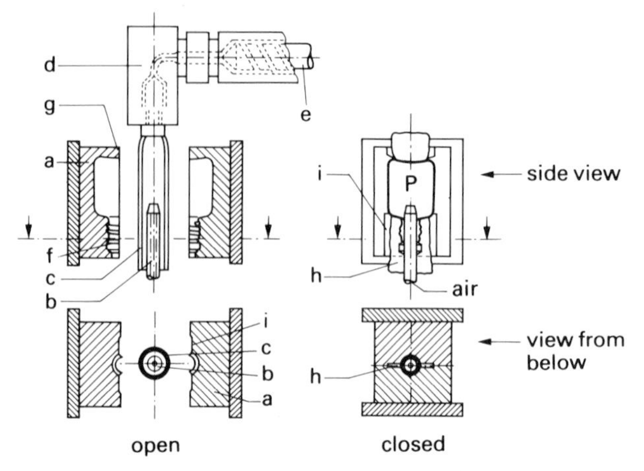

Fig. 4.14 Production of a parison of melt for the blow moulding of a container. (a) Mould; (b) calibrating mandrel; (c) parison; (d) parison head; (e) extruder; (f) neck insert; (g) pinch-off blade; (h) pinch-off; (i) clearance for waste (from Holtzmann, *Industrial and Production Eng.*, 1980, **4**, 36)

the shear stress under these conditions must be less than 100 kN m^{-2}. The hanging extrudate must not stretch excessively under its own weight while it is being extruded. The tensile stress at the top of the extrudate is

$$\sigma = \rho g L \tag{4.31}$$

For a length $L = 0.2$ m, a density $\rho = 750$ kg m^{-3} and $g = 9.8$ m s^{-2} the stress $\sigma = 1.5$ kN m^{-2}. To avoid significant stretching the tensile strain rate must be less than 0.2 s^{-1}, which means that the tensile viscosity must exceed 7500 N s m^{-2}. When these two conditions are imposed on a typical shear flow curve in Fig. 4.15 it is clear that the melt must be highly non-Newtonian. This explains why a similar bottle making process cannot be used for Newtonian silicate glasses. The molecular weight must be $M_W > 150\,000$ for polyethylene to provide suitable parison stability.

When the mould halves close, a weld is made at one end of the container by pressing the two layers of melt together. Inflation of the parison from the other end causes rapid extensional flow until the melt contacts the cooled aluminium mould. A uniform thickness parison produces a bottle of non-uniform wall thickness; this is wasteful of polymer and may cause weakness at highly stressed corners. Consequently the blow moulding machines have been developed that allow the parison thickness to be programmed. A hydraulic actuator moves the conical die interior vertically to restrict or open up the die width. The parison thickness is controlled in 20 or more steps and the resulting container has a nearly uniform wall thickness.

The moulds need only resist an air pressure of 5 bar, compared with melt pressures of several hundred bar in injection moulding. Consequently aluminium moulds are adequate, and the high thermal conductivity of this metal aids the cooling process. The conduction cooling from one side only is

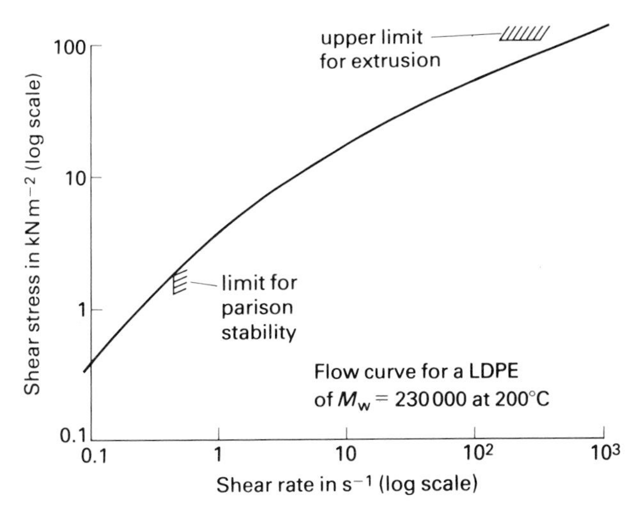

Fig. 4.15 Shear stress versus shear strain rate for a polyethylene used for blow moulding. The limits shown are for the stability of the hanging parison, and for the parison to have a smooth surface

relatively slow for large containers; cooling from both sides would be 4 times as fast. Interior cooling by injecting liquid CO_2 has been tried to increase the productivity of the machines.

The *stretch blow moulding* process involves the injection moulding of a preform, and the subsequent stretching of the preform. There are no weld lines in the injection moulded preform (Fig. 4.16) which is gated at the base. This means that in the stretching operation higher pressures can be used without the risk of the parison of melt splitting at the weld line. There are two versions of the process: either an injection moulding is removed from the mould at about 100 °C and almost immediately blown in a second mould, or the moulding is cooled into the glassy state and stored prior to stretching process. The latter will be described as it is used for high output rates. The crystallisation kinetics of PET suit the process (Fig. 2.23). A glassy preform can be injection moulded with a wall thickness of up to 4 mm; if the mould is kept at 10 °C the inner layers of the polymer cool fast enough for the crystallinity to be negligible. High molecular weight PET, with $M_N \cong 24\,000$, is used so that the rate of crystallisation is suitably low.

When the preform is heated to 100 °C it is in the rubbery state. The preform is stretched in length by a rod inserted through the neck, and inflated by an air supply at 25 bar to the dimensions of the mould. The stretched rubbery PET crystallises when the extension ratio exceeds 2 (Fig. 7.18) so it crystallises before the expanding preform hits the wall of the mould. This stabilises the shape of the expanding preform, so that the diameter expansion spread along the preform (rather like blowing up a cylindrical rubber balloon). Fig. 4.16 shows that the neck is not expanded at all, and there is a lower degree of expansion in the base. This can be revealed by placing a PET bottle in an oven at 120 °C for about 10 min. The neck and base, which were glassy, will

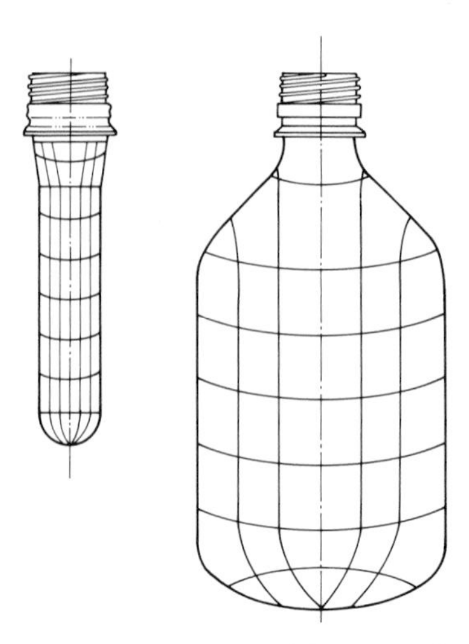

Fig. 4.16 The preform and final bottle in the stretch blow moulding of PET. The grid shown on the preform is measured on the blown bottle to find the extension ratios (from T. S. Chung, *Polym. Plast. Technol. Eng.*, 1983, **20**, 147)

crystallise in spherulitic form and be opaque, while the wall, which had formed crystals smaller than the wavelength of light, will shrink noticeably while remaining transparent. Section 10.4.2 discusses light scattering in semi-crystalline polymers. The high air pressure acting on the final dimensions of $r = 45$ mm, $t = 0.5$ mm causes a hoop stress of 200 MN m^{-2} just before the bubble reaches the mould wall. This high stress can easily stretch the polymer which is semi-crystalline at this stage, so the crystals in the bottle wall are high oriented (Section 2.4.7).

4.5.3 Thermoforming

The secondary process of thermoforming starts with extruded sheet, and converts it into curved parts with a non-re-entrant shape (e.g. margarine tubs, baths and curved panels). The first stage in the process is to heat an appropriate sized sheet into the melt state. Thicker sheets, such as the 6 mm PMMA for baths, are pre-heated in ovens, but sheets of 2 mm and less are heated over the mould by electric radiant heaters. In the basic version of the process the melt is sucked down into a female mould. The draw ratio of the mould cavity is defined by

$$\text{draw ratio} \equiv \frac{\text{mould depth}}{\text{mould width}}$$

Forming trials are carried out with cavities of a range of draw ratios, and the largest draw ratio cavity that can be successfully filled is used as a measure of the formability of the plastic. The maximum draw ratio increases with increasing temperature, as the melt viscosity falls, but there are limits to the draw ratio that can be achieved with any particular polymer (Fig. 4.17). For PVC the best conditions are with the melt at 60 to 100 °C above the T_g of 68 °C. At temperatures exceeding 180 °C thermal degradation becomes rapid in PVC. The heating conditions used depend on the thickness of the sheet; for a sheet less than 0.5 mm thick there is hardly any temperature gradient across the sheet, but for thicker sheets the heating rate is limited by the need for all the polymer to be above the forming temperature while the surface is not over-heated. Heat must not be supplied to the surface by absorbed radiation faster than it can be conducted through the thickness. Fig. 4.18 shows the temperature profiles in 1 mm sheet, calculated for a radiant temperature of 500 °C; a heating time of 12 s is required. Cycling the heaters on and off every 5 s or so is a possible way of reducing the temperature differential across a thicker sheet. The time for deforming the melt bubble is low and the thin walled products solidify almost immediately they contact the mould surface. Hence the overall cycle times is dominated by the heating time if it is necessary to heat the sheet over the mould.

One disadvantage of using a concave mould for thermoforming is the *excessive thinning* that occurs at bottom corners of any box-like product. These corners will be vulnerable in use, and should really be thicker than the rest of the box. The film of melt is sucked down as a bubble into the mould cavity when the air beneath the polymer sheet is evacuated. Once part of the melt contacts the sides of the cold mould, and its outer surface rapidly cools to a temperature at which it can no longer extend, the stretching is limited to the remaining lower part of the bubble. If the sides of the mould are nearly vertical the wall thickness of the moulding decreases exponentially, and the thinnest part will be at the bottom corner of the container. A draw ratio of 1

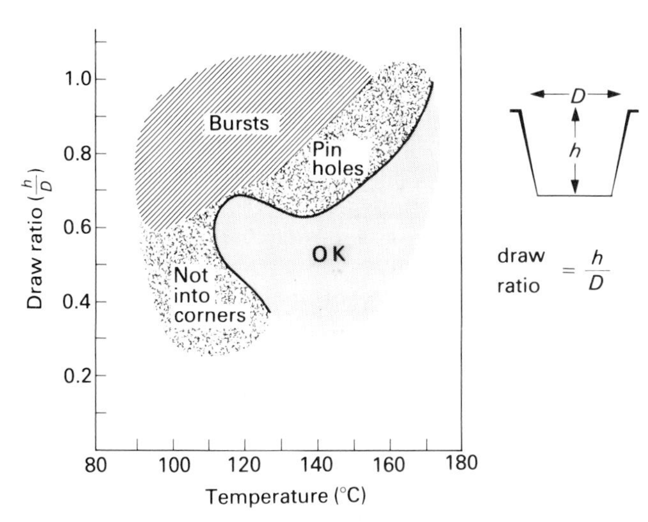

Fig. 4.17 The draw ratio and melt temperature conditions that give acceptable mouldings in a certain PVC (from P. F. Bruins (ed.), *Basic Principles of Thermoforming*, Gordon and Breach, 1973)

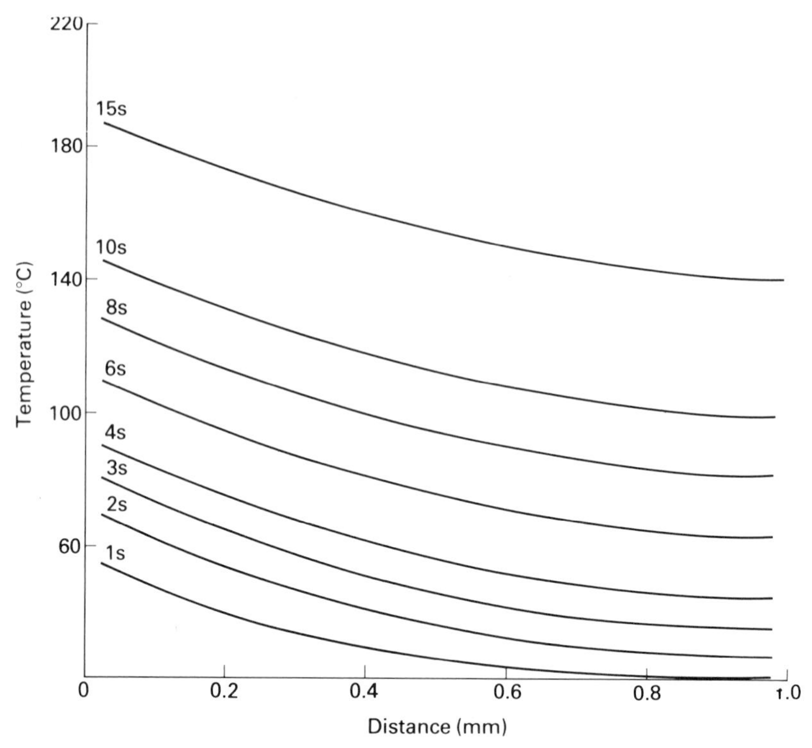

Fig. 4.18 Calculated temperature profiles in a 1 mm plastic sheet after different times of exposure to a radiant source at an effective temperature of 500 °C

corresponds to the melt thinning to about 25% of its original thickness, which is close to the limit at which holes form.

The simplest method of manufacturing containers with thicker bases is to invert the thermoforming process (Fig. 4.19) and to use a male mould for the interior dimensions of the container. A positive air pressure may be used to inflate the sheet of melt upwards before the male mould is raised into the interior of the bubble, or the mould movement itself can just stretch the sidewalls of the container. Subsequent evacuation of the remaining air between the bubble and the mould will complete the forming. Vending machine drink cups and containers for margarine are made in this way, and the base thickness is somewhat greater than the wall thickness. It is easy to make very thin containers because there are no flow ratio restrictions on the wall thickness, as in injection moulding. However containers where the height is several times the width must be made by blow moulding.

The low pressures in thermoforming mean that cast aluminium moulds are adequately strong, and their high thermal conductivity is an advantage over the steel required for injection moulds. As these moulds can be cast to shape, and there is no requirement for high forces to move parts of the mould, the thermoforming process has low capital costs compared with injection moulding. However there is no independent control of the product thickness and it

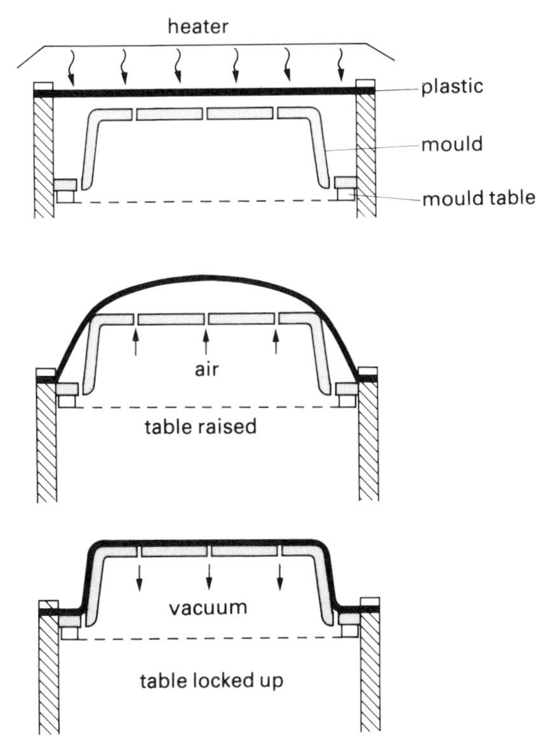

Fig. 4.19 Thermoforming machine showing a male mould being raised into the melt bubble before the remaining air is evacuated

is impossible to mould reinforcing ribs into the products. The product walls are likely to buckle elastically if compressive forces are applied. Corrugations in the walls of thermoformings can be used to increase the second moment of area of the cross section, and hence the critical force for buckling. The only way to make a really stiff product would be to thermoform separately the upper and lower skins of the product, and then to inject a polymer foam into the cavity to make a sandwich structure.

4.6 INJECTION MOULDING

4.6.1 The cycle of operations

Any three-dimensional shape that can be extracted from a steel cavity when it opens into two or more parts can be made by injection moulding. Features such as internal threads or intersecting holes can be made using unscrewing or sliding cores attached to the main mould. Hollow or re-entrant parts can be made by assembling or welding together several injection mouldings. It is preferable to keep the product of constant wall thickness so that the mould can be filled efficiently, and the wall thickness is best kept between 0.5 and 10 mm for efficiency of production.

Fig. 4.20 shows the main parts of an injection moulding machine. To minimise the cycle time several of the operations in the cycle are carried out simultaneously. Thus the solidification of one moulding occurs at the same time that a new batch of melt is being prepared. The method of melting is the same as in extrusion, and the operating diagram (Fig. 4.9) of the extruder can be applied. As the screw rotates it moves backwards against a set back pressure, which determines the rate of melt accumulation at the front of the screw. A higher back pressure slows down the rate of melt accumulation, but gives better mixing and generates more viscous heating. At the beginning of the next cycle the mould closes hydraulically and the extruder moves forward until the nozzle contracts the fixed mould half. The screw has a non-return valve at its tip, so that when it moves forward it acts as a piston to inject the melt. The injection pressure is controlled in two or more stages so that there is a high pressure during mould filling and a lower holding pressure during the feeding of the initially full mould. The rates of melt flow can be set, and there are various control options for the mould filling (see Section 4.6.2).

The machine size is given in terms of two main parameters, the maximum *shot size* injected in a single forward movement of the screw, and the maximum *clamping force*. The shot size is usually quoted in terms of the maximum mass of polystyrene that can be injected, whereas the clamping force ranges from 100 kN to 100 MN. The clamping force restricts the

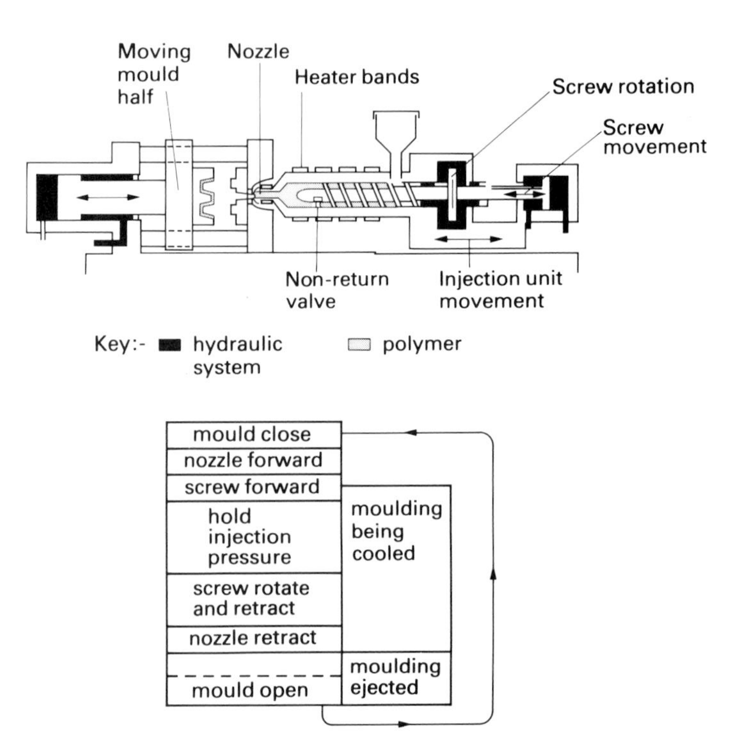

Fig. 4.20 An injection moulding machine with hydraulic mould closing, showing the cycle of operations

maximum projected area A of a moulding on to the mould parting plane (Fig. 4.21). A typical average melt pressure p in the mould is 20 MN m^{-2}, and the condition

$$pA < F_{\text{lock}} \tag{4.32}$$

must be obeyed if the mould is not to open and flash be formed at the edges of the moulding.

The mould block, because it must resist high pressures without distortion, and resist wear over 10^5 cycles or more, is usually made from a forged block of low-alloy steel, that can be air-hardened after machining. The blocks act as thick walled pressure vessels, and the tensile stresses in the walls are rather high. If the melt pressure is 50 MN m^{-2} and the ratio of cavity diameter to wall thickness is 4:1 then the average hoop stress in the wall is from equation (7.16) equal to 100 MN m^{-2}. Corners in the mould cavity act as stress concentrating features, so to avoid localised yielding at such locations, the mould must be made from a steel of yield stress exceeding 300 MN m^{-2}.

The mould must contain a number of features that add to its complexity and cost. There is a tapered sprue that leads through the fixed mould half to a series of runners that distribute the melt. At the entries to the cavity there are gates, a restriction that controls the flow into the mould, and makes the removal of the sprue and runners easy. Gating systems are designed so that the flow orientation in the mould is favourable, and so the flow paths to the farthest points in the mould are approximately equal. Thus any part with axial symmetry is preferentially gated in the centre. The cross sectional area of the gate influences the time for which more melt can be packed under pressure

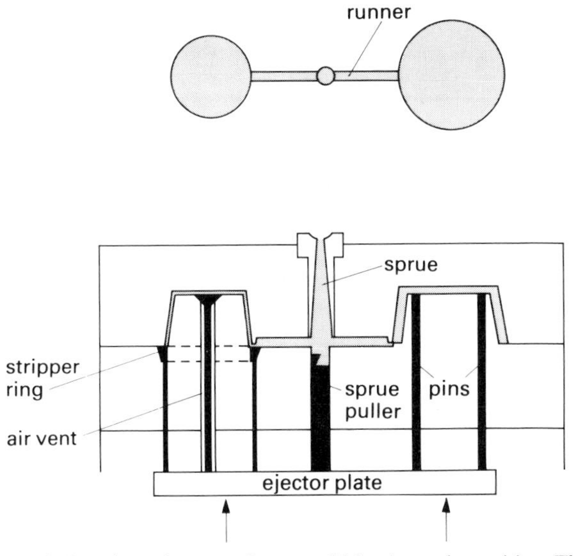

Fig. 4.21 Plan and elevation of a two-plate mould having twin cavities. The upper diagram shows the projected area of the moulding. Alternative methods of ejecting the cup shaped moulding are shown; the stripper and air vent are used if the moulding is very thin

into the cavity; the gate will usually be the first part of the moulding to solidify completely. Provision must be made for the air that the cavity initially contains to be vented—a gap of 25 μm between the mould blocks is sufficient. There are four guide pins at the corners of the mould to keep the mould halves correctly located.

The mould needs to be kept at a constant temperature by recirculating water or oil through cooling channels. Polyethylene when injected has a heat content of about $700\,\mathrm{J\,g^{-1}}$ of which only half is left in the ejected warm moulding. The mould temperature is not necessarily low to maximise the cooling rate. Other requirements such as keeping the orientation or residual stress to acceptable levels may mean higher temperatures are used, for example 90 °C for polycarbonate. The cooling channel system is a compromise between achieving a uniform mould temperature and the complexity of the machining required.

Ejection of the solidified moulding depends on a series of ejector pins whose heads are set flush into the surface of the moving mould half. The pins are mounted in an ejector plate which is actuated mechanically or hydraulically after the mould has opened a small distance; there are also ejector plate guide pins which automatically retract the ejector plate when the mould closes. When the still-warm moulding is released from the mould it shrinks by ~1% (Section 5.3.1). This means that prior to ejection there is a ~1% tensile strain in the moulding. This will cause the moulding to grip any male part of the cavity, and the frictional forces at the plastic/steel interface will resist ejection. Consequently there must be a 1° to 2° taper on the inner walls of box-like products (Fig. 4.21), so that the moulding becomes loose after a small ejector movement. Deep thin-walled containers are troublesome to eject; the forces from pin ejectors may be unacceptably high, causing distortion and damage. Consequently a stripper ring can be used to distribute the force around the rim of the container. There may also need to be a means of breaking the vacuum on the flat base of the container. Further details of mould construction are given in the books listed in *Further Reading*.

4.6.2 Analysis of mould filling

The filling of a cold steel mould with a polymer melt is a process in which melt flow and heat transfer interact strongly. The highly viscous melt takes a number of seconds to fill the mould, and during this time a skin of solid polymer builds up at the mould walls. The temperature profile through the moulding has an intermediate peak due to the viscous heating that occurs in the high shear stress regions. The mould cavity must be filled completely without any flash occurring. The problem is worse with multi-cavity moulds because the flow to the different cavities must be balanced.

The rheological information available from materials suppliers is usually shear flow curves at a range of temperatures, which can be approximated in the relevant high shear rate range by the power law relationship. Data may also be given for the filling of a spiral test mould. This has a rectangular cross section, and the maximum flow length is recorded for different fixed injection pressures to give the data shown in Fig. 4.22. The flow length is nearly

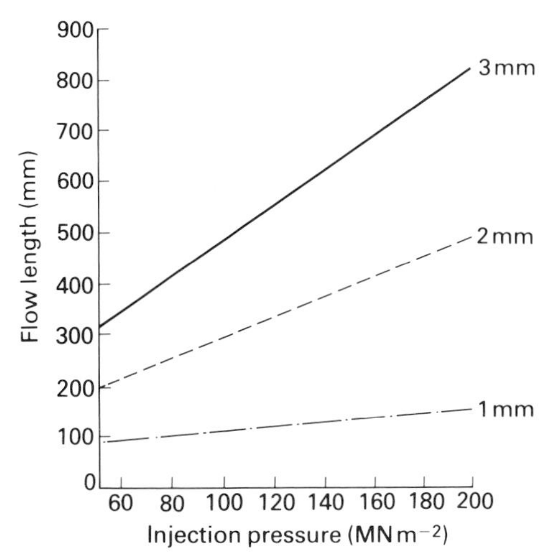

Fig. 4.22 Spiral test moulding data for a thermoplastic. The flow length versus injection pressure data are for three moulding thicknesses (from *Lexan Noryl Valox* booklet, General Electric Plastics, 1981)

proportional to the cavity thickness, at a given injection pressure. The simplest guide rule for the filling of a mould is that the thermoplastic selected should have a flow ratio that is greater than the flow ratio of the mould, which is defined by

$$\text{flow ratio} \equiv \frac{\text{maximum flow path from the gate}}{\text{cavity thickness}}$$

For the data in Fig. 4.22 and an injection pressure of 1000 bar, the polymer flow ratio is 100 to 150 depending on the cavity thickness.

Most injection mouldings have regions where the melt is flowing radially outwards from a gate, as well as regions where the flow lines are nearly parallel. Computer modelling is often used to predict whether a mould can be filled satisfactorily before the mould cavity is machined. The most realistic predictions of the melt front shape use 3-D finite element modelling. The simpler 2-D layflat models give accurate predictions of when the extremities of the mould are filled, and the basis of the 2-D model is easier to explain. Fig. 4.23 shows how an instrument panel moulding has been approximated as a series of rectangular slots and radial flow segments in series with one another. The mould filling problem is solved at a number of discrete times; let us assume that the shaded region of the mould is already filled and that the volume flow rates Q at the gates are known. Because of material continuity the average melt velocities $\bar{V}$ are known at each of the segments; in a continued radial flow the velocity $\bar{V}$ is inversely proportional to the radial distance if the thickness remains constant. The apparent shear rate $\dot{\gamma}_{aw}$ at the segment wall can be calculated from equation (B.11) in Appendix B. The shear flow curve at the appropriate melt temperature is then used to find the

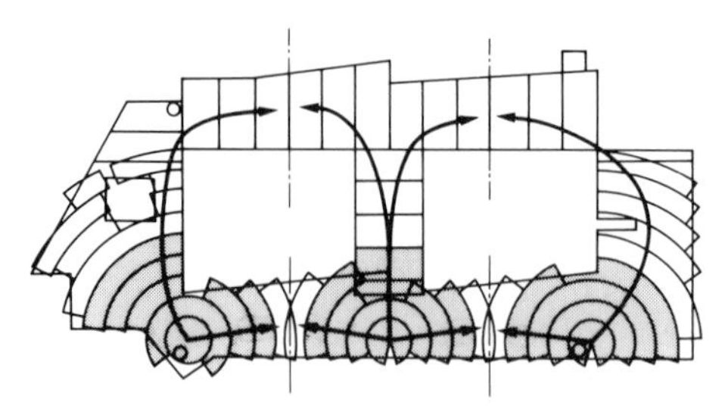

Fig. 4.23 Division of a moulding into radial flow segments and rectangular flow segments for the purpose of mould filling analysis. The grey segments are filled with melt (from *Plastics Machinery and Equipment*, Mouldflow Pty., 1980)

shear stress for that $\dot{\gamma}_{aw}$ value. Finally the pressure difference Δp across the segment of length L and thickness h is given by

$$\frac{\Delta p}{L} = \frac{2\tau_w}{h} \tag{4.33}$$

If this process is carried out for all the full mould cavity segments, and the sum of the Δps calculated, the injection pressure is known. The injection moulding machine often has a set injection flow rate Q, so long as the injection pressure does not exceed a set maximum value. The computer program checks whether the injection pressure is possible, then updates the melt temperatures, and melt channel thickness h at the end of each time interval.

Many injection moulding machines are computer controlled, so that the optimum process parameters for each mould/polymer combination is stored for future use. The control of one polymer variable is affected by a number of the machine settings. For example the melt temperature prior to injection is influenced by at least 4 inputs: the screw speed, the back pressure on the screw, the barrel and nozzle heater temperatures. There is no simple way of measuring the melt temperature at the nozzle; any thermocouple projecting into the melt would be sheared off by the high flow stresses. Therefore there is open loop control on the melt temperature, with the value fluctuating both during and between cycles.

The simplest way of ensuring that the mould cavity is filled is to use limit switches on the screw travel. These actuate after the correct volume of melt has been injected and reduce the injection pressure to a lower holding value. However, variations in the melt leakage past the screw non-return valve, or at the nozzle, will cause the part mass to vary, with a consequent variation in the dimensions of the product. A better method of controlling the changeover to holding pressure is to place a melt pressure transducer in the mould. The mould pressure rises rapidly once the mould is full so the injection pressure can be switched to the holding level once the mould pressure reaches a set level. Subsequent feeding of the moulding will occur at a nearly constant pressure until the gate freezes; thereafter the cooling and contraction of the moulding causes the mould pressure to fall.

The capital cost of the mould is a major consideration for injection moulding. Firstly it limits the application of the process to products made on a scale of more than 10 000. Secondly the productivity of the process depends on minimising the cooling stage which dominates the cycle time. Rather than invest in a second mould to reach the required production rate, it is preferable to reduce the cycle time. This means that product wall thicknesses must be minimised (see equation (4.6) for the effect that this has on the cooling time), and wherever possible kept constant throughout the moulding. The design features that must be used to achieve adequate product stiffness are discussed in Chapter 13.

4.7 REACTION INJECTION MOULDING

Reaction injection moulding (RIM) is being used on an increasing scale to make large automotive panels, and is an example of a process where a chemical crosslinking occurs in the mould. The majority of mouldings are made from polyurethanes and polyureas, the chemistry of which was described in Chapter 3. Other systems used include a block copolymer between a crystalline polyamide (nylon 6) and a rubbery polyether (polypropylene oxide). In principle any polymerisation reaction that can be substantially completed after about 30 s in the mould is a candidate for RIM.

The main features of the RIM process are shown in Fig. 4.24. The two components are kept in temperature controlled tanks, with pumped recirculation when injection is not taking place. For the polyurethane system one tank contains the isocyanate (usually MDI) and the other a mixture of the polyol, chain extenders, catalyst, blowing agent and mould release (and possibly reinforcing additives). There is an amine catalyst to accelerate the initiation of the polymerisation and an organotin catalyst to accelerate the gelling (when the liquid gels its viscosity becomes infinite and it acts as a rubbery solid). The surfactant enables the polymer to wet the mould surface to obtain a better surface finish. A low boiling point liquid such as $CClF_3$ can be used as a blowing agent if the mouldings are required with low density; this boils at 25 °C and generates a foam when the exothermic reaction takes place. Otherwise air or nitrogen is dissolved in the polyol. The two components are low viscosity liquids with viscosities of the order of $1\ N\ s\ m^{-2}$. When they are pumped in accurately metered amounts to the mixing head and the valve is opened, the pressure of 100–200 bar causes the two streams of liquid to meet head on at a $100\ m\ s^{-1}$ velocity in a small ($<5\ cm^3$) chamber. The Reynolds number of the flow must exceed 200 for efficient turbulent mixing on a scale less than 0.1 mm to take place, so that the diffusion distances for the chemicals is low enough for rapid reaction. For rubbery polyurethanes the liquids pass through an after-mixer before they enter the mould; this further improves the mixing. The pressure drop in mixing generates heat so that the mixture is at about 50 °C as it enters the mould; at this temperature the components are highly reactive.

The design of gates and the method of mould filling are more akin to those in the gravity casting of metals than to those in the injection moulding of

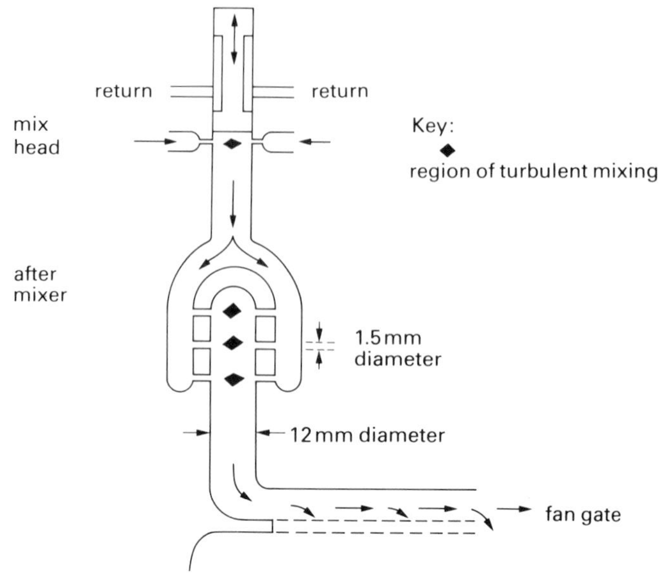

Fig. 4.24 Impingement mixing and mould gating used in the reaction injection moulding process

thermoplastics. The moulds are fed at the lowest point, and a laminar flow into the mould is required so that no air bubbles become entrapped. Consequently long film gates are often used; the liquid flowing through the gate as a 1 to 2 mm thick film at about 1 m s^{-1}. The mould must be filled in a small proportion of the gel time; in 1 to 2 s for a 10 s gel time. The strongly exothermic reaction (Fig. 4.25) causes gas bubbles to be generated. The aluminium or steel mould is controlled at about 60 °C so the surface of the moulding never heats above this. The poor thermal conductivity, however, means that the centre can reach 150 °C. Consequently the polyurethane cures first in the interior. The high polymerisation shrinkage is compensated for by the foaming of the core. Air or nitrogen, which had been dissolved into the holding tanks under pressure, will cause the liquids to froth as they enter the mould, but this gas will redissolve when the mould is full and the pressure rises to 5 bar. Later the polymerisation shrinkage will cause the pressure to drop and the bubbles to reappear. The final microstructure has a solid skin of density 1100 kg m^{-3} and a foamed core of density 800 to 950 kg m^{-3}. The overall cycle times are currently 30 to 90 s. Mould release chemicals such as 1 to 2% of zinc stearate plus a fatty acid are incorporated into the constituents, but it is still necessary to spray a layer of mould release agent into the mould every 10 to 30 mouldings. The advantage *vis-à-vis* conventional injection moulding is that the liquid pressure in the mould, and hence the mould clamping force for a given moulding projected area, is reduced by more than 95%. The capital costs of both the machine and the mould are thus considerably smaller than for injection moulding. There is an overall energy saving as the polymerisation is carried out in the mould, cutting out all the

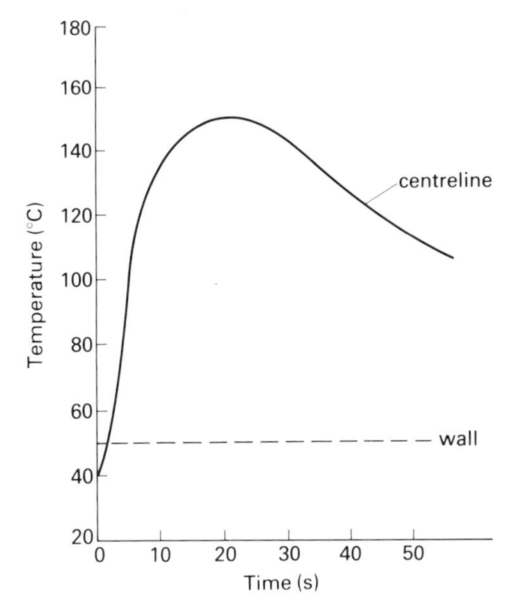

Fig. 4.25 Temperature history at different positions in a polyurethane RIM moulding 5 mm thick, injected at 40 °C into a mould at 50 °C

extrusion, granulation and processing operations needed with thermoplastics.

The range of applications for RIM has been increased by the introduction of 20% by weight of chopped glass fibres into the polyol component. Apart from changes in the pump type to cope with the higher viscosity liquid and the abrasive wear there is little modification in the process. The flexural modulus is increased by a factor of 3 or so, so that with suitable polyol systems it is comparable with that of thermoplastics. Chapter 13 discusses the application of these materials.

5

Effects of processing

5.1 INTRODUCTION

This chapter explains the features in plastic products that are the effects of processing. The cooling rate in the latter stages of processing, and the stress in the melt as solidification occurs, are the main causes of these features. They can be divided into microstructural phenomena, such as crystallinity changes, and macroscopic phenomena, such as dimensional changes. Residual stresses will be characterised as a macroscopic feature even though there is a variation in the stress with position the product. For a few polymers such as PVC there are effects that come from the initial melting and mixing stages—these will be dealt with at the end of the chapter. Chapter 4 dealt with how the choice of process affects the design of the product. Chapter 13 will consider how redesign can optimise the mechanical properties of mouldings.

5.2 MICROSTRUCTURAL CHANGES

5.2.1 Effects of cooling rate

In all plastics processes the cooling rates are relatively fast. Nevertheless the rate of cooling varies with position in the product, with the outer layers being cooled at a rate that can be orders of magnitude larger than the rate at the centre. To illustrate this Fig. 5.1 shows the rate of cooling through the glass transition temperature at different positions in a sheet. The sheet is cooled from both sides by a medium that has an effectively infinite heat transfer coefficient. In contrast with the temperature profiles in Fig. A4 in Appendix A, the cooling rate varies more strongly near the surface of the sheet.

If a polymer is capable of crystallising then the effect of changing the cooling rate may be dramatic, especially if the maximum rate of crystal growth (Section 2.4.4) is not too high. In this case it may be possible to cool thin sections of the polymer fast enough to avoid any significant crystallisation. Thus injection moulded PET bottle preforms are glassy—this allows greater extensions to be achieved when the preforms are reheated to above the glass transition temperature T_g and biaxially stretched in the injection–

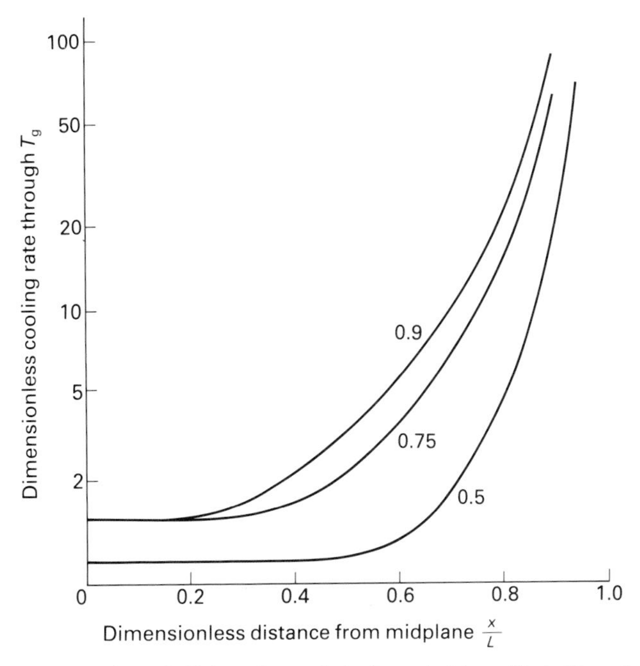

Fig. 5.1 Cooling rate through T_g in a sheet of plastic cooled from T_m to T_b on both sides, as a function of the distance from the midplane, for different values of $(T_g - T_b)/(T_m - T_b)$

blow moulding process. This should be contrasted with PBT where the crystallisation rates are higher so it is not possible to change significantly the average crystallinity of a 3 mm thick moulding, although the crystallinity of a 100 μm thick surface layer can be reduced.

For some polymers metastable crystals are formed at high rates of cooling. In polypropylene a hexagonal form of crystal is formed just below the rapidly cooled skins of injection mouldings. This metastable form will recrystallise as the stable triclinic form if the moulding is annealed at a sufficiently high temperature. Even if there is no change in the form of crystallisation the degree of crystallinity increases with the time spent in the crystallisation temperature range. Fig. 5.2 shows the variation in density through the 22 mm thick wall of an extruded HDPE pipe. The crystallinity is 5% higher at the bore of the pipe because of the lower cooling rate.

When a glass forming polymer is cooled, the density of the glass slightly increases as the cooling rate into the glassy state is decreased. Slower cooling gives the polymer more time to relax towards an equilibrium glassy state; this may be associated with changes in the local conformation of the polymer chains. This slow approach towards an equilibrium state continues if the polymer is held at a temperature not more than 50 °C below T_g. Section 2.3.6 described this as an increase of the fictive temperature of the glass, but it is also referred to as *ageing*. Ageing can occur at 20 °C for polymers like PVC with a low T_g. Fig. 5.3 shows the stress strain curve of a PVC specimen

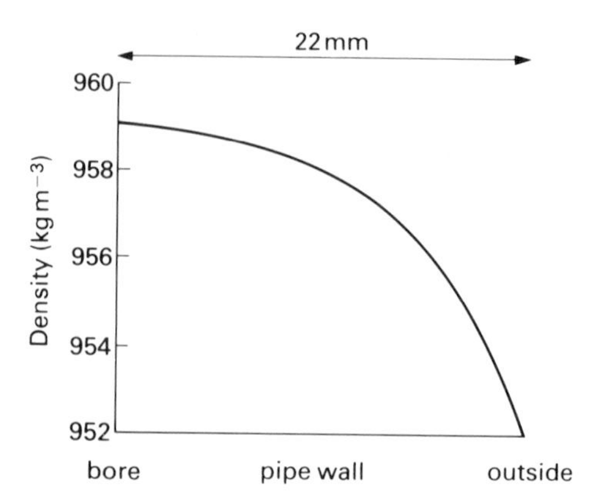

Fig. 5.2 Variation of the density, hence of the crystallinity, through the thickness of a 22 mm wall HDPE pipe

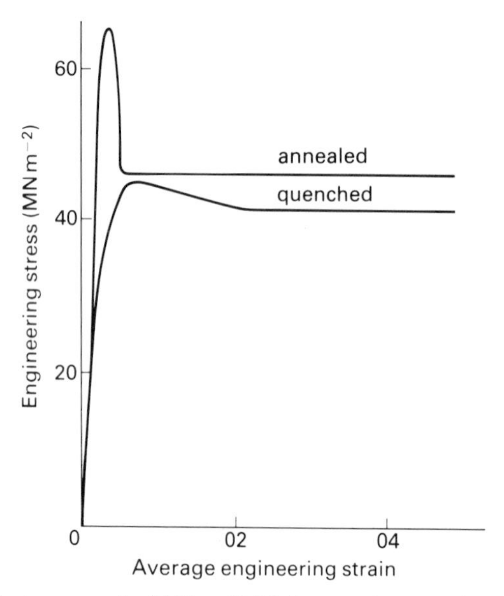

Fig. 5.3 Stress–strain curves for PVC at 23 °C, for samples annealed at 60 °C, and quenched from 90 °C

1.7 mm thick immediately after it has been quenched from 90 °C into cold water, and after annealing at 65 °C then at 40 °C. There is a considerable increase in the yield stress as a result of the annealing; most of this would also occur if the PVC is kept at 20 °C for several years. The quenching treatment will also produce some residual stresses (Section 5.3.3) which could affect the yielding behaviour. In injection mouldings the molecular orientation (see the next section) will also change from the interior to the surface of the test piece, so it is often difficult to isolate the effect of the cooling rate.

5.2.2 Melt stress effects for glassy polymers

The shear and extensional flows in polymer processing have an effect on the microstructure of the melt. On the molecular level, as a result of the entanglements between molecules, elongation of parts of the molecules occurs (Fig. 2.9), giving rise to tensile stresses in the melt. If there are rigid additives present, such as glass fibres, these can be aligned relative to the flow. Alternatively initially isotropic but deformable inclusions, such as rubber spheres, can be both elongated and aligned in the flow. If this non-equilibrium microstructure persists into the solid stae, then there is said to be *orientation* in the product. Orientation is greatest where the polymer solidifies while the stresses that shape the melt are still acting. Thus blown film is likely to have strong orientation, as is the *skin of an injection moulding* which solidifies while the mould is still being filled. The *core* of an injection moulding is defined as the region that solidifies after the mould cavity is full. There is a pressure gradient down the mould while it is being filled (Fig. 5.4a), consequently there is a shear stress in the melt which is a maximum at the mould wall. Immediately the mould is full the pressure becomes uniformly high and the shear stress drops to zero.

When the *molecular orientation* in polystyrene is measured using the optical birefringence technique (Fig. 5.4b) it is found that the values are highest in the skin, and that the skin is thickest near the gate of the moulding, where flow continues for longest. The lower orientation at the very surface is due to the 'fountain flow' as the mould fills, with relatively unstressed melt from the centre of the channel coming into contact with the mould wall.

In the packing stage of injection moulding when the mould is full there is very little flow except that due to feeding near the gate. With conventional liquids the melt stresses would become zero immediately the flow ceases, but this is not the case for polymer melts. Fig. 5.5 shows a simple viscoelastic model (see Chapter 6) which can be applied to the relaxation of molecular orientation once flow has ceased. The spring represents the temporary shear modulus G of the entanglement network in the melt, and the damper represents the melt viscosity. In this series model the shear stress τ is constant in both elements, and the shear strain rates $\dot{\gamma}$ can be added giving

$$\dot{\gamma} = \frac{\tau}{\eta} + \frac{\dot{\tau}}{G} \tag{5.1}$$

If there is steady shear flow at a strain rate S until time $t = 0$ and no flow thereafter, equation (5.1) can be solved to give

$$\tau = \eta S \exp(-Gt/\eta) \tag{5.2}$$

The quantity η/G represents the *relaxation time* of t_0 the melt, the time in which shear stress decays to 1/e (34%) of its original value. The magnitude of the relaxation time depends on the molecular mass and the melt temperature, both of which affect the viscosity. In reality the stress relaxation behaviour is more complex than this with a range of relaxation times (see Section 6.2). The molecular orientation in the core of an injection moulding will be greatest

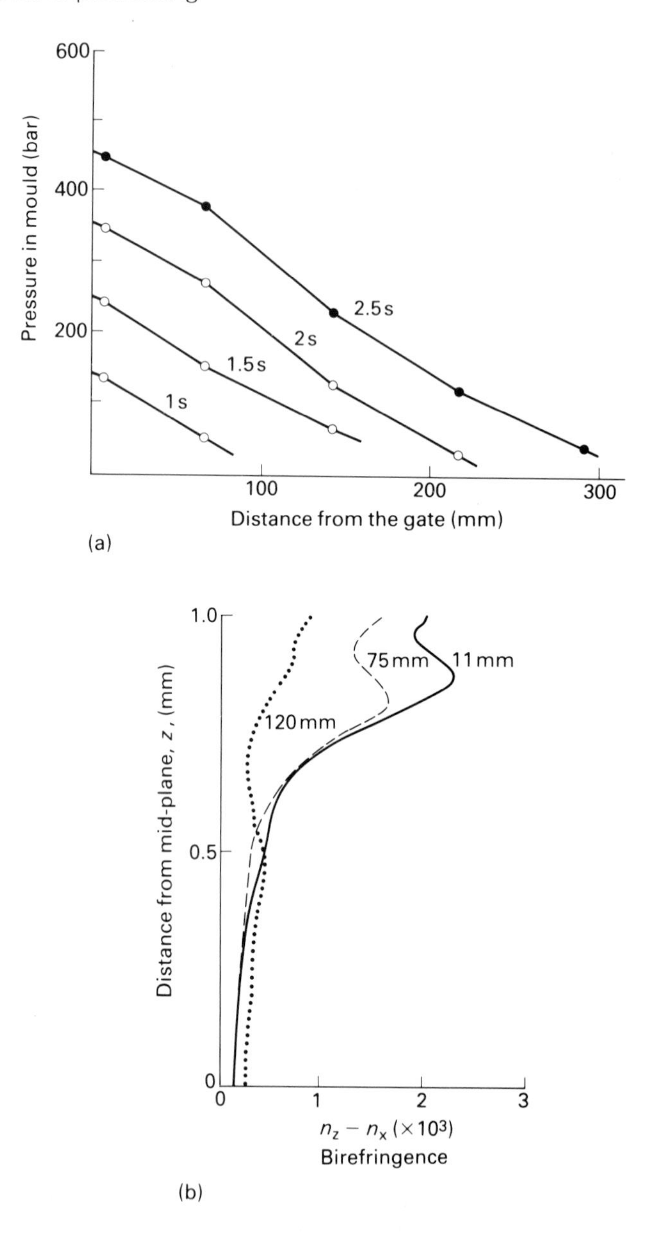

Fig. 5.4 (a) Pressure distribution in a large injection mould at various times during filling (from Wales, *Polym. Eng. Sci.* 1972, **12**, 360). (b) Refractive index difference versus position through the thickness of a polystyrene moulding at different distances from the gate, in a mould 127 mm long. The x and z axes used are defined in Fig. 5.11

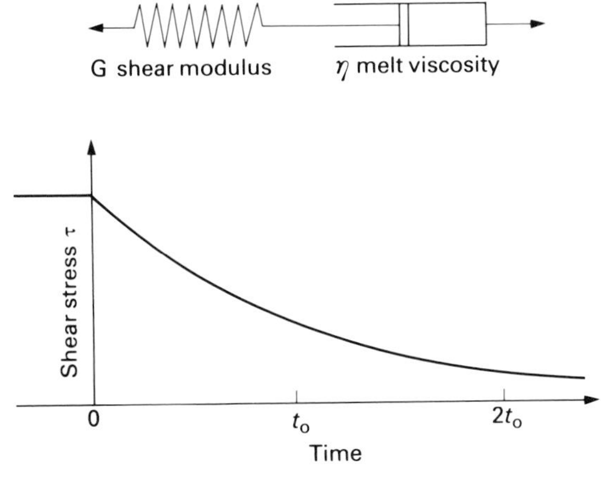

Fig. 5.5 Maxwell viscoelastic model and its prediction of stress relaxation in a melt after the cessation of steady shear flow; t_0 is the relaxation time

near the skin than at the centre (Fig. 5.4b) because there was less time for relaxation before solidification occurred.

In many processes (blow moulding, thermoforming) solidification only commences when the melt contacts the mould wall and flow stops. The time for the tensile stresses to relax will depend on the thickness of the product, so there is likely to be high orientation in a thin walled thermoformed cup, made from a highly viscous glassy polymer.

A high level of molecular orientation is usually regarded as a bad feature, the reason being that the strengthening in the direction of orientation is more than offset by the weakening in the perpendicular direction. Although the orientation could be useful if it were in the direction of the main structural loads, impacts on the surface will seek out the weakest direction. If a spherical indenter presses on the surface the stress field is symmetrical about an axis normal to the surface, and there are surface tensile stresses just outside the contact area. The cracks that develop in the less tough glassy plastics (PS and PMMA) can be used to map out the flow pattern in a large moulding (Fig. 5.6) because they align with the flow direction. There is a particular weakness near the gate of the moulding where the orientation is largest. If feeding is allowed to occur under a high holding pressure the problem will be exacerbated.

5.2.3 Melt stress effects for semi-crystalline polymers

If crystallisation occurs in an oriented melt, then non-spherulitic microstructures can form, with a preferred orientation of the crystals (Section 2.4.7). Fibrous nuclei, sometimes visible by optical microscopy, can form in an oriented melt. Fig. 5.7a shows several fibrous nuclei in a polyethylene injection moulding that are aligned with the flow direction. They are believed to contain fully extended polymer chains. On either side of these dark nuclei

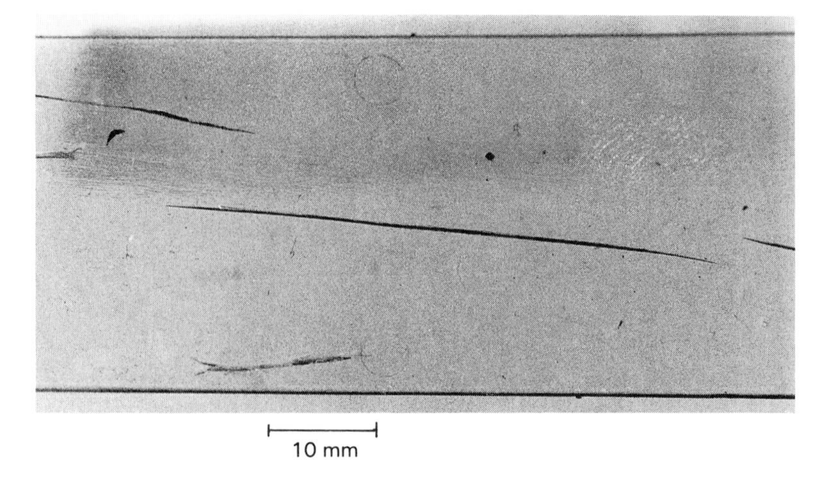

Fig. 5.6 Cracks in a polystyrene injection moulding due to surface indentations. The cracks lie along the flow direction

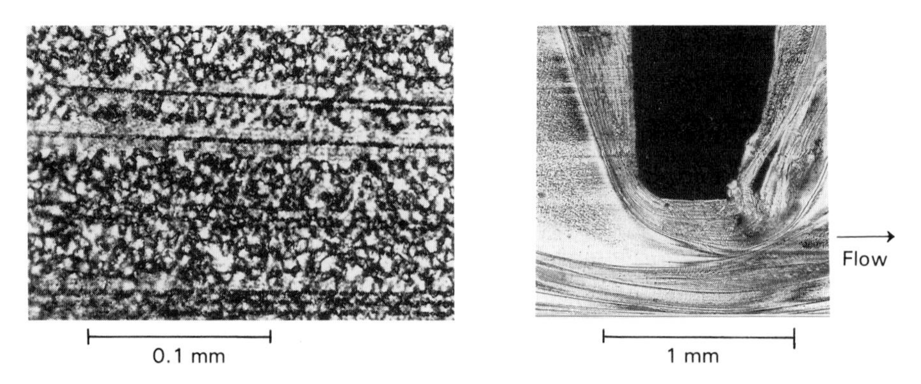

Fig. 5.7 (a) Polarised light micrographs of fibrous nuclei in a polyethylene injection moulding surrounded by parallel lamellar crystals, then by spherulites. (b) Oriented skin of a polypropylene moulding, forming the hinge of a box where there is a thickness restriction

is a bright layer, which is where lamellar crystals have grown from the nucleus. The **c** axes of all the lamellar crystals are parallel to the fibrous nucleus, and the microstructure of platelet crystals skewered by a rod-like nucleus has been described as a 'shish-kebab'. The rest of the microstructure consists of small spherulites. An oriented skin can form at the surface of an injection moulding. Fig. 5.7b shows a polypropylene injection moulding, in which the 0.3 mm thick skin of the moulding has a completely different oriented microstructure to the core. This microstructure is used in the manufacture of *moulded hinges* between the two halves of a box. The mould is gated so that the melt flows from the one half of the box to the other through a constriction that is only 0.5 mm thick. This ensures that the whole of the hinge has the highly oriented microstructure, and consequently the hinge is very strong in bending. Splitting of the hinge parallel to the orientation when it is flexed has no effect on the strength of the hinge, and it reduces the stiffness for subsequent flexure.

In an extrusion blow moulded polyethylene container with a 1 mm wall, 3 s will elapse before crystallisation takes place at the inner surface (equation 4.6). This is time for the melt tensile stresses to relax, so the microstructure will be spherulitic, even though the shape of the spherulites may be somewhat distorted. In contrast there will be high orientation in the wall of a stretch blow moulded PET bottle (Section 2.4.7) because the crystallisation occurs while the preform is stretching.

In a semi-crystalline polymer there can be orientation in both the amorphous and the crystalline phases. One method of assessing orientation is to heat the product into melt state and observe the shape change (Fig. 5.8). The entropic elastic forces are given free rein in the melt, and as the molecules retract to their equilibrium coiled shapes the melt changes shape. Just because entropic stresses exist temporarily in the melt does not necessarily mean that there were residual elastic stresses in the product. Section 5.3.4 explains that the average residual stress across the cross-section is zero so residual stresses could not cause the shape change in Fig. 5.8.

5.2.4 Weld lines

Polymer melts cool very slowly in contact with air, so it is possible for a weld to be formed when two sections of melt are pressed together. This occurs at the base of a blow moulded container. In injection moulding welds occur in larger mouldings when the melt from neighbouring gates (Fig. 4.23), or when the melt stream parts and recombines. Extruded sections can be joined

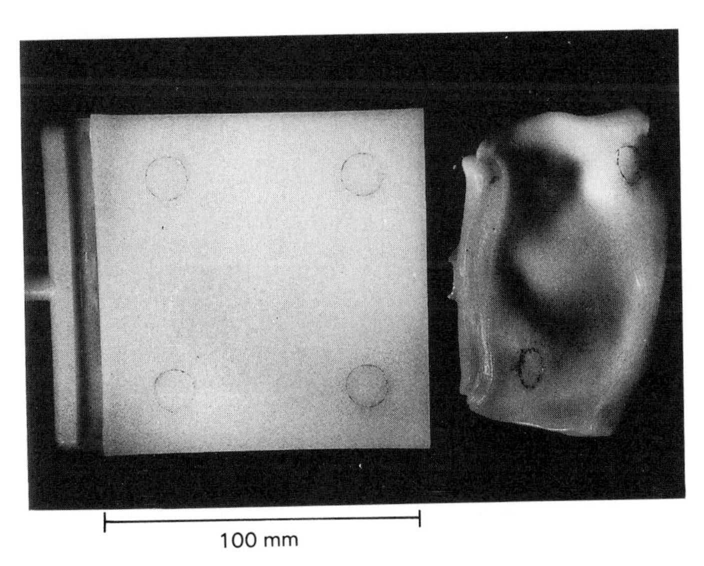

Fig. 5.8 An edge gated polyethylene injection moulding before and after melting in a bath of silicone oil

together by hot plate welding, for example to butt weld HDPE pipes in a trench or to make PVC window frames. In this case the flat surface of the end of the extrusion is pressed against a metal heating plate with a non-stick coating for a few minutes to prepare a layer of melt 2–3 mm thick. When the two extrusions are then pressed together a weld is formed. The common feature of these processes is that two flat or slightly convex melt surfaces come together under pressure, and there is some outwards flow (Fig. 5.9). The orientation produced will be at right angles to the original flow direction. When the melt is backed-up by soft semi-solid polymer, as in the welding of extrusions, there will be some shear deformation of this semi-solid material as the melt flows outwards to form a bead at the free surface.

Weld lines can be points of weakness especially in semi-crystalline plastics. The transverse crystal axis orientation has already been described in the last section. While this is not ideal, it is not as serious as regions where there is lack of fusion. If the surface of the melt has become contaminated, or cooled to a temperature at which interdiffusion of the two melt layers is slow, then a low strength region can form. It is possible that air can be trapped when two opposing melt streams meet in an injection moulding. Therefore the strength of products should be checked in areas that contain weld lines. When the tensile stress strain curve is measured for a specimen that has a weld line perpendicular to the tensile stress, it is often found that it fails just after the yield point. The entanglement network across the weld line is not strong enough to survive the process of necking that occurs without a weld line.

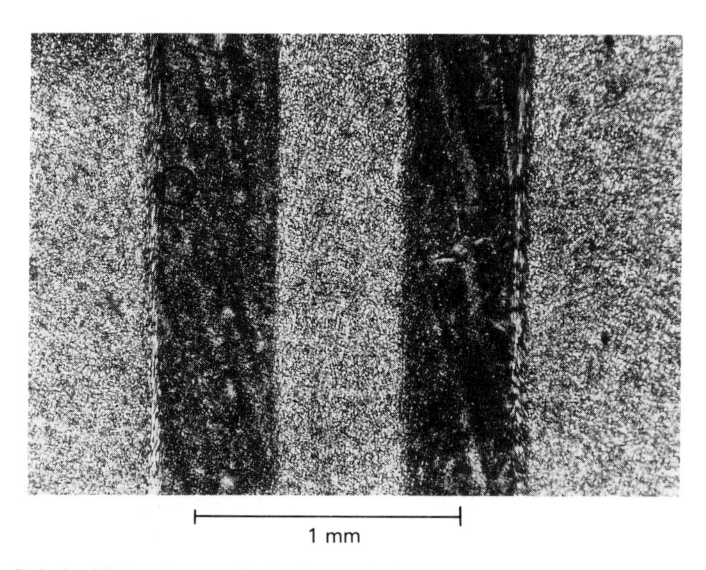

Fig. 5.9 Polarised light micrograph of a butt weld in a polyethylene pipe. The arrow shows the direction of melt flow during weld fusion

5.3 MACROSCOPIC EFFECTS

5.3.1 Shrinkage and distortion

The reproducibility of the dimensions of injection mouldings is extremely good, but if they are to be assembled to other parts without excessive elastic stresses being used, the absolute dimensions must be within specification. This means that the lengths L between the assembly points of the mouldings at 23 °C must fall between close limits. Before the equivalent length M in the mould is machined, it is necessary to have an estimate of the linear shrinkage S defined by

$$S = [100(M - L)/M]\% \tag{5.3}$$

To understand the shrinkage values it is necessary to consider both the microstructure of the polymer and the processes that occur in moulding. Polymers have high thermal expansion coefficients because the increase in thermal vibrations with increased temperature is only weakly resisted by the van der Waals bonding between the polymer chains. Thermal expansion is anisotropic if there is molecular orientation present, with the expansion coefficient being smaller in the direction that has the greatest fraction of covalent bonding; however, in the following it is assumed that the polymer is isotropic. If the polymer is partly crystalline there is an increase in specific volume when the crystals melt. In the pressure–volume–temperature (p–V–T) data for polyethylene in Fig. 5.10a the crystalline phase finishes melting at 130 °C at 1 bar pressure. Pressure is included as a variable because of the large pressure variations that occur during solidification in the injection moulding process. The bulk modulus is low because of the weak van der Waals forces between the chains. This partly compensates for the high thermal expansion coefficient. We can use Fig. 5.10 to estimate the volume shrinkage that would occur if a polyethylene melt was injected at 200 °C and 300 bar ($V = 1.286 \times 10^{-3}$ m^3 kg^{-1}) and there was no packing stage, so that this volume of melt contracted to a V of 1.035×10^{-3} m^3 kg^{-1} at 20 °C and 1 bar pressure. The volume shrinkage of 19.5% equates to an isotropic linear shrinkage of 7.0%. In reality the length shrinkage of injection moulded semi-crystalline polymers is in the range 1 to 3%, whereas the thickness shrinkage can be considerably higher. The p–V–T data for amorphous polymers (Fig. 5.10b) only show a change in slope at the glass transition, so we expect much lower shrinkage values for these materials. Fig. 5.10 shows how the T_m of polyethylene and the T_g of polystyrene shift to higher temperatures as the pressure is increased. This is another reason for injecting the melt at a temperature well above these values.

Fig. 5.11 shows a flat section of a moulding, in which the thickness in the z direction is much smaller than the other dimensions. The heat flow down the temperature gradient is only in the z direction from the hot melt to the water cooled mould. A solid skin develops adjoining the mould surfaces; it forms a closed box containing a molten core. This low modulus skin is easily sucked inwards by the contracting core once the pressure in the mould has fallen to zero. Apart from the shaded edge regions the skin does not resist the

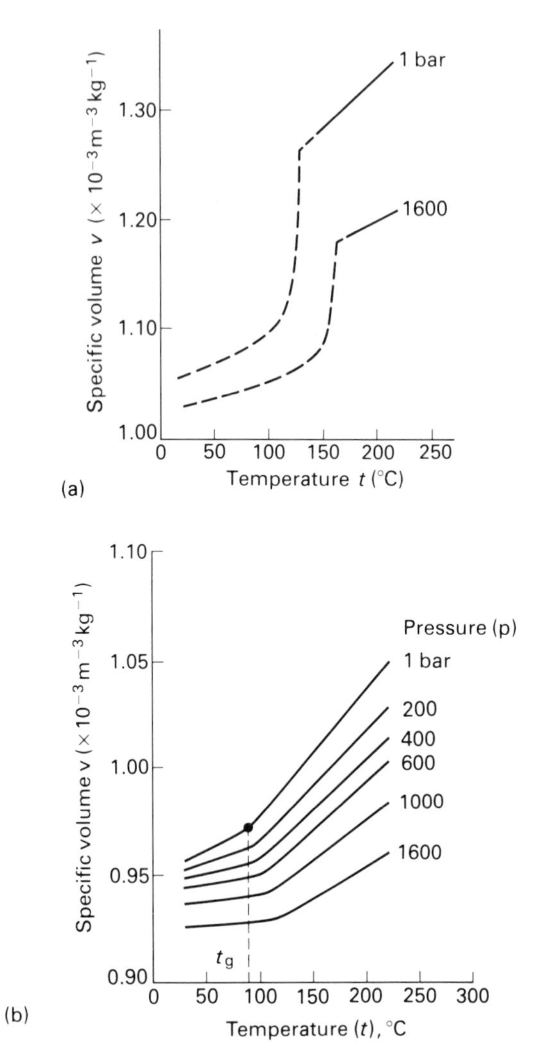

Fig. 5.10 Pressure–volume–temperature relationships for (a) polyethylene and (b) polystyrene (from Wang, *Polym. Plast. Technol. Eng.*, 1980, **14**, 88, Marcel Dekker, New York, and Menges, *Polym. Eng. Sci.*, 1977, **17**, 760, Soc. Plastics Eng. Inc.)

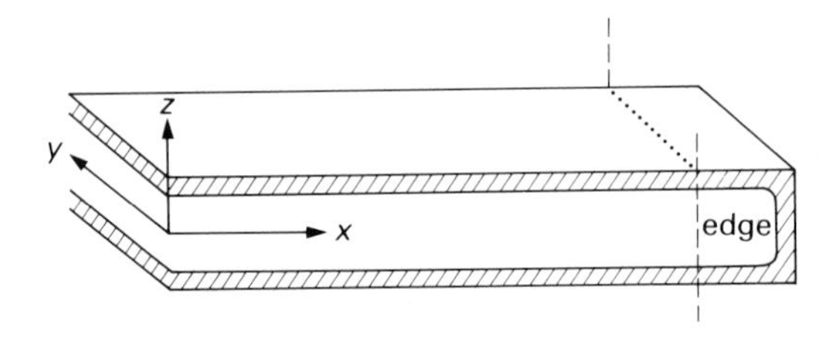

Fig. 5.11 The solid skin on a partially solidified injection moulding

contraction of the core in the z direction. However, the skins and the core are thin layers connected in parallel in the x and y directions, having the same length L. Therefore the skins reduce the overall shrinkage in the x and y directions, and as a consequence there are residual stresses in these directions in the cold moulding (see Section 5.3.3).

The shrinkage of a moulding can be altered by changing the amount of feeding of the mould, either by increasing the holding pressure or the length of the holding time, or by enlarging the gate. Fig. 5.12a shows how the changeover from the high hydraulic pressure for filling the mould to the lower hold pressure is felt in the mould cavity. The mould pressure only rises rapidly once the mould is full, and it then decreases during the solidification process, as the melt channels connecting the injection unit to the mould gradually constrict with time. If the holding pressure is removed before the gate or feeders freeze then there is a rapid pressure drop as some melt flows out of

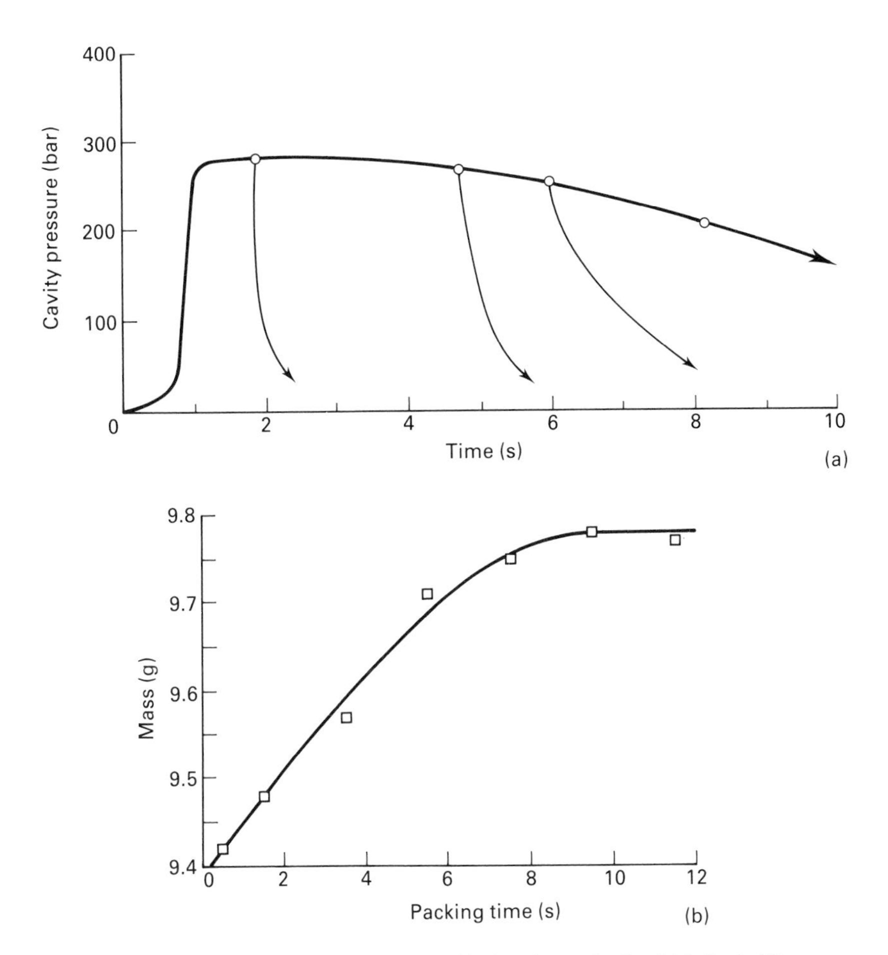

Fig. 5.12 (a) Variation of the cavity pressure with time for cycles in which the holding pressure is switched off at various times. (b) The corresponding variation of the mass of a polyethylene moulding

the mould. Once the gate region has solidified no further feeding can occur, and Fig. 5.12b shows how the mass of a particular moulding reaches an asymptotic value. When the corresponding shrinkage values are examined, it is found that most of the extra material injected in the feeding phase has been used in reducing the thickness shrinkage. In contrast the length and width shrinkage decrease less, because of the constraining influence of the solid skins.

If there is orientation in a moulding, the anisotropy of shrinkage can cause distortion. The lid of a box often has a central gate, and if the shrinkage is higher in the radial than in the circumferential direction this can be accommodated by the lid warping. This is a topological solution to the problem of a disc of radius r, for which the circumference is greater than $2\pi r$. Such a disc cannot exist in a plane, but it can exist on the surface of a rumpled sheet, so the circumference rises above the plane in two or more sectors. Other forms of distortion can occur if the moulding contains regions that solidify earlier than others—if the rim of a box lid is thicker than the rest of the moulding, it is likely to shrink more, and this can cause distortion as the product seeks to avoid residual stresses developing.

It is usual to design moulds so that the melt flows from thicker sections into thinner sections. This is to make sure that the extremities of the moulding can be fed properly. A cooling calculation for a two-dimensional heat flow illustrates one of the problems than can arise when an island of melt is cut off from the main supply during solidification. Fig. 5.13a shows the isotherms relevant to the cooling of a reinforcing rib on an injection moulding. The rib, which is two-thirds as thick as the plate that it reinforces, solidifies first, and an island of melt is eventually left at the intersection of the rib and the plate. Contraction of this last island of melt can cause sink marks in the plate surface if the moulding is relatively thin, and the skin pulls in, or it can lead to a central shrinkage cavity in a 10 mm thick moulding. A sudden increase in cross section in the direction of flow increases the probability of a shrinkage cavity forming, which would substantially weaken the moulding. Fig. 5.13b shows a section through an injection moulded gear. A shrinkage cavity has formed where the flange intersects the rim that supports the gear teeth.

5.3.2 Surface roughness

There are a number of types of surface roughness on plastics products that can occur as a result of melt processing. These are especially noticeable on any surface of the product that has solidified in contact with air. Thus if one looks down the bore of an extruded pipe towards a light source it is often possible to see surface grooves that mark the positions where the melt passed the spider legs of the die; there is a slight difference in the elastic recovery of the melt in the region where the two melt streams welded together. A more serious roughness may be found on the interior of blow moulded containers. This is due to a form of flow instability as the polymer melt is rapidly extruded from the die. A periodic slipping of the melt at the die wall produces ridges at right angles to the flow direction (Fig. 5.14a). In addition to being unsightly these could be the site for the initiation of a crack. The roughness becomes

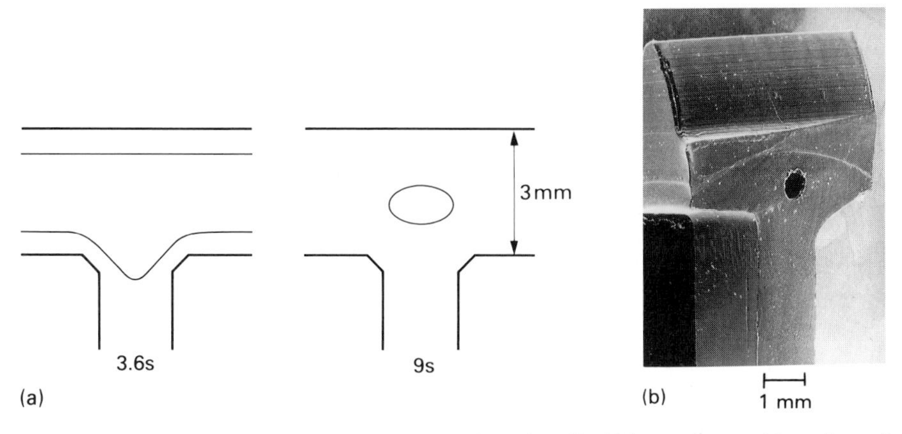

Fig. 5.13 (a) 100 °C isotherms around the junction of a rib thickness 2 mm with a plate of thickness 3 mm, during cooling. (b) Shrinkage cavity in a polypropylene gear, at the intersection of the flange and the gear

more severe as the molecular mass of the polymer increases, or the extrudate velocity increases.

It might be expected that the surface of injection mouldings would be a replica of the surface of the steel mould. However, waves may be visible on the moulding surface, running at right angles to the flow direction (Fig. 5.14b). The waves are prominent if the mould is filled slowly. They form on the skin of the polymer during mould filling. There is a tensile stress parallel to the flow direction in core of the melt (Section 4.3.3). As there is no net tension on the advancing melt front, this puts the skin of the moulding ino compression and the skin buckles.

5.3.3 Residual stresses in extrudates

Residual stresses occur both in extruded products and in injection mouldings. Extruded products are easier to analyse because the pressure does not vary during solidification, and the extrudate dimensions are not externally constrained during solidification. Consider a flat area in an extruded product, similar to that shown in Fig. 5.15. For modelling purposes the polymer is divided into a number of thin parallel layers. It is assumed that the flow stresses in the molten polymer are negligible so that the layers of polymer are stress free as they solidify. The solidification temperature T_s is taken to be equal to T_g for a glassy polymer, or to the temperature at which crystallisation is 75% complete. As the ith layer solidifies it acquires a *reference length* L_i defined as the length at which the layer is stress free at temperature T_s. The L_i value is equal to the current length L of the extrudate at the time of solidification.

The total strain in any solid layer is the sum of the thermal strain and the elastic strain. If the plate is at a temperature T its thermal strain is $\alpha(T - T_s)$ where α is the linear thermal expansion coefficient. The elastic strain in the x

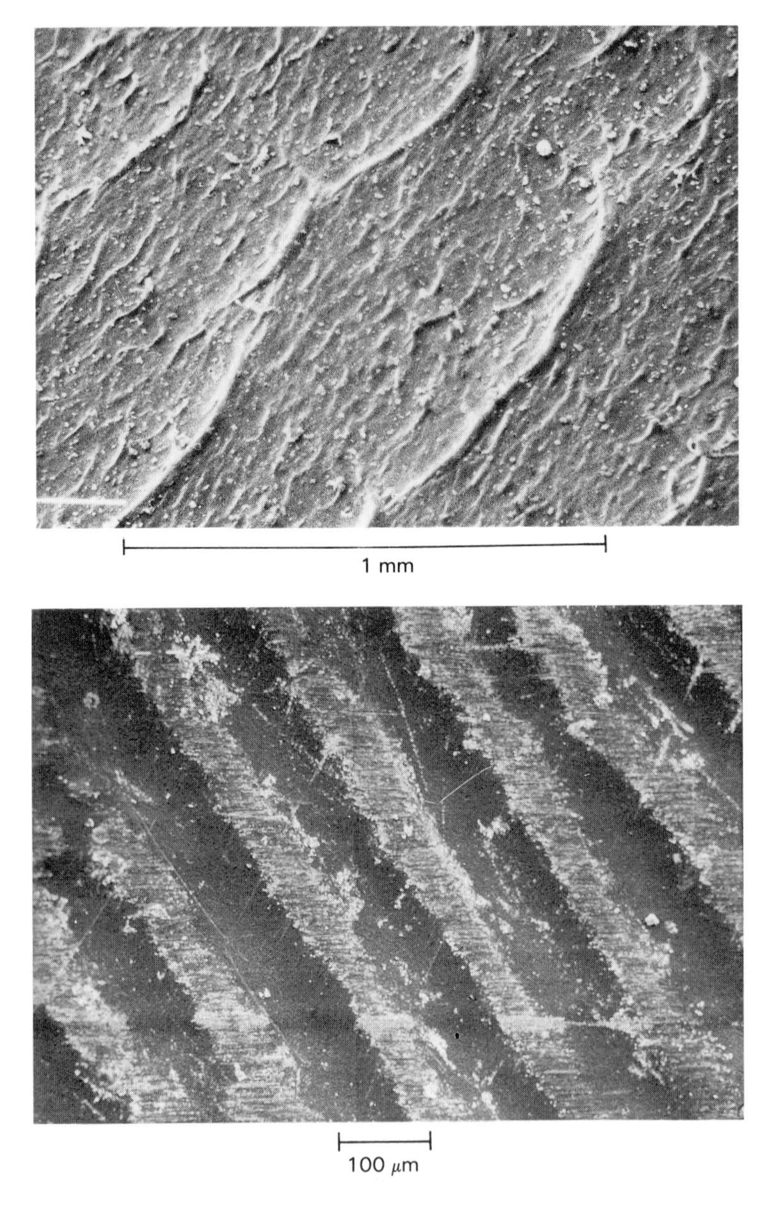

Fig. 5.14 Scanning electron micrographs of surface defects on (a) the inner surface of a blow moulded HDPE bottle and (b) a polyethylene injection moulding. In both micrographs the ridges run at 90° to the flow direction

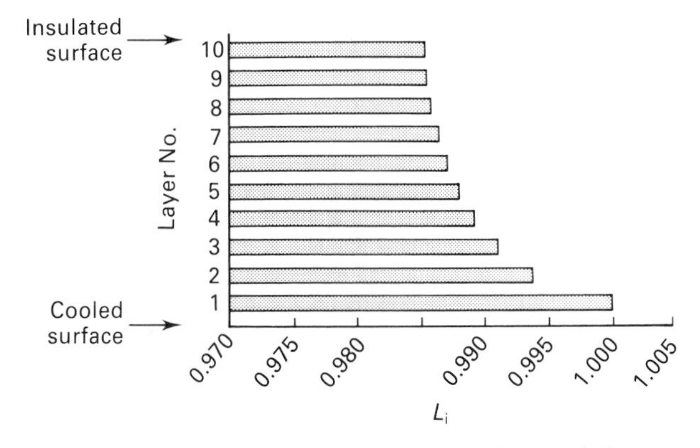

Fig. 5.15 The layers into which a 6 mm thick MDPE extrudate, cooled on one surface only, is divided for residual stress analysis. Each has the reference length at which it would be stress free at the solidification temperature T_S

direction due to a biaxial stress system

$$\sigma_{xx} = \sigma_{yy} = \sigma_i, \qquad \sigma_{zz} = 0$$

in the ith layer is $\sigma_i(1-\nu)/E$. Consequently the total strain is

$$e_i = \alpha(T - T_s) + \frac{\sigma_i(1-\nu)}{E} \tag{5.4}$$

There is no external force on the plate in either the x or y directions, so the total internal force on a cross section

$$w\Delta z \sum_{i=1}^{n} \sigma_i = 0 \tag{5.5}$$

The width w and thickness Δz of the layers are equal. Therefore the compressive stresses in the outer layers with high L_i values must be balanced by tensile stresses in the interior, and the net stress on the section is zero. The cooling of the layers can be calculated using the finite difference methods of Appendix A. The equilibrium length of the extrudate can be calculated by using equations (5.4) and (5.5). The value of L_i varies from a maximum at the surfaces which solidified first, to a minimum at the centre which solidified last. Fig. 5.16a shows how the surface and the centre stresses vary with Biot's modulus (Section 4.2.3). The stresses are proportional to

$$\sigma^* = E\alpha(T_s - T_b) \tag{5.6}$$

so the magnitude of the residual stresses can be reduced by using a higher cooling bath temperature T_b or reducing the value of Biot's modulus. If the extrudate is only cooled from one side, as occurs for pipes, the residual stresses vary from compressive values at the outer to tensile values at the inner surface (Fig. 5.16b). If a sheet is cooled unequally from two sides, the product will bend until internal stresses have a zero net bending moment. Therefore if a moulded part is bent this suggests that the cooling has been uneven.

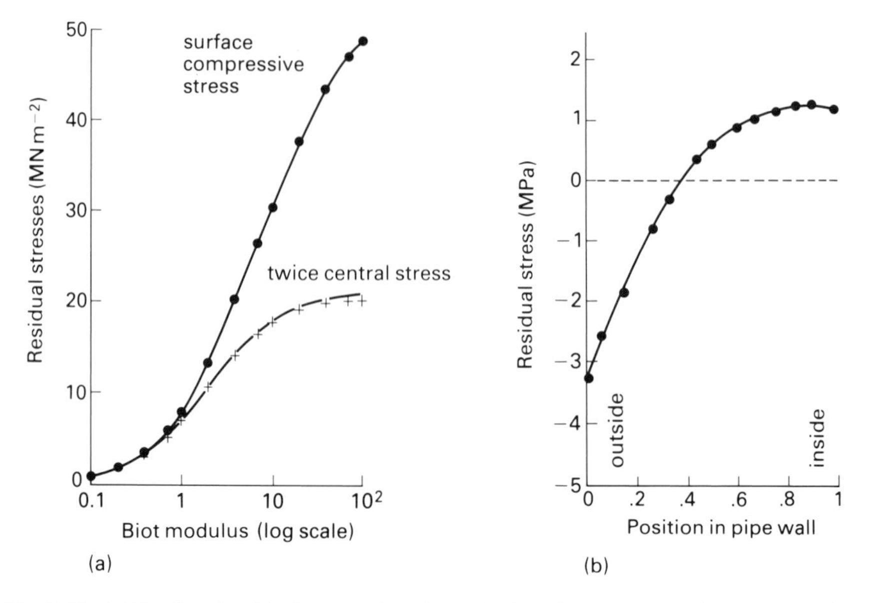

Fig. 5.16 (a) Predicted residual stresses in a sheet of polycarbonate versus Biot's modulus (from Mills, *J. Mater. Sci.*, 1982, **17**, 558). (b) Measured residual stresses in an extruded polyethylene pipe wall (from *Plastics Pipes V*, Plastics and Rubber Inst. Conference, 1982)

5.3.4 Residual stresses in injection mouldings

For injection moulding, varitions in the cavity pressure p during solidification leads to variations in the reference lengths L_i of layers in the moulding. The latter must be redefined as the length at which the layer is stress free at atmospheric pressure and temperature T_s. If the layer solidifies at a pressure p_i then when the moulding is at a pressure p a term $(p-p_i)/3K$ must be added to the right-hand side of equation (5.4), where K is the bulk modulus of the melt. The complex shape of the mould means that the solidifying moulding is restrained in length to the dimension L of the mould. Hence the left hand side of equation (5.4) is zero while the moulding is in the mould, and after the moulding cools to room temperature the stress distribution in the mould is

$$\sigma_i = \frac{E}{1-\nu}\left(\alpha(T_s - T) - \frac{p_i}{3K}\right) \tag{5.7}$$

Once the moulding is released from the mould the average residual stress on the cross-section falls to zero according to equation (5.5). There are many possible residual stress distributions across the thickness of an injection moulding, because of the different possible pressure histories in the mould.

The shrinkage of thermoplastics can be turned to advantage if metal components are to be attached to a moulding. The most common case is a metal insert, often with an internal thread (Fig. 5.17a) to allow the repeated assembly and dismantlement of a structural component. The inserts are usually knurled on the outside to prevent the possibility of rotation or pull-out

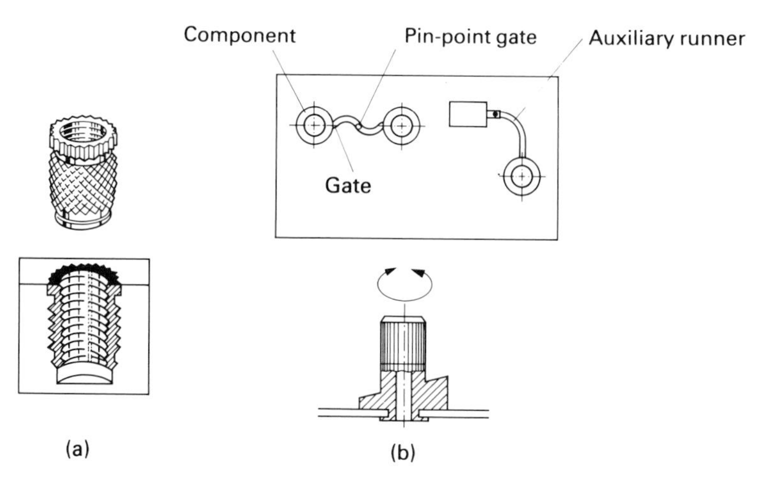

Fig. 5.17 (a) A metal insert before and after insertion into a plastic part. (b) Injection moulded outserts on a steel plate and a cross section of an outsert knob (from *Lexan Noryl Valox* booket, General Electric Plastics, and *Outsert Moulding with Hostaform* booklet, Hoechst, 1978)

under high loads. If the insert is placed in the injection mould and the plastic moulded around it, it will not allow the plastic to contract so it will be held in place by the residual tensile hoop stresses in the plastic. There is the disadvantage that a moulded-in insert will cause a weld line to form by dividing the melt flow (see Section 5.5), and in some plastics the residual tensile strains are high enough to cause environmental stress cracking. For these materials it is preferable to heat the inserts and press them into the moulded component. Localised heating can be provided by mounting the insert on the tip of an ultrasonic vibrator and pressing it into the hole in the moulding.

'Outsert' mouldings are used when it is desirable to combine the dimensional stability and stiffness of a metal baseplate with the assembly advantages of moulded plastics bearings, springs, axles, etc. For example a video recorder can use a number of outsert mouldings to construct the main load bearing framework for the moving parts. The galvanised steel baseplate, containing suitable punched holes, is placed in the injection mould. The mouldings are gated individually because of the high shrinkage of the plastic, or local clusters of components are fed through curved runners to prevent shrinkage stresses betwen the components (Fig. 5.17b). The plastic is prevented from shrinking in the direction perpendicular to the baseplate, so it has a residual tensile stress in the central region, balanced by compressive stresses at the periphery that grip the baseplate.

5.4 MIXING AND FUSION

5.4.1 Mixing

Often the reason for a failure in a moulding will be found to be the poor dispersion of pigments and other additives, associated with high orientation. Polyolefins in particular are weakly bonded to pigments. Elongated strings of pigment act in the same way as bands of inclusions in wrought steel products, so a fracture can initiate from one of these strings of pigment when a bending impact occurs.

The two main types of mixing relate to the nature of the phase that is to be dispersed. *Distributive mixing* occurs when fine solid particles or liquids are distributed evenly throughout the melt, whereas *dispersive mixing* is required to break up agglomerates of particles. In a shear flow, as in the barrel of an extruder, the distributive mixing increases with the total shear strain. Fig. 5.18 shows how the layer thickness S of regions of initially pigmented and natural polymer decreases according to

$$S = S_0/\dot{\gamma} \tag{5.8}$$

The validity of equation (5.8) fails when the layer thickness approaches the diameter of any solid particles in the melt. As the total shear strain received by polymer passing down an extruder screw of length L equal to 40 times the diameter exceeds 500, the distributive mixing is good. Hence granules of coloured masterbatch polymer are adequately dispersed by the time that they leave the extruder.

Extrusion is not adequate to disperse agglomerated powders such as carbon black of a high 'structure factor'. What is necessary is higher stress mixing such as can be produced by batch internal mixers that contain intermeshing blades. An alternative is the cavity transfer mixer that fits on to the front of an extruder screw and which has a cutting and folding action on the melt.

5.4.2 The processing of PVC

The effects of processing on PVC are more complex than those of most other polymers because the suspension polymerisation particle structure. This

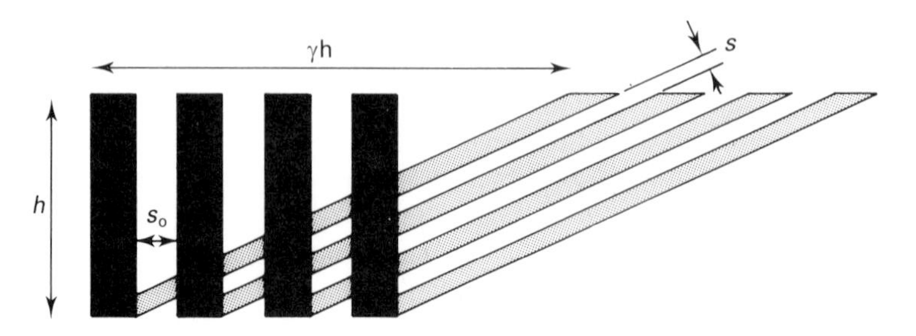

Fig. 5.18 The spatial disposition of layers of coloured polymer before and after a shear strain in a flow causing distributive mixing

particle structure would be irrelevant to the product properties if the grains melted into a homogeneous melt on heating. However, PVC is slightly crystalline, with about 10% crystallinity measured by wide angle X-ray diffraction and this crystallinity binds the particles together. The crystals melt in the range 200 to 240 °C; lower temperatures are used in melt processing because thermal degradation must be avoided. Transmission electron microscopy reveals a microdomain substructure inside the primary particles, so the crystals must be smaller than 10 nm in size.

A typical method of manufacturing PVC extrusions is to process a powder blend through a twin screw extruder (Fig. 5.19). The counter-rotation of the two intermeshing screws inside a barrel with a figure of eight cross section is an effective way of pumping a powdery type of melt that can slip against the barrel wall. Its action is closer to that of a positive displacement pump than is the single screw extruder; the C shaped segments of material are passed from one screw to the other. The initial stage of processing occurs once the PVC is heated to above its T_g of 80 °C; the grains are compacted together to increase the bulk density from about 500 to 1200 kg m^{-3} (Fig. 5.20). At this stage the various solid additives are at the grain boundaries. Next a densification of the grains occurs with the density increase starting at the surface and spreading to the interior as the porosity is eliminated. The grains then become deformable and elongate in the direction of flow. Finally the grain boundaries fuse and a melt containing only primary particles is left. The breakdown is not the same if the powder is compounded in an internal mixer, in which large kneading blades impart a much higher level of shear at an early stage in the process, and break up the grains into fragments.

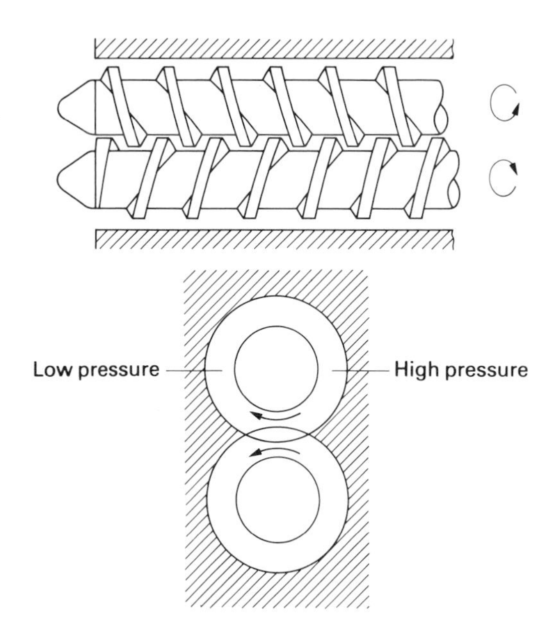

Fig. 5.19 Intermeshing counter-rotating screws in a twin screw extruder

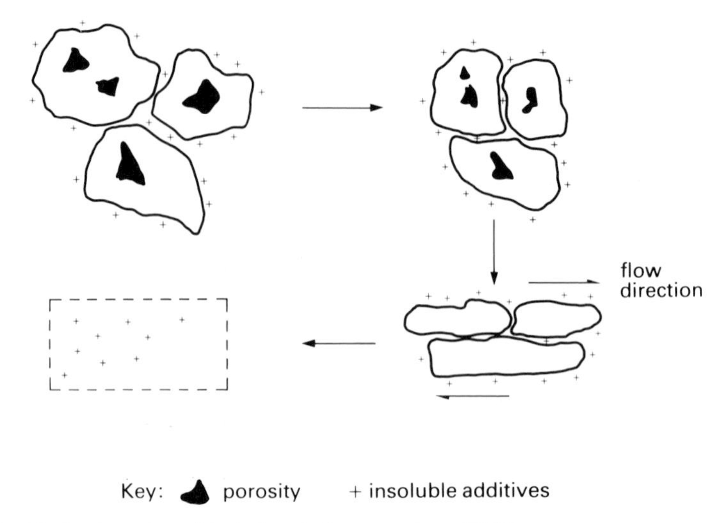

Fig. 5.20 Mechanism of particle fusion during extrusion of PVC

There are various methods of assessing the state of fusion of the PVC powder. If the grain boundaries are intact they are very weak bcause of the coating of stabilisers etc. Consequently immersing the PVC in a liquid such as methylene dichloride causes the grains to separate and gives a white appearance. Fig. 5.21 shows the appearance of PVC pipe after such a test. The wall of the pipe has been machined away by varying amounts to expose the different layers. Fusion is only complete at the inner and outer surfaces which have experienced the highest levels of shear in passing through the die. A pipe with this level of fusion will have inferior mechanical properties. Often the fusion of PVC particles is referred to as gelation, although no crosslinked network develops. The usage is probably due to the changes in the melt rheology as the grain boundaries disappear. As PVC is mixed in an internal mixer the viscous dissipation causes a steady rise in the PVC temperature. The PVC can be removed at various temperatures and the pressure to make it flow through a very short cylindrical die measured. There is a minimum pressure for extrusion at about 160 °C when the inter-grain voids and internal porosity have disappeared but when the melt is still flowing as a collection of subgrains. When the internal mixing is taken to 180 °C the viscosity rises, as does the melt elasticity. The extrudate shows severe melt fractures as a result. The formation of a continuous entanglement network throughout the melt is responsible, hence the reference to gelation. The remaining crystallinity at this temperature may help to make the entanglement network more effective as the crystals physically bind together neighbouring molecules.

PVC pipes of the type shown in Fig. 5.21 are commonly used for water distribution, in which case they are subjected to an internal pressure, which can vary according to the demand for water. The fluctuating stresses that are generated in the pipe wall are an example of fatigue loading. Fatigue stresses, if high enough, will eventually cause the initiation of small cracks. If weak

Fig. 5.21 The wall of a PVC pipe after exposure to methylene chloride for 10 min. The white areas are where there is incomplete fusion of the particles (courtesy of Dr. J. Marshall, Manchester Polytechnic)

grain boundaries remain in the PVC they can be the site where a crack the size of a grain starts. Subsequent fatigue cycles will then cause the crack to grow slowly until failure occurs. The initiation of a crack will take place at the weakest place in the microstructure. Therefore the average level of fusion is less important than the homogeneity of the fusion process. In a pipe there will be positions of weakness where the melt has passed on either side of the die spider legs. Also the hoop stresses in the pipe are slightly larger at the bore than at the outer surface of the pipe. Consequently cracks tend to start at the bore of the pipe from a point of weakness.

6

Viscoelastic behaviour

6.1 INTRODUCTION

Although most metals and silicate glasses behave as elastic materials at ambient temperature, this is not the case for plastics where the mechanical behaviour is time dependent. They are referred to as viscoelastic materials because their behaviour is between that of a viscous fluid and an elastic solid. The most dramatic viscoelastic effects occur when the polymer is at a temperature at which it is in transition between two states; for example between the glassy and rubbery liquid state.

The treatment of viscoelastic models is limited to a demonstration of the theoretical framework, showing how the different phenomena are connected. This is balanced by engineering design calculations of the deflections of viscoelastic structures, using modifications of methods for elastic materials. The re-design of plastic products to compensate for the low elastic moduli is dealt with in Chapter 13.

Table 6.1 lists a number of simple experiments which demonstrate viscoelastic behaviour. The loading on a plastic product may be one of these simple kinds. For example a person sitting on a plastic chair for an hour will subject it to creep loading, whereas a plastic spring deflected by a fixed amount will undergo stress relaxation. Other more complex types of loading can be analysed in terms of combinations of these simple cases. It must not be thought that time dependent phenomena only occur at low stresses, as we shall see in subsequent chapters that yielding and fracture phenomena are also time dependent.

Table 6.1 Simple cases of viscoelastic behaviour

	Variation with time of	
Test	Strain	Stress
Creep	Increases	Constant
Stress relaxation	Constant	Increases
Cyclic	Sinusoidal	Sinusoidal but phase shift
Tensile	Constant rate of increase	Usually increases

6.2 LINEAR VISCOELASTIC MODELS

6.2.1 The Voigt model for creep

We shall start by using simple linear viscoelastic models for the phenomena in Table 6.1. Although the models are inadequate at high stress levels they are an aid to understanding the deformation processes, and they are the basis for more complex treatments.

Table 6.2 Elements in linear viscoelastic models

Type	Symbol	Equation	Constant
Spring		$f = kx$	k
Dashpot		$f = c\,\mathrm{d}x/\mathrm{d}t$	c

Mechanical analogues of viscoelastic behaviour are constructed using the linear mechanical elements shown in Table 6.2. They are linear because the equations relating the force f and the extension x only involve the first power of both variables. The elements can be combined in series or parallel as shown in Fig. 6.1 and Fig. 5.5.

The convention to remember for these models is that elements in parallel undergo the same extension. Thus in the series or *Maxwell model* the elements experience the same force, and in the parallel or *Voigt model* they experience the same extension. The total force f across the Voigt model can be written as

$$f = kx + c\frac{\mathrm{d}x}{\mathrm{d}t} \tag{6.1}$$

Such a combination of mechanical elements occurs on the rear suspension of many cars, where a helical spring surrounds a shock absorber, and both connect the wheel to the car body.

The viscoelastic version of the Voigt model is produced by imagining the elements to be contained in a unit cube so that the force on the unit area end face is the tensile stress σ, and the extension per unit length is the tensile strain e. This conversion is not perfect, as the Voigt mechanical model only operates along one axis, whereas the viscoelastic material behaviour is isotropic. Hence the mechanical model must be imagined to aligned with the stress direction inside a 1 m^3 black box. The mechanical analogue is an aid to understanding the behaviour under different types of loading.

The differential equation of the Voigt viscoelastic model has the same form as equation (6.1) and is

$$\sigma = Ee + \eta\frac{\mathrm{d}e}{\mathrm{d}t} \tag{6.2}$$

The renamed constants are a Young's modulus E and a viscosity η. *It is not*

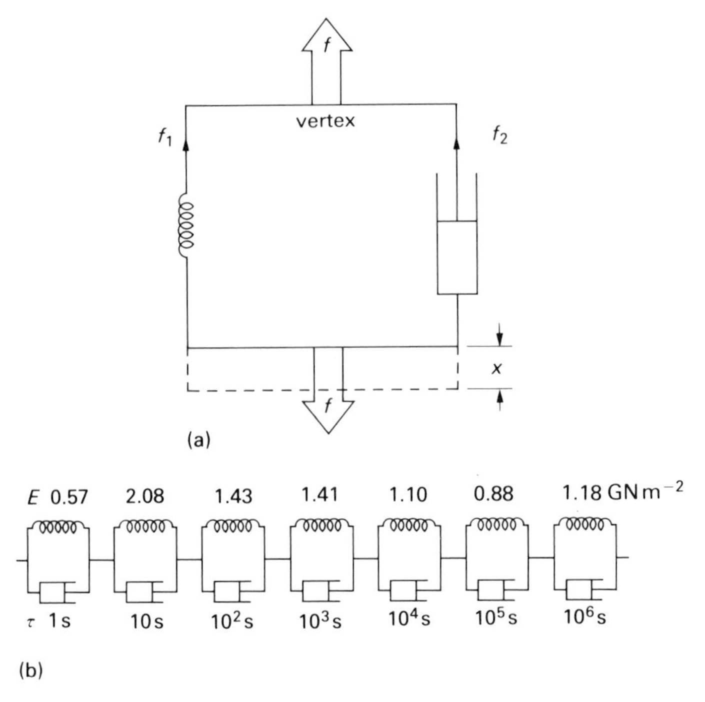

Fig. 6.1 Viscoelastic models employing (a) a single Voigt element and (b) multiple Voigt elements connected in series. The values of the moduli and retardation times are used to model the creep of HDPE in Fig. 6.3

possible to link the values of these constants directly to the modulus of the crystalline phase and the viscosity of the amorphous interlayers in a semi-crystalline polymer. Hence the Voigt model is an aid to understanding creep, rather than a mechanism of microstructural deformation.

When the Voigt model is subjected to creep loading, this means that

$$\sigma = 0 \qquad \text{for } t \leq 0$$

$$\sigma = \sigma_0 \qquad \text{for } t > 0$$

Making this substitution for $t > 0$, and dividing equation (6.2) by η we obtain

$$\frac{\sigma_0}{\eta} = \frac{Ee}{\eta} + \frac{de}{dt}$$

This equation can be integrated, once both sides have been multiplied by $\exp(Et/\eta)$, giving

$$\frac{\sigma_0}{E} \exp\left(\frac{Et}{\eta}\right) = e \exp\left(\frac{Et}{\eta}\right) + A$$

The constant of integration A is found to equal σ_0/E when the initial condition, that $e = 0$ when $t = 0$, is substituted. Therefore the solution is

$$e = \frac{\sigma_0}{E}\left[1 - \exp\left(-\frac{Et}{\eta}\right)\right] \tag{6.3}$$

The physical response of the Voigt model is made more obvious if we introduce the *retardation time* $\tau = \eta/E$. Fig. 6.2 shows that 63% of the creep strain occurs in the first τ seconds and 86% within twice the retardation time.

6.2.2 Creep compliance of the generalised Voigt model

In order to compare the prediction of the model with experimental creep data we need to define the *creep compliance*

$$J(t) \equiv \frac{e(t)}{\sigma_0} \tag{6.4}$$

where the parentheses indicate that both J and e are functions of the time t since the creep load was applied. For the Voigt model the creep compliance

$$J(t) = \frac{1}{E}\left[1 - \exp\left(-\frac{t}{\tau}\right)\right] \tag{6.5}$$

The qualitative response of the Voigt model to creep loading is correct but Fig. 6.3 shows that the quantitative agreement with the creep of a polyethylene is poor. Better quantitative agreement can be obtained by combining, in series, Voigt models with different retardation times as in Fig. 6.1b. The series connection means that there is the same force (stress) in each Voigt model, so the creep compliances of each can be added to give

$$J(t) = \sum_{i=1}^{n} \frac{1}{E_i}\left[1 - \exp\left(-\frac{t}{\tau_i}\right)\right] \tag{6.6}$$

The real creep response of polyethylene in Fig. 6.3 can be adequately reproduced by using equation (6.6) with retardation times at intervals of factors of 10, i.e. $\tau_1 = 1$ s, $\tau_2 = 10$ s, $\tau_3 = 100$ s, etc. Thus polyethylene can be said to have a *spectrum of retardation times*. The retardation time spectrum can be found by curve fitting the creep response; it can then be used to predict other forms of viscoelastic behaviour.

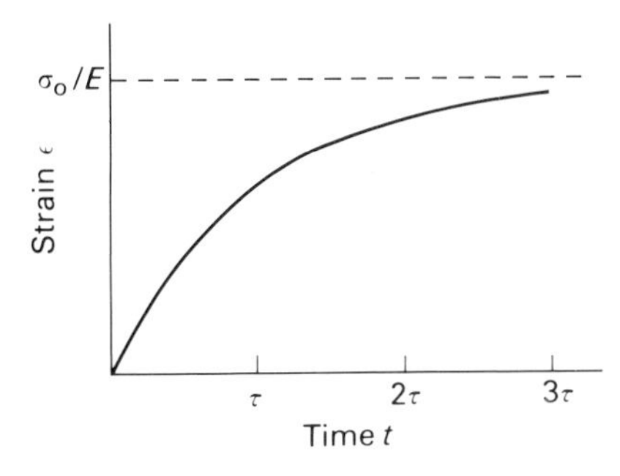

Fig. 6.2 Response of a single Voigt element to a creep stress σ_0 applied at time $t = 0$, shown on linear scales

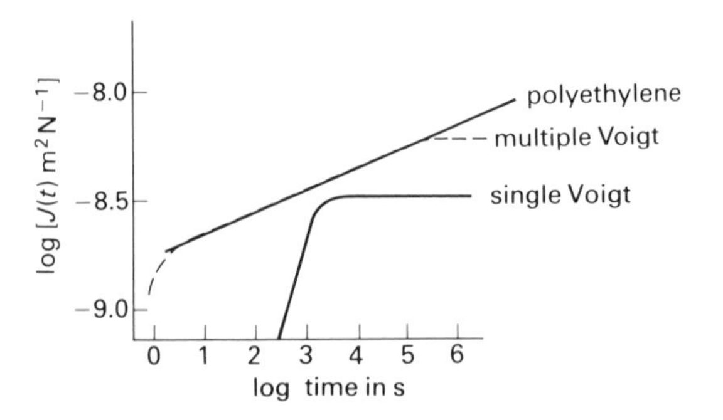

Fig. 6.3 The shear creep compliance of an HDPE at 29 °C versus time on logarithmic scales. The dashed curve is for the multiple Voigt element model of Fig. 6.1. The response of a single Voigt element having $E = 300$ MN m^{-2}, $t = 1000$ s is also shown

6.2.3 Prediction of stress relaxation

The generalised Voigt model (Fig. 6.1b) will be used to predict stress relaxation. A similar analysis of the cyclic loading response is dealt with in Section 6.4.1. The Maxwell model (Fig. 5.5), or a combination of Maxwell models in parallel, can be used to model stress relaxation, but the purpose here is to show the predictive power of a known viscoelastic model. A single Voigt element will not exhibit stress relaxation; when a constant extension is imposed there is a constant force in the spring, and a zero force in the stationary dashpot. However, when the model in Fig. 6.1b is used the extension can redistribute itself from one Voigt element to another, allowing stress relaxation. In order to obtain a solution we need a rule for time-varying loads on a viscoelastic model.

When a combination of loads is applied to an elastic material the stress (and strain) components caused by each load in turn can be added. A similar concept applies to linear viscoelastic materials; the *Boltzmann Superposition Principle* states that if a creep stress σ_1 is applied at time t_1, and the further creep stress σ_2 is applied at time t_2, the total creep strain at times $t > t_2$ is

$$e = \sigma_1 J(t - t_1) + \sigma_2 J(t - t_2) \tag{6.7}$$

The result is shown schematically in Fig. 6.4 for the case when σ_2 is negative. It can be seen that the creep strain is equal to a particular value e_0 at two different times. This suggests an approximate method of predicting stress relaxation. Further negative stress increments $\sigma_3, \sigma_4, \ldots$ can be made at times $t_3, t_4, \ldots$ and their values chosen so that the strains, calculated from equations (6.6) and (6.7), are $e(t_4) = e_0$, $e(t_5) = e_0$, etc. A computer can be used to calculate the necessary stress increments at sufficiently short time intervals, and hence predict the stress relaxation curve. In the limit that the applied stress is changing continuously at a rate $d\sigma/d\tau$ equation (6.7) becomes the convolution integral

$$e = \int_0^t \frac{d\sigma}{d\tau} J(t - \tau)\, d\tau \tag{6.8}$$

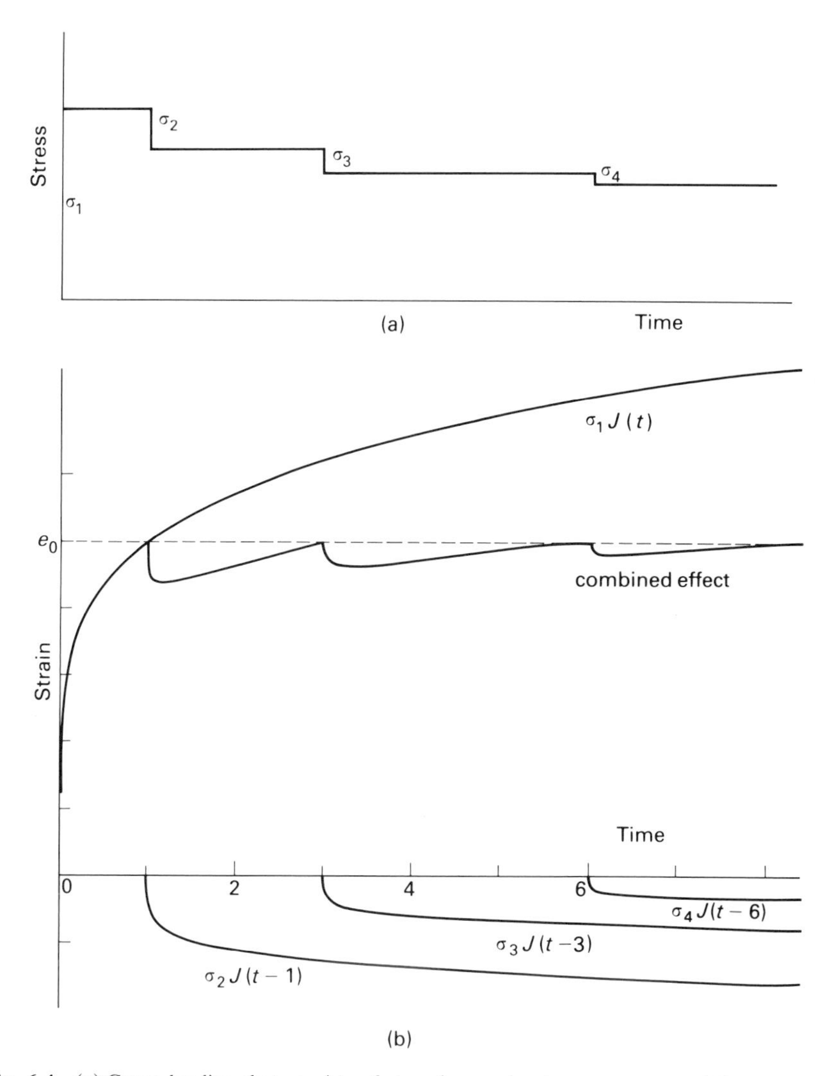

Fig. 6.4 (a) Creep loading that consists of step changes in stress σ_1, σ_2, σ_3 at times t_1, t_2, t_3. (b) The creep strains due to the separate application of the stresses, and their combined effect according to Boltzmann's superposition principle

This is a statement of the exact relationship to which the computer program approximates. We now define the *stress relaxation modulus* $G(t)$ as

$$G(t) \equiv \frac{\sigma(t)}{e_0} \tag{6.9}$$

where e_0 is the imposed fixed strain. Aklonis and MacKnight (1983) show by using Laplace transforms that the relationship between the creep compliance and the stress relaxation modulus is

$$\int_0^t G(\tau)J(t-\tau)\,\mathrm{d}\tau = t \tag{6.10}$$

This relationship has a simple solution when the creep compliance is varying relatively slowly with time; that is when

$$J(t) = J_0 t^n$$

where n is a constant less than 0.15. In this case

$$G(t) \cong \frac{1}{J(t)} \tag{6.11}$$

with an error of less than 4%. For a more rapidly varying creep compliance, equation (6.10) has to be solved numerically. Stress relaxation data are far less common than creep data, and this approximation is normally used for design purposes when stress relaxation data is not available.

6.2.4 The temperature dependence of viscoelastic behaviour

Any of the viscoelastic measurements described so far can be repeated at a series of increasing temperatures. If graphs of the viscoelastic function versus log time or log frequency are compared there is found to be a similarity in shape for measurements made at neighbouring temperatures. This is particularly the case if the temperature range spans the glass to rubber transition for amorphous polymers. For example Fig. 6.5a shows the shear creep compliance of polystyrene over such a temperature range, plotted on logarithmic scales. The data are taken at a low stress level, so it is in the linear viscoelastic region. It was found empirically that the data could be made to superimpose if each curve was shifted horizontally by a factor $a(T)$ where T is the absolute test temperature; this is equivalent to multiplying all the creep times by a constant factor. A small vertical shift factor $\rho T/\rho_0 T_0$, where ρ is the density and the subscripts refer to the reference temperature, was used to conform with the predictions of rubber elasticity theory, but this vertical shift is often ignored.

This *time–temperature superposition* for linear viscoelastic data is the basis of data gathering for academic studies of microstructure–property relationships. The *master curve* at a temperature of 100 °C is shown in Fig. 6.5b. This covers the times from 1 s to 10^{10} s and can be used for the extrapolation of linear viscoelastic behaviour. In terms of the linear viscoelastic models, such as Fig. 6.1b, it means that all the retardation times τ_i have reference values τ_{i0} at temperature T_0, and a common temperature shift factor $a(T)$ so that

$$\tau_i = \tau_{i0}\, a(T) \tag{6.12}$$

whereas the elastic moduli E_i must stay constant. This implies that the temperature dependence is connected to the viscosity η of the dashpots. In Section 2.3.6 on the glass transition the temperature dependence of the melt

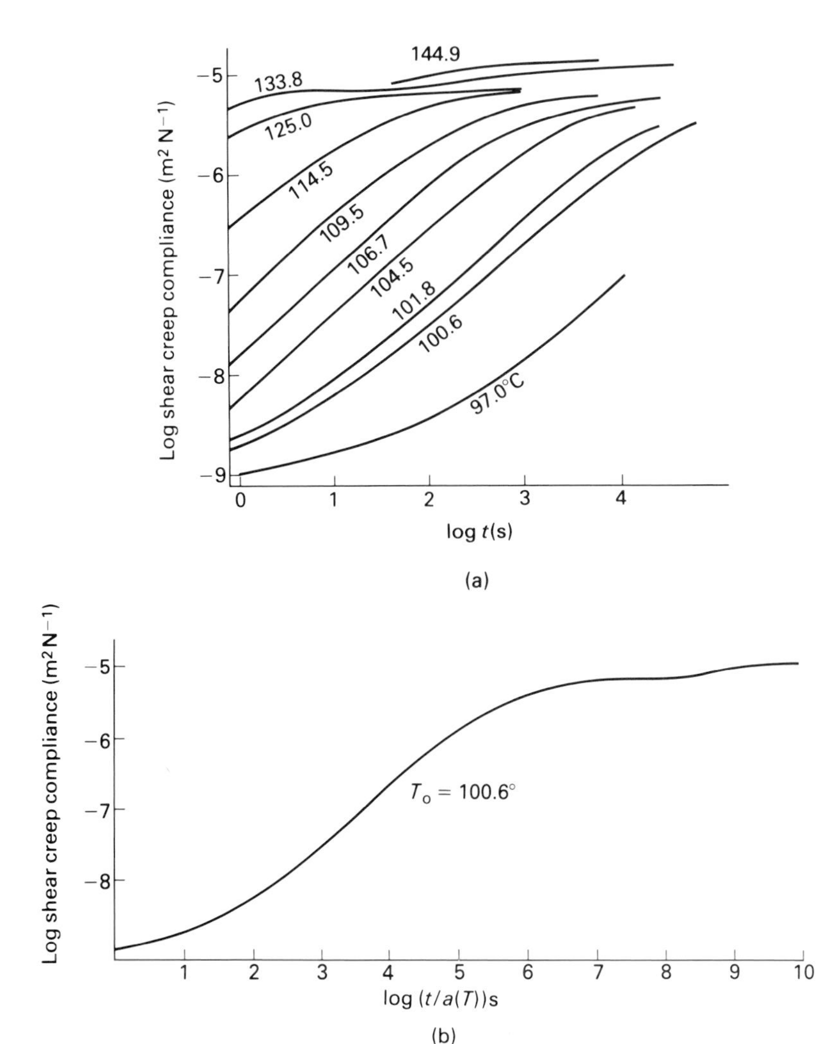

Fig. 6.5 (a) The shear creep compliance of polystyrene versus the creep time at various temperatures. (b) Master curve at 100 °C constructed by shifting the curves in (a) horizontally until they superimpose

viscosity is discussed, and it is mentioned that different molecular weight melts have the same temperature dependence of viscosity. This could imply that the different retardation times in the viscoelastic response are due to different molecular weight entities in the amorphous phase.

After time–temperature superposition was found to work for amorphous polymers, it was tried for semi-crystalline polymers. However these may start to recrystallise into more stable morphologies if heated within 50 °C of T_m, and residual stresses may start to relax. Nevertheless data for polyethylene for the temperature range 20 °C to 80 °C will usually superimpose.

6.3 CREEP DESIGN

6.3.1 Creep data

There is a conflict between the demand for creep data under a wide variety of conditions and the testing time necessary to generate such data. For most plastics there will be tensile creep data for times up to perhaps 1 year at 5 to 10 different stress levels (Fig. 6.6) that are chosen to span the range of stresses that do not cause rapid failure. There may also be creep data at selected elevated temperatures.

There is a lack of creep data from test rigs where there are biaxial stresses, yet the use of plastic pressure pipes for the distribution of gas and water is widespread. For a pipe of mean diameter D_m and wall thickness t under internal pressure p, the hoop strain in an elastic material is given by

$$e_H = \frac{pD_m}{2t} \frac{(2-\nu)}{2E} \tag{6.13}$$

For a viscoelastic material, it is assumed that equation (6.13) becomes

$$e_H(t) = \sigma_H(1-\nu/2)J(t) \tag{6.14}$$

Poisson's ratio ν is assumed not to be time dependent, but this assumption can only be checked by comparing creep data obtained under tensile and biaxial loading. If the stress σ_x in a tensile creep experiment, and the stress σ_H in a pipe creep experiment are equal, the equation above predicts that the hoop strain e_H under biaxial stresses will be smaller than the strain in a tensile creep experiment by a constant factor $(1-\nu/2)$.

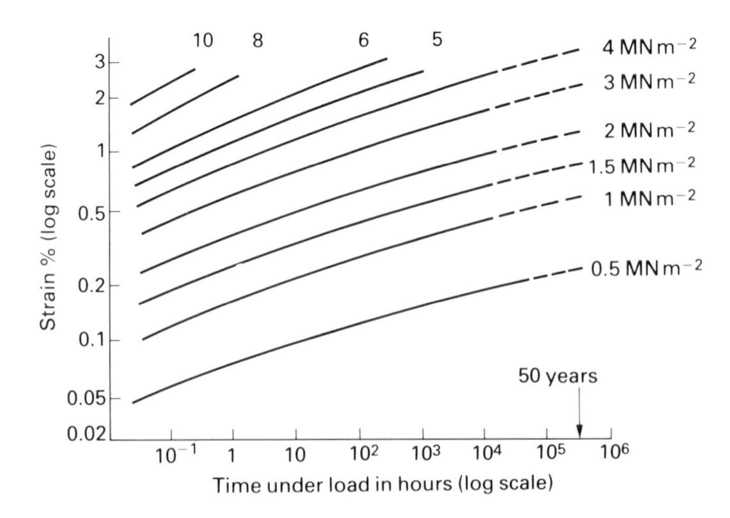

Fig. 6.6 Tensile creep data for HDPE at a range of creep stresses (from *Pipes: Hostalen GM5010 T2, GM7040G*, 1980, Courtesy of Hoechst AG)

6.3.2 Creep deflection calculations

A simple design problem where creep data is required is shown in Fig. 6.7. A cantilever arm with a float at one end is used to operate a water valve in a cold water tank. The maximum load at the free end is 5 N (due to the buoyancy of the float), the operating temperature range is 0 °C to 30 °C, and the deflection of the free end of the arm should not exceed 30 mm after 1 year. It has been decided to injection mould the arm from polyethylene, and the cross section of the arm must be designed. We shall follow through the stages in such a design. A simple conservative design method is used, called the *pseudo-elastic* approach.

Firstly, a cross-plot of the tensile creep data is carried out for the *design time* t_D of 1 year. It may be necessary to extrapolate the creep curves to the time required; for this reason it is normal to plot log strain versus log time, so that the creep data are approximately linear, and extrapolation is straightforward. This *isochronous stress–strain curve* (Fig. 6.8) serves several purposes; it can be used for interpolation to other creep stress levels, and it allows us to classify the type of viscoelastic behaviour. If the plot is linear to within 5% up to a certain stress level then we can say there is *linear viscoelastic behaviour* up to this stress level and the creep compliance $J(t)$, defined by equation (6.4), is not a function of stress. For the data in Fig. 6.8 there is *non-linear viscoelastic behaviour* above a tensile stress of 1 MN m^{-2} and the creep compliance becomes stress dependent

$$J(\sigma, t) \equiv \frac{e(\sigma, t)}{\sigma} \tag{6.15}$$

Secondly we use a modification of the elastic material analysis to calculate the deflection in the product. For the point force F on the cantilever beam of length L in Fig. 6.7 the deflection Δ at the free end of an elastic beam is

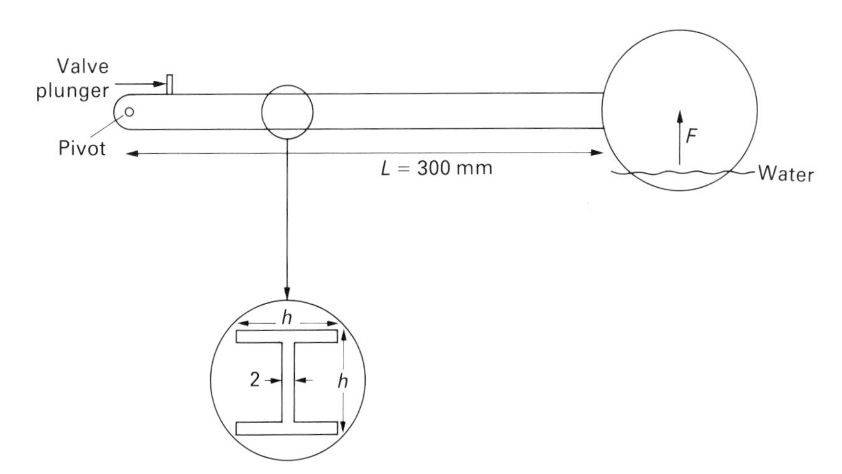

Fig. 6.7 Cantilever beam used with a spherical float to operate a valve in a cold water tank. The inset shows a possible cross section for the beam

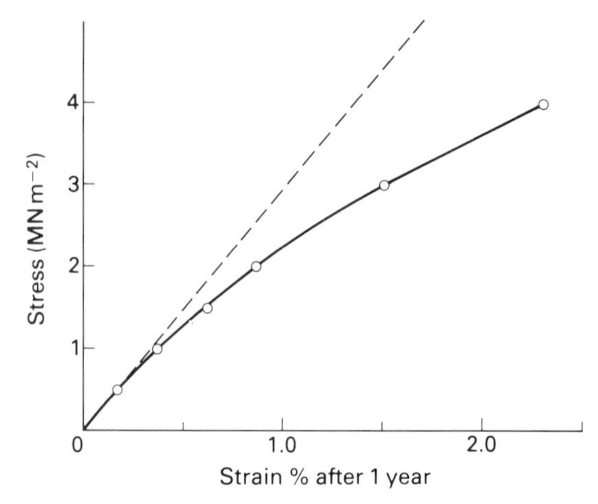

Fig. 6.8 Isochronous stress–strain curve at a time of 1 year constructed from the creep data in Fig. 6.5. The broken line represents linear viscoelastic behaviour

$$\Delta = \frac{FL^3}{3EI}$$

where E is Young's modulus, and I the second moment of area of the beam cross section. We replace $1/E$ in this equation by the creep compliance to obtain

$$\Delta(t_D) = \frac{FL^3 J(\sigma_{max}, t_D)}{3I} \tag{6.16}$$

Fig. 6.7 shows a possible cross section for the beam; the 2 mm thick section can be injection moulded with a short cycle time, and the beam shape is efficient in I to mass ratio. An initial cross section of the beam can be designed using the linear viscoelastic compliance of $J(t_D) = 3.3 \times 10^{-9}\,\text{m}^2\,\text{N}^{-1}$ from Fig. 6.8 in equation (6.16). The deflection limit of 30 mm means that the second moment of area is $I = 5.0 \times 10^{-9}\,\text{m}^4$ which can be met if the height of the beam is $h = 18$ mm.

Thirdly, we have to make a stress analysis of the product. We use the analysis for an elastic material as an approximation, and calculated the maximum stress σ_{max} in the product using

$$\sigma_{max} = \frac{y_{max} M_{max}}{I} \tag{6.17}$$

where y_{max} is the maximum distance from the neutral axis and M_{max} is the maximum bending moment. Substituting the value of I, the maximum y value of 9 mm and the maximum moment of 5×0.3 N m at the left hand end of the beam gives the stress $\sigma_{max} = 2.65\,\text{MN m}^{-2}$.

Fig. 6.9 shows the variation of the longitudinal stress σ_z and strain e_z through a beam. For a long slender beam the strain e_z varies linearly with the

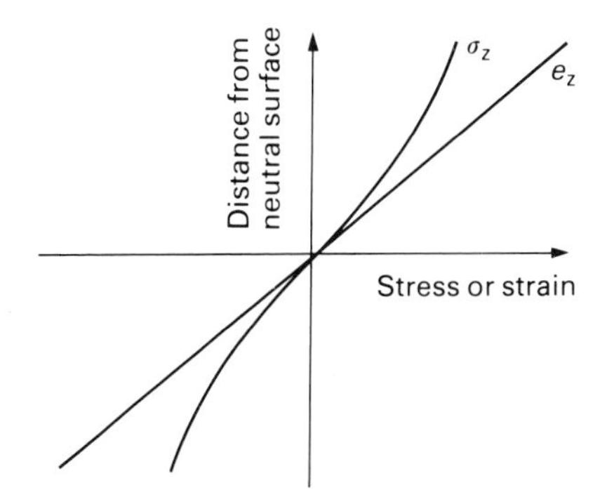

Fig. 6.9 Variation of the longitudinal strain with position in a bent beam, and variation of the longitudinal stress for a non-linear viscoelastic material

distance y from the neutral surface, regardless of whether the material is elastic or viscoelastic. For linearly viscoelastic materials the stress variation is linear so the concept of second moment of area remains valid. However, for the usual case of non-linear viscoelasticity the stress variation through a beam is approximately the shape of the isochronous stress–strain curve and the maximum stress at the upper and lower surfaces will be smaller than that calculated by the elastic equation.

Fourthly the compliance is calculated for the design time and the maximum stress in the product. The creep compliance for this stress level and the design time is calculated from the isochronous curve (Fig. 6.8) as $5.8 \times 10^{-9}\ m^2\ N^{-1}$. This compliance is higher than before, and when it is substituted in equation (6.16) the deflection of 51 mm is found to be too large. The cross sectional dimensions are then increased slightly to meet the new I target and the calculations repeated. This will give a slightly too low deflection, and the process can be repeated if the safety margin on the design is considered too large.

The pseudo-elastic approach, by assuming that the stress distribution follows the elastic analysis, will overestimate the real deflection. There are other sources of error as well. One of these is the assumption in the mechanics analysis that the geometry of the product does not change under load, and the compressive loads are not high enough to cause buckling. However, since plastics can have creep strains of 1 or 2%, and they are comparatively thin, large deflections can easily occur. Consequently it is often necessary to construct a prototype of a product and test its performance, rather than rely on a mechanics analysis alone.

6.3.3 Recovery and intermittent creep

The most important difference between creep in plastics and creep in metals is that only in plastics is the creep strain recoverable when the load is removed. The linear Voigt viscoelastic models predict that the creep strains are 100% recoverable. In practice the fractional recovered strain, defined as

$(1-e/e_{max})$ where e is the strain during recovery and e_{max} is the strain at the end of the creep period, exceeds 0.8 when the recovery time is equal to the creep time. Fig. 6.10 shows that the recovery is quicker for low e_{max} and short creep times, i.e. when the creep approaches linear viscoelastic behaviour.

In reality it is likely that a product will be loaded intermittently. It is impossible to produce data for all kinds of loading history, but experiments with a regular loading and unloading cycle show the general trend. Fig. 6.11 shows how the creep strain changes with the cumulative creep time for both continuous loading, and a 6 hour per day loading cycle. For metals it is expected that the two curves would coincide, but here the intermittent creep loading produces only slightly more strain than a single 6 hour cycle. In the continuous experiment creep time accumulates at four times the rate of the intermittent loading, so the intermittently loaded sample has half the creep strain after 1 year that the continuously loaded polymer has after 3 months. Consequently an intermittently loaded product is less likely to fail from excessive creep strain.

6.4 CYCLIC DEFORMATION

6.4.1 Linear viscoelastic analysis

Vibrations occur in most rotating parts, and many structures will resonate at a particular frequency. We start by examining a simple situation, a sinusoidally

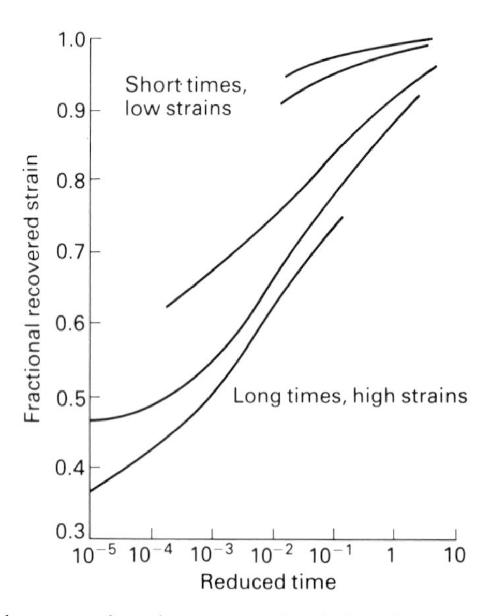

Fig. 6.10 Fractional recovered strain versus reduced time (= recovery time/creep time) for acetal copolymer at 20 °C and 65% humidity. The data from different creep times and stresses do not superimpose (from *Thermoplastics and Mechanical Engineering Design*, ICI Plastics Division, booklet G117)

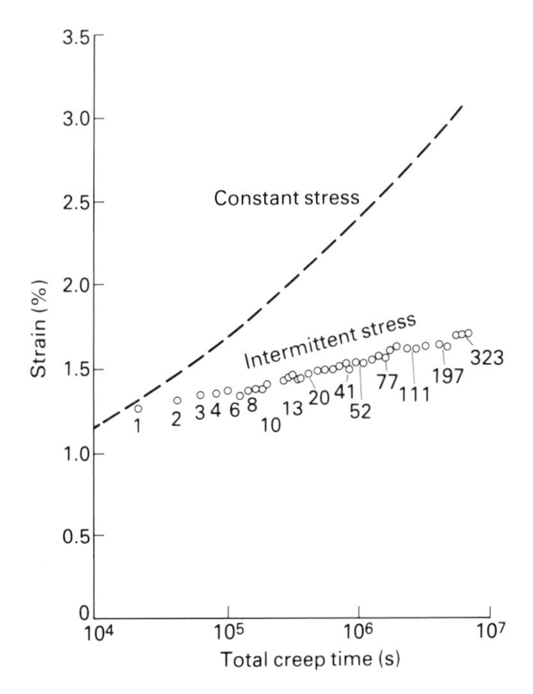

Fig. 6.11 Comparison of continuous and intermittent creep (6 hours per day) of polypropylene at 20 °C and 10 MN m^{-2} stress. The maximum strain at the end of each intermittant creep period is plotted against the cumulative creep time (from *Plastics and Mechanical Engineering Design*, ICI Plastics Division, booklet G117)

varying strain at a single angular frequency ω

$$e = e_0 \sin \omega t \tag{6.18}$$

The response of a material in the linear viscoelastic region is that the stress also varies sinusoidally, but leads in phase by an angle δ

$$\sigma = \sigma_0 \sin (\omega t + \delta) \tag{6.19}$$

At higher strains in the non-linear region other harmonics can be generated and more complex stress–strain responses occur. If we plot equation (6.19) versus equation (6.18) for the first half of the stress–strain cycle (Fig. 6.12a) we see that when the strain has its maximum value e_0 the stress is $\sigma_0 \cos \delta$. The energy input per unit volume of material in the first quarter of the strain cycle is

$$W = \int_0^{e_0} \sigma \, \mathrm{d}e$$

Substituting equations (6.18) and (6.19) and writing θ for ωt gives

$$W = \sigma_0 e_0 \int_0^{\pi/2} (\sin \theta \cos \delta + \cos \theta \sin \delta) \cos \theta \, \mathrm{d}\theta$$

so

$$W = \sigma_0 e_0 \left(\frac{1}{2} \cos \delta + \frac{\pi}{4} \sin \delta \right) \tag{6.20}$$

If the calculation is repeated for the second quarter cycle to obtain the energy output per unit volume the result is the same as equation (6.20) but with a minus sign for the sin δ term. Therefore the first term in equation (6.20) is the maximum energy stored elastically in the cycle, and the second term is the energy dissipated in a quarter cycle. Therefore we can write

$$\frac{\text{energy dissipated per cycle}}{\text{maximum stored elastic energy}} = \frac{\pi \sin \delta}{0.5 \cos \delta} = 2\pi \tan \delta \tag{6.21}$$

Tan δ is a useful parameter of energy dissipation, but it is not the only one used. Many published data are in terms of the complex compliance J^* and the complex Young's modulus E^* and shear modulus G^*. These are defined as

$$E^* \equiv \frac{\sigma}{e} \qquad G^* \equiv \frac{\tau}{\gamma} \qquad J^* \equiv \frac{e}{\sigma} \tag{6.22}$$

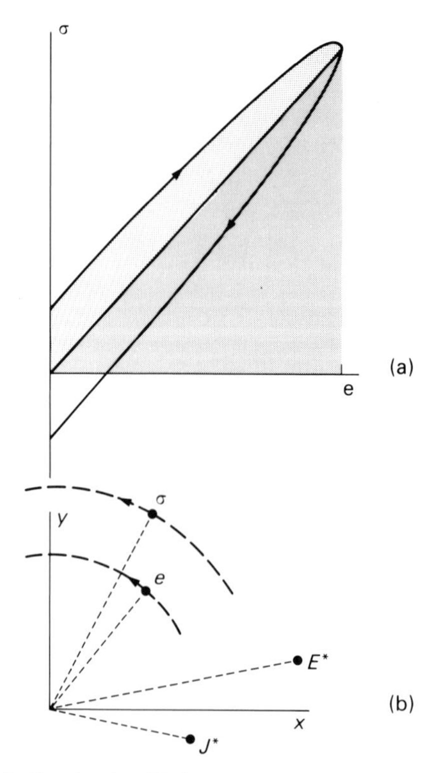

Fig. 6.12 (a) The first half cycle of oscillation, showing the maximum stored elastic energy, and the energy dissipated in the first quarter cycle. Drawn for tan δ = 0.2. (b) Positions of the corresponding stress, E, strain, e, Young's modulus, E^*, and compliance, J^*, in an Argand diagram

The interrelation between these quantities is best understood by writing the stress and strain in complex number form

$$e = e_0 \exp i\omega t$$

$$\sigma = \sigma_0 \exp i(\omega t + \delta)$$

This does not mean that the strain is a complex number, but is a convenient mathematical shorthand, with the modulus $|e|$ representing the amplitude, and the argument Arg e representing the phase of the sinusoidal variation. On an Argand diagram (Fig. 6.12b) the quantities e and σ trace out circles as time varies, but E^* and J^* have the fixed values

$$E^* = \frac{\sigma_0}{e_0} \exp i\delta \qquad J^* = \frac{e_0}{\sigma_0} \exp -i\delta \tag{6.23}$$

Data for the complex modulus are usually presented as graphs of the real and imaginary parts, given the symbols E' and E'', against frequency or temperature. A common way of characterising the temperature dependence of the moduli of polymers is to measure the complex moduli over a range of temperatures. The complex shear modulus G^* can be measured by imposing oscillations on a long slender rod of polymer, with an inertial mass hanging from it. Fig. 1.14 showed how G' varies with temperature for plasticised PVCs and Fig. 3.14 for a polyurea. The tan δ value increases in magnitude as the slope of G' against temperature becomes steeper, and it will reach a maximum at a transition temperature, so this is a method of identifying the transition temperatures.

Expanding equations (6.23) shows that

$$\begin{aligned} E' + iE'' &= \sigma_0/e_0(\cos \delta + i \sin \delta) \\ J' - iJ'' &= e_0/\sigma_0(\cos \delta - i \sin \delta) \end{aligned} \tag{6.24}$$

The imaginary part of J^* is called $-J''$ so that J'' is a positive quantity. The imaginary parts of equations (6.24) show that

$$\sin \delta = \frac{e_0}{\sigma_0} E'' = \frac{\sigma_0}{e_0} J'' \tag{6.25}$$

and this can be substituted in equation (6.20) to give the energy dissipated per unit volume per cycle as

$$W_{\text{dis}} = \pi e_0^2 E'' = \pi \sigma_0^2 J'' \tag{6.26}$$

Thus the imaginary parts of the complex moduli and compliances determine the rate of energy dissipation.

Finally we can examine how the linear viscoelastic models of Section 6.2 behave with a sinusoidal strain input. When the strain variation of equation (6.18) is substituted in the constitutive equation (6.2) of the Voigt model, the stress is given as

$$\sigma = Ee + \eta i\omega e \tag{6.27}$$

Therefore the complex compliance of the Voigt model is

$$J^* = \frac{e}{Ee + i\eta\omega e} = \frac{1}{E(1 + i\omega\tau)}$$

so

$$J' - iJ'' = \frac{1}{E(1 + \omega^2\tau^2)} - \frac{i\omega\tau}{E(1 + \omega^2\tau^2)} \tag{6.28}$$

where τ is the retardation time η/E. The single Voigt model is a poor representation because the tan δ value increases linearly with the frequency ω. However, the model of Fig. 6.1b, with a number of Voigt elements in series, is effective. The stress on each element is the same so the complex compliances of the elements can be added. This means that both J' and J'' are the sums of terms of the form shown in equation (6.28).

6.4.2 Vibration damping

There is likely to be vibration with rotating or reciprocating machinery, and plastics can assist in vibration damping. Any panel or thin shell structure that is part of the machine can be excited to resonate in bending at a relatively low frequency. Unacceptable levels of noise may be generated because of the large area of the panel. If the product can be redesigned then metal panels may be replaced with plastic. The tan δ value rises from 0.001 (metals) to 0.01–0.02 for structural plastics, and this may be enough to reduce the vibration levels. However, there may need to be a metal panel for stiffness reasons, or to contain an environment like hot oil that would attack plastics. In this case a layer of plastic can be applied to one side of the metal panel, or the plastic can be sandwiched between the main panel and a thin metal skin (Fig. 6.13).

The mode of action of these two designs is different. In the former, the neutral surface of the metal panel hardly changes with the addition of the low modulus polymer, and the strain in the polymer alternates from tensile to compressive as the panel bends. Consequently the energy absorbed by the polymer per cycle is proportional to its E'' value, and the strain in the polymer is only slightly greater than that in the metal. In the second case we have a sandwich beam (Section 3.3) where the deflection can either be by extension of the metal skin or by shear of the core (Fig. 6.13b). For a beam of length L and width W the tensile stiffness (force/extension) of a skin of thickness t and Young's modulus E_s is $E_s wt/L$. Similarly for a core of shear modulus G_c and thickness b the shear stiffness of the core is $G_c wL/b$. When the core shears

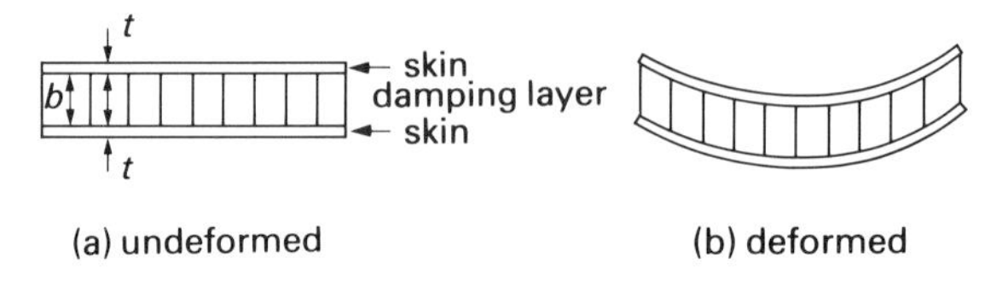

Fig. 6.13 (a) Damping the vibration in a panel using a high loss plastic layer plus a stiff metal skin. (b) The intended shear deformation mode in the plastic when the panel bends

the shear strain increases linearly from zero at the centre to a value Γ at the ends of the beam. It can be shown, by minimising the stored elastic energy of the beam, that the value of Γ is

$$\Gamma = \frac{3E_s bt}{G_c LR} \tag{6.29}$$

if the metal panel has a constant radius of curvature R. Consequently the average shear strain in the polymer can be high if the metal skin has a much higher tensile stiffness than the core shear stiffness, and the skin is relatively long.

Next we need to consider the efficiency of a vibration damping device as a function of frequency. Fig. 6.14 shows a simple device with only one degree of freedom in terms of its motion. The mass can either represent the mass of a machine resting on a separate mounting spring, or it can be the self-mass of a vibrating viscoelastic panel. The transmissibility T can be defined, either in terms of the ratio of the amplitudes of the applied force F_0 and the transmitted force F_T when there is a sinusoidal force input to the mass, or of the ratio of the amplitudes of displacement when there is a sinusoidal displacement x_0 input to the support

$$T = \frac{|F_T|}{|F_0|} = \frac{|x_T|}{|x_0|} \tag{6.30}$$

The differential equations for the forces, in terms of the spring constant k, the dashpot constant c and the time derivatives $\dot{x}$ and $\ddot{x}$ of the spring displacement x, are

$$\frac{F_T}{F_0} = \frac{kx + c\dot{x}}{kx + c\dot{x} + m\ddot{x}} \tag{6.31}$$

When the sinusoidal displacement $x = \exp(i\omega t)$ is substituted in equation (6.31) it becomes

$$\frac{F_T}{F_0} = \frac{k + ic\omega}{k + ic\omega - m\omega^2} \tag{6.32}$$

Next the modulus of the complex quantities in equation (6.32) is taken, yielding

$$T = \left[\frac{k^2 + c^2\omega^2}{(k - m\omega^2) + c^2\omega^2}\right]^{0.5}$$

This result can be re-expressed in terms of the natural frequency $\omega_0^2 = k/m$ of the undamped system, and the characteristic damping time $\tau = c/k$ of the spring dashpot combination, giving

$$T = \left[\frac{1 + \omega^2\tau^2}{(1 - \omega^2/\omega_0^2) + \omega^2\tau^2}\right]^{0.5} \tag{6.33}$$

This is shown in Fig. 6.14 for various values of $\omega_0\tau = c/(mk)^{0.5}$. If strong

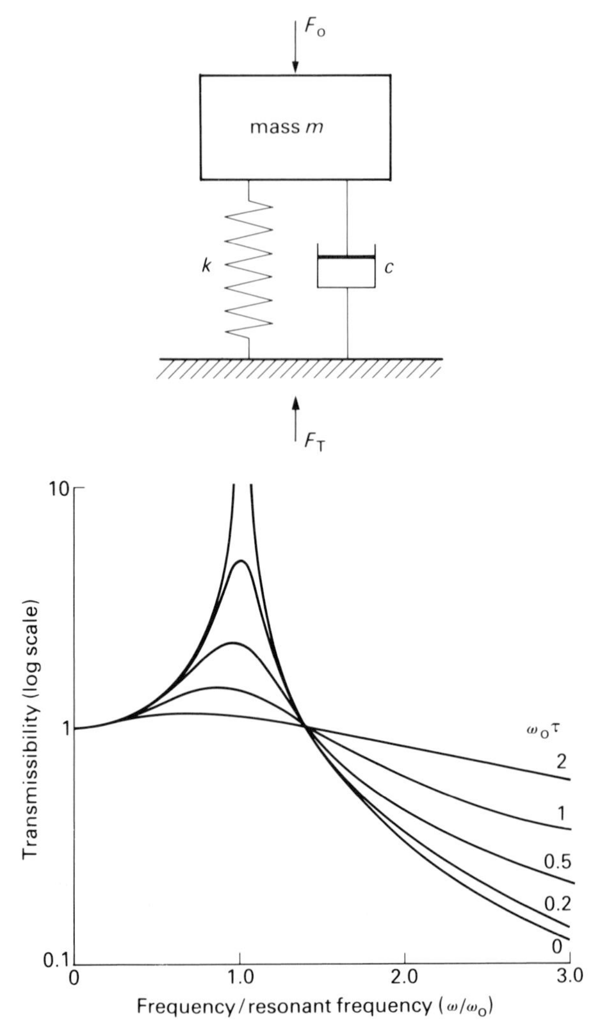

Fig. 6.14 The isolation of a vibrating machine of mass m by a damped spring, considering one mode of vibration. The transmissibility is given as a function of the vibration frequency and the parameter $\omega_0 t$

damping of the resonant frequency ω_0 is required then the value of $\omega_0\tau$ must be >1. On the other hand if the aim is to isolate the vibration of a machine from the floor then the natural frequency of the mounting must be less than 40% of the vibration frequency and the best performance occurs with little or no damping. In the latter case rubber and steel laminated bearings (Chapter 3) have low resonant frequencies and the moderate tan δ values of the rubbers (0.07 to 0.30) prevent severe resonance on start-up.

Some vibration damping applications require high tan δ values. If the damping is to be achieved at a fixed temperature then a polymer that is in the centre of its glass to rubber transition will have a high tan δ over a reasonably wide frequency range. On the other hand if vibrations are to be damped in a product like a steel car door panel that has a wide operating temperature range, it is not so easy to achieve a good performance. A blend of incompatible polymers will produce a material that has separate tan δ peaks at the two T_g s, but neither will be as high as in the homopolymers. Proprietary mixtures of this kind, used in conjunction with an outer metal skin as in Fig. 6.13, provide the best damping performance.

6.4.3 Viscoelastic heating as a cause of fatigue failure

The energy dissipated by viscoelastic deformation is liberated as heat, and in some circumstances this is unwelcome. Consider the fatigue loading of a plastics product, such as occurs every time a gear rotates. Fig. 6.15 shows how the compressive contact stresses between gear teeth are localised near the surface. The contact moves down and up the face of the tooth as the gear rotates. If the contact force is F and the contact width is b on a tooth of face width w the Hertzian contact stress normal to the tooth surface in the contact region in Fig. 6.15 is

$$\sigma_N = \frac{2F}{\pi b w} \tag{6.34}$$

There will also be tensile stresses tangential to the surface due to the bending action of the force F. This tensile stress is calculated by treating the tooth as a stubbly cantilever beam and using equation (6.17). It has its maximum magnitude at the root of the tooth as R in Fig. 6.15. Two heat inputs to the

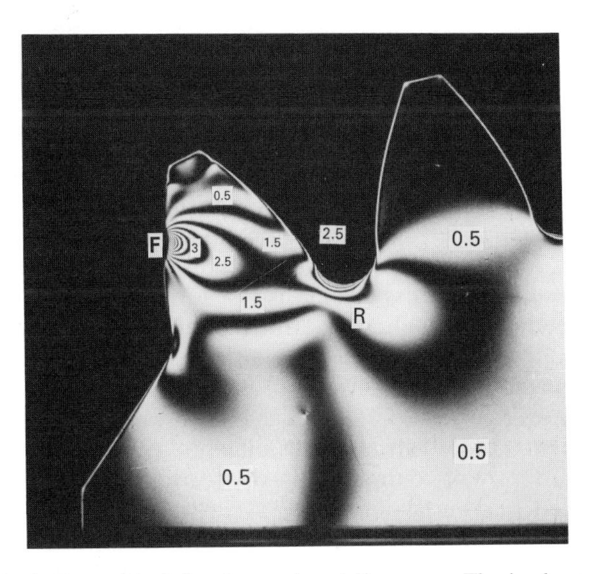

Fig. 6.15 Photoelastic model of the stresses in rotating gears. The isochromatic fringes are contours of maximum shear stress (from E. J. Hearn, *Mechanics of Materials*, 2nd ed., Pergamon Press, 1985)

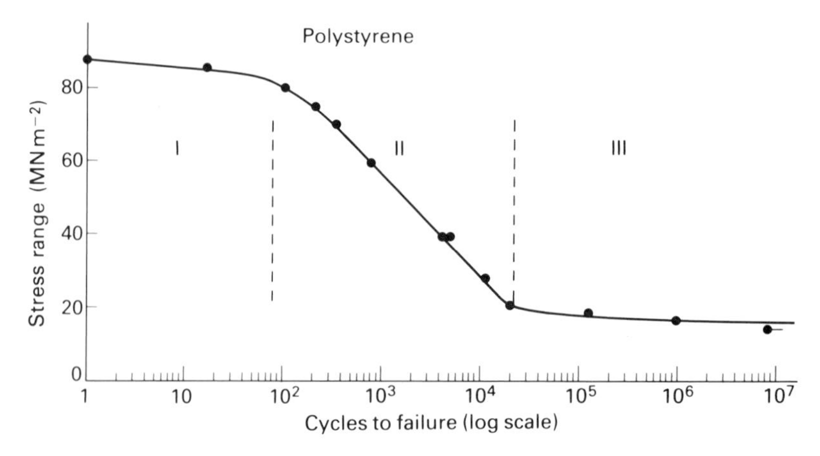

Fig. 6.16 Fatigue life of polystyrene in tension–compression cycles at 0.1 Hz, plotted as stress range versus cycles to failure (from S. Rabinowitz *et al.*, *J. Mater. Sci.*, 1973, 8, 14, Chapman and Hall).

tooth face are from the viscoelastic response to the stress σ_N and the bending stresses, which has to be dissipated by conduction to the interior and convection to the air. Another heat input is from the sliding action of the meshing tooth surfaces which occurs on either side of the pitch circle (which is roughly half way up the tooth).

In contrast to the behaviour of metal gears, the temperature rise which occurs in plastics has a significant effect in lowering the modulus and often increasing the tan δ value. Under severe conditions the temperature rise can be self-reinforcing, so that thermal runaway occurs until the gear teeth melt and distort.

In the usual fatigue test the polymer is cycled between a tensile stress S and a compressive stress of equal magnitude $-S$. The original reason for this was that it was easy to carry out experiments in which a cylindrical beam was rotated while a bending moment was maintained. Fig. 6.16 shows the regimes in a fatigue S–N curve, in which the cycles to failure N are plotted against the magnitude S of the applied stress. The low cycle failures are due to thermal runaway, but at lower stress levels crack initiation and growth becomes the failure mechanism. The stress regime is not the same as that seen by the rotating gear teeth as it contains a tensile part to the cycle. The low 0.1 Hz frequency of the test will mean that the temperatures will not rise as fast as in gear action in which the frequency can easily be 50 Hz.

Viscoelastic heating can be used as a means of rapid localised heating for joining components. In the ultrasonic welding process a vibration at 20 kHz is concentrated into a small volume of plastic by using a horn (Fig. 6.17). The energy dissipated per unit volume per cycle is given by equation (6.26). Consequently for a PVC with an E'' value of 10 MN m^{-2} it only requires a strain amplitude of 0.04 to produce a power of 1 GW m^{-3} which will heat the polymer at 1000 degrees per second. Welding times of a fraction of a second can be achieved so long as the power can be concentrated into a sufficiently

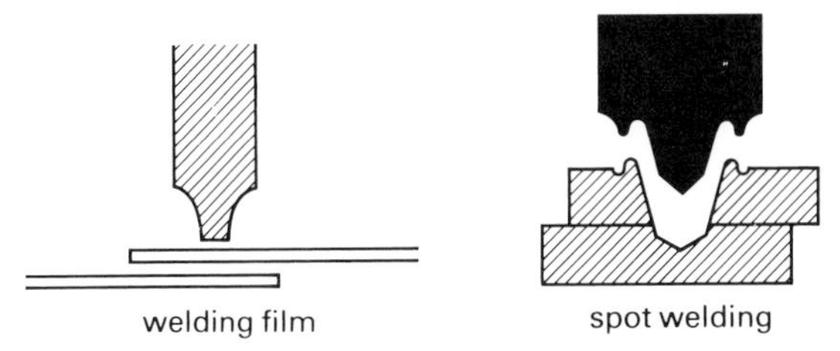

Fig. 6.17 Use of viscoelastic dissipation for welding using ultrasonic vibration and a horn to concentrate the vibrations

small area. This means that only relatively thin sheets of polymer can be joined.

7

Yielding

After a brief survey of the molecular mechanisms of yielding, an overview will be given of the phenomena that occur when plastics yield under different stress states. The analyses for metals will be reviewed and modified where necessary to allow for the different deformation mechanisms in plastics. For instance it is rare that plastic products yield in compression by becoming squatter and thicker; rather they buckle. The rate dependent effects, that were evident in the study of viscoelasticity, will be shown to affect the values of yield stresses used for design purposes. Orientation hardening is another pecularity of polymers, which leads to enhanced strength products in quite different ways to the strengthening of alloys. Finally the yielding mechanisms in polymers that occur on a micron scale can differ in nature to those occurring on a bulk scale. One of these phenomena, crazing, will lead on the study of fracture in the next chapter, while the study of yielding in polymer foams will illustrate the effects of product thickness on the dominant deformation mechanism.

7.1 MOLECULAR MECHANISMS OF YIELDING

We begin by making a brief survey of the molecular mechanisms of yielding. There has been relatively little progress in establishing these mechanisms, in contrast with the situation for metals where the details of dislocation behaviour are well established. Part of the problem is the complex microstructure of semi-crystalline polymers, and part is the lack of experimental techniques that allow observation of processes on a sufficiently small scale. The deformation processes are expected to be strongly influenced by the anisotropy of the microstructure on a scale $\sim$5 nm; segments of polymer chains tend to pack parallel to one another, and the covalent bonding along the chains is much stronger than the van der Waals bonds between chains. Therefore it is expected that deformation processes that minimise the breaking of covalent bonds will be favoured.

In contrast with metals there is less need to understand the yielding mechanisms, because it is not possible to change the yield stress significantly by minor additions of a second element (as with carbon in steel), or by mechanical working (as the dislocation structures build up in complexity) or by heat treatment (to precipitate a second phase).

7.1.1 Glassy polymers

Without a regular crystal lattice the concept of a dislocation ceases to have a meaning. However, the idea of yielding proceeding by a small sheared region propagating through the material is appealing. For some glassy polymers the initial stages of compressive yielding are seen to proceed inhomogeneously by the propagation of *shear bands*. These are layers the order of 10 μm thick in which there is a shear strain of ~1 unit (Fig. 7.1). The patterns which they create will be used in Chapter 8 to aid the stress analysis of fracture processes. Many glassy polymers yield homogeneously in compression, as the conditions for strain localisation (a reduction in the yield stress—see Section 7.2.1) are not met.

It is difficult to know exactly what happens to the polymer chains during the yielding process. The following facts have been established.

(a) Conformational changes occur in the polymer chains. Spectroscopic methods of measuring the populations of trans and gauche isomers show that

Fig. 7.1 Pattern of shear bends in a block of polystyrene that has been indented by a flat indenter

in PET the population of gauche isomers increases as unixial deformation proceeds.
(b) Some polymer chains are broken in the yielding process. At the broken ends of the chains there are short lived free radicals that can be detected using the technique of electron spin resonance.
(c) The entanglement network, created by the inter-twining of neighbouring polymer chains, survives the yielding process. This can be demonstrated by heating a necked tensile specimen of a glassy polymer to a temperature just exceeding T_g, whereupon it slowly regains its original shape (the tensile specimen should be cut from compression moulded or extruded sheet because injection mouldings will have further molecular orientation due to the mould filling flow). The entanglement network is reverting to its maximum entropy unstrained state once the molecules are hot enough to be able to change their shapes.

7.1.2 Semi-crystalline polymers

In semi-crystalline polymers the crystal lamellae must play a part in the yielding process; they form too large a fraction of the microstructure and are too well connected by interlamellar links to be able to act as rigid inclusions in a deformable matrix. The problem is that the lamellae are too thin to be observed individually during yielding. Consequently most of the information on crystal deformation has been obtained by X-ray diffraction of deformed samples which only gives a statistical picture of crystal orientation. The types of deformation mechanism that occur in crystalline metals also occur with polymer crystals: slip, twinning and stress-induced phase changes. Fig. 7.2 shows a side view of a lamellar crystal that is undergoing 'chain-slip'—that is the slip plane contains the covalently bonded polymer chains and the slip direction is parallel to the crystal **c** axis. If this process occurs inhomogeneously on a number of parallel slip planes it will eventually cause the lamellar crystals to break up into blocks without the polymer chains breaking. This happens at tensile strains of 200 to 600% when the polymer is drawn into a fibre (Fig. 7.17).

For semi-crystalline polymers that are above their glass transition temperatures the volume fraction crystallinity has the largest effect on the yield stress. Fig. 7.3 shows that there is a linear variation of yield stress with crystallinity for the range of commercial polyethylenes. If this line is extrapolated it reaches a zero yield stress at a crystallinity of 27%; however, it is not possible to have a continuous microstructure of spherulites at such low crystallinities. In contrast with metals, where a small grain size increases the yield stress because the grain boundaries are obstacles to dislocation movement, the spherulite size has little effect on the yield stress. Average spherulite sizes larger than 50 μm are usually avoided because the polymer tends to become brittle, with fracture occurring at the spherulite boundaries. Such large spherulites will only be formed if crystallisation is slow or nucleating agents are absent. The problems occur either because weak low molecular mass fractions migrate away from the growing spherulites, or because the volume contraction on crystallisation causes voids to form at spherulite boundaries.

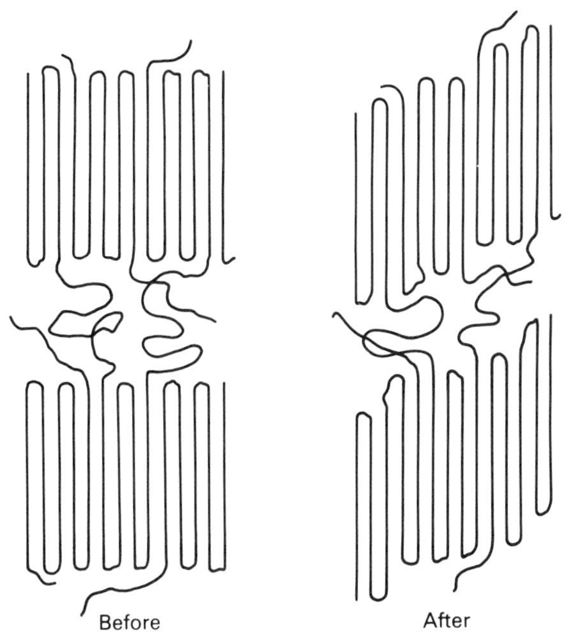

Fig. 7.2 'Chain-slip' in lamellar crystals, seen edge-on. The slip direction is parallel to the **c** axis and occurs on many parallel slip planes

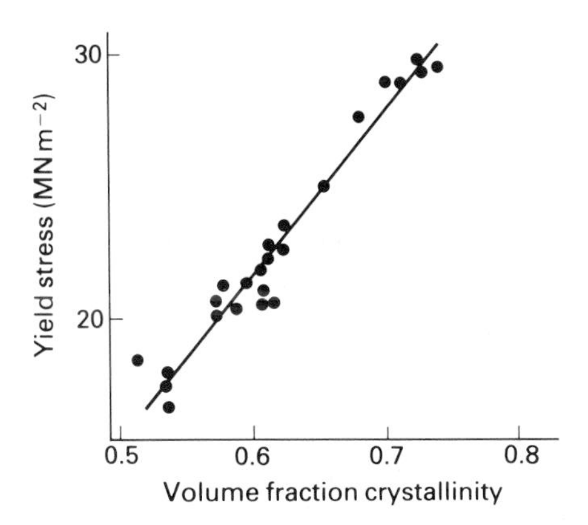

Fig. 7.3 Variation of the yield stress of polyethylene at 20 °C with the volume fraction crystallinity

7.2 YIELD UNDER DIFFERENT STRESS STATES

7.2.1 Tensile instability and necking

Most mechanics-of-solids textbooks analyse the necking instability that occurs in a tensile test. The analysis will be extended here because a number of useful products are made by the stable propagation of a neck in a polymer

(textile fibres, and the 'Tensar' soil stabilising grids shown in Fig. 7.4).
The standard analysis starts with two assumptions.

(a) The yield processes do not cause the density of the material to change. This is true for most plastics, but if crazing occurs in semi-crystalline or rubber toughened plastics the density will decrease. Hence the theory will not apply to polymers that craze (Section 7.5.1).
(b) The slope of the tensile true stress σ versus true strain ε curve decreases as the true strain increases. This is no longer true at high strains for most polymers.

The analysis proceeds by considering the relationship between the true tensile strain, defined by

$$\varepsilon \equiv \ln\left(\frac{L}{L_0}\right) \tag{7.1}$$

where L is the current and L_0 the initial length, and the cross sectional area A, while the tensile specimen is extending uniformly (Fig. 7.5). The constant volume of the element of initial cross section area A_0 means that

$$\frac{A_0}{A} = \frac{L}{L_0}$$

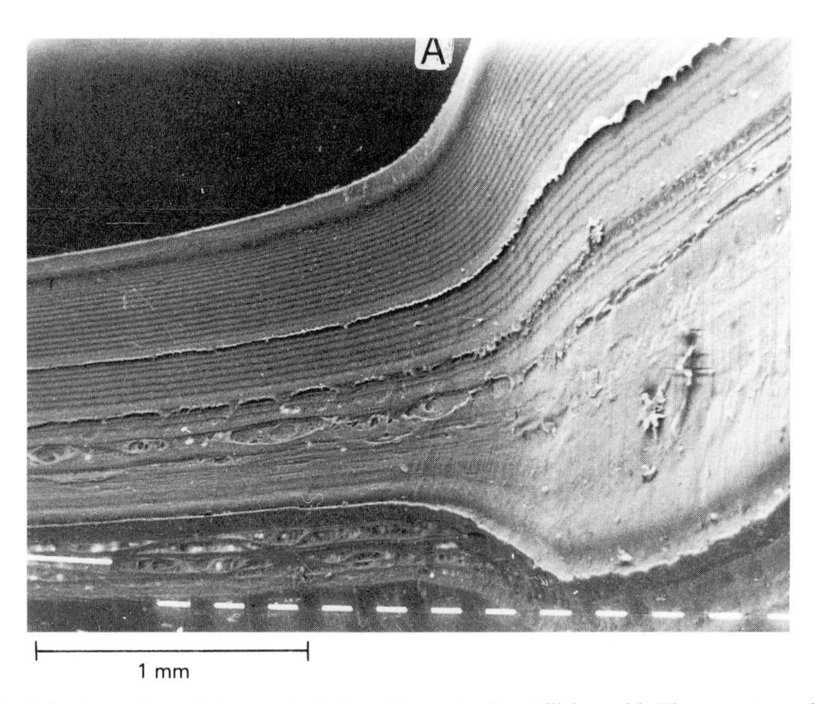

Fig. 7.4 A tensile neck in a polyethylene 'Tensar' soil stabilizing grid. The curvature of the directions in which the principal stresses act causes an increase in the average yield stress on the section AA

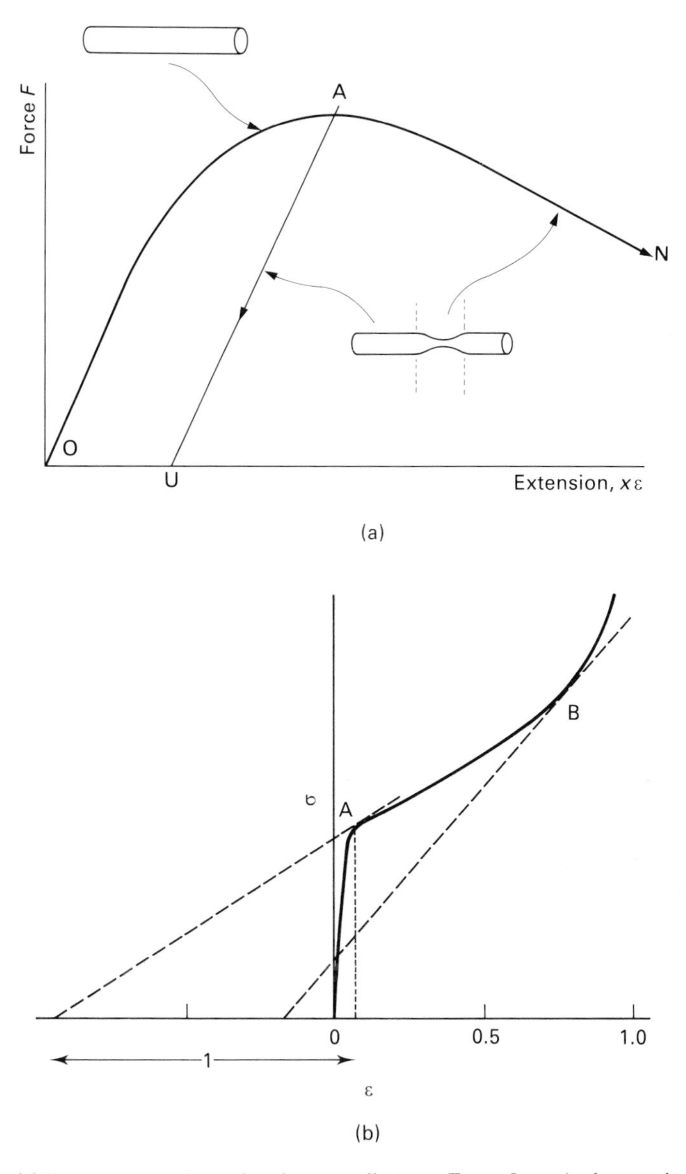

Fig. 7.5 (a) Force versus elongation in a tensile test. From 0 to A the specimen extends uniformly, but beyond A the end parts unload elastically along the path AU, while the necked portion proceeds along path AN. (b) The true stress versus true strain relationship for a polymer that cold draws, showing the points A and B at which a neck respectively forms and stabilises

where L is the current length. Taking natural logarithms and using equation (7.1) gives

$$\ln\left(\frac{A_0}{A}\right) = \varepsilon$$

Differentiaton leads to the result

$$\frac{d\varepsilon}{dA} = -\frac{1}{A} \tag{7.2}$$

A neck will begin to form in part of the specimen, leading to non-uniform strain, when there are two possibilities for the next strain state, as shown in Fig. 7.5a. They are the elastic unloading of part of the specimen along path AU, and the further plastic straining along path AN. These two possibilities can only occur simultaneously when A is at the maximum in the force extension curve, that is

$$\frac{dF}{dx}\frac{dF}{d\varepsilon} = 0 \tag{7.3}$$

As the force is the product of the cross sectional area A and the true stress σ, defined as F/A, this condition becomes

$$A\frac{d\sigma}{d\varepsilon} + \sigma\frac{dA}{d\varepsilon} = 0$$

Substitution of $dA/d\varepsilon = -A$ from equation (7.4) then gives

$$\frac{d\sigma}{d\varepsilon} = \sigma \tag{7.4}$$

This relationship, known as *Considere's criterion*, is usually as far as the analysis is taken. It shows that the ultimate tensile strength occurs for metals, not when the true tensile stress reaches a critical value, but when the rate of work hardening $d\sigma/d\varepsilon$ falls to a critical value σ. It can then no longer compensate for the cross sectional area decrease, and necking occurs followed by failure. It is usual to define the tensile yield stress of polymers as the engineering stress F_{max}/A_0 calculated from the maximum in the force extension curve. The reason for not using a definition of a 1% offset from the initial straight line (as for metals) is that non-linearity in polymers is a result of viscoelasticity rather than yielding. The formation of a neck is the first sign of permanent deformation.

For polymers which *cold draw* as in Fig. 7.4, the condition (7.4) can be applied twice. When $d\sigma/d\varepsilon$ falls below σ a neck initiates and the strain in the bar becomes inhomogeneous. This condition is true between the points A and B on the true stress versus true strain graph (Fig. 7.5b), then beyond B the slope $d\sigma/d\varepsilon$ is once more greater than σ, owing to orientation hardening of the deformed molecules (Section 7.4). There is a region of instability between A and B, so the neck rapidly converts the polymer from the strain state A to B. The neck then propagates the length of the tensile bar until there is no more material in state A, whereupon further homogeneous straining occurs at true strains higher than that at B.

The analysis can be taken further by splitting the total differential $d\sigma/d\varepsilon$ into a number of contributions

$$\frac{d\sigma}{d\varepsilon} = \frac{\partial\sigma}{\partial\varepsilon} + \frac{\partial\sigma}{\partial\dot{\varepsilon}}\frac{d\dot{\varepsilon}}{d\varepsilon} + \frac{\partial\sigma}{\partial T}\frac{dT}{d\varepsilon} + \frac{\partial\sigma}{\partial G}\frac{dG}{d\varepsilon} + \ldots \tag{7.5}$$

In an incipient neck the strain rate $\dot{\varepsilon}$ must increase, and the yield stress increases with increasing strain rate (equation 7.18); hence the second term in the right hand side of equation (7.5) is a stabilising influence. The yielding process generates heat in the neck; all the work input is converted to heat, which is only slowly conducted down the specimen or convected into the surrounding air. As the yield stress is always a decreasing function of temperature T the third term in equation (7.5) is a destabilising influence. Finally there is a geometrical factor G associated with the shape of the neck. Because of the curvature of the lines of tensile stress in the neck (Fig. 7.5) it can be shown that for a material obeying Tresca's yield criterion the average yield stress needed across a flat strip specimen of thickness $2a$ is

$$\bar{\sigma} = 2k\left(1 + \frac{a}{3R}\right) \tag{7.6}$$

where $2k$ is the yield stress in a straight sided specimen. Initially the radius of curvature R is negative at the edge of the neck, so the fourth term in equation (7.5) is a destabilising factor. When the neck is fully formed the fourth term is a stabilising factor at a point such as A in Fig. 7.5.

This analytical approach shows that at least four factors are involved in the stability of necks in polymers. Computer modelling, rather than analytical treatment, is necessary to deal with the large geometrical changes and the interactions between factors that occur in high-strain necks. Fig. 7.6a shows how the predicted temperature rise in the neck in a polycarbonate tensile specimen changes with the crosshead speed in a tensile test. The temperature increases are not large enough to dramatically reduce the yield stress; however they can be in low T_g glassy plastics like PVC, or in polyethylene where the strain in the neck is much higher. Careful control of heat transfer is necessary if the necking process is to be used commercially to make highly oriented products. If the necking process occurs too rapidly the heat generated in the neck can soften the polymer to the point where *thermal runaway* occurs. Fig. 7.6b shows such a failure in a thin walled polyethylene liquid container that was dropped from 5 m on to a hard surface.

7.2.2 Yielding in bending

Any ductile material can be bent until it stays permanently deformed. The mechanics-of-solids analysis of the formation of a *plastic hinge* assumes that one side of the beam or sheet yields in tension at a stress T and the other side yields in compression at a stress $-T$. The material is assumed not to work harden after yielding (this is a good approximation for polymers for strains up to 50%). The longitudinal strain e varies linearly through the sheet according to

$$e = \frac{y}{R} \tag{7.7}$$

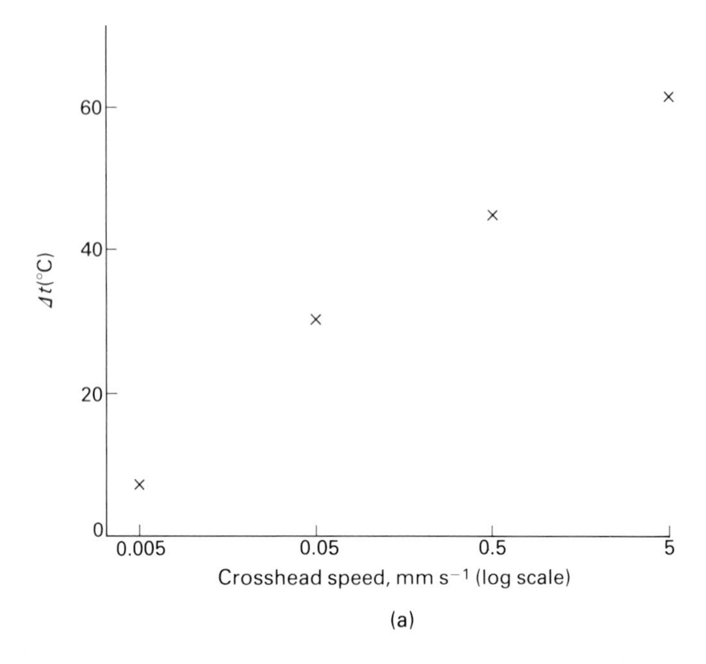

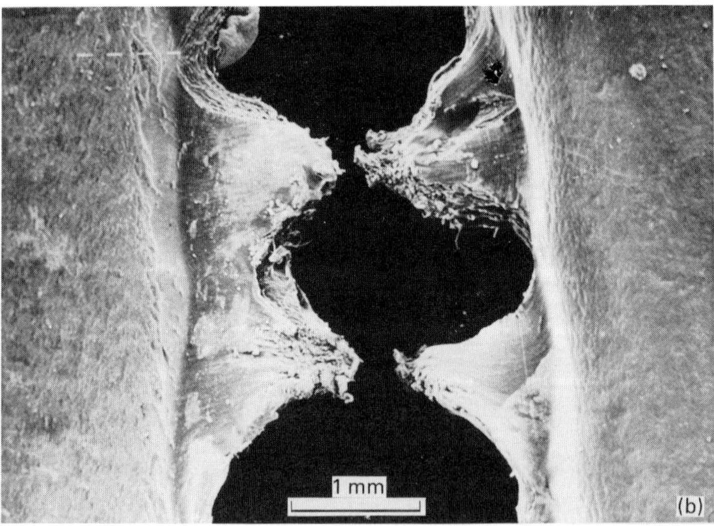

Fig. 7.6 (a) Predicted maximum temperature rise in a 50 mm long polycarbonate tensile specimen versus the extension rate (from Mills, *Br. Polym. J.*, 1978, **10**, 1). (b) Impact fracture in the wall of a polyethylene container. The temperature rise in the neck has made the necking unstable

where y is the distance from the neutral surface and R is the radius of curvature. Hence the yielding will spread inwards from the surfaces of the sheet. Fig. 7.7a shows a partly yielded beam in 3-point bending, where the bending moment M increases linearly from the ends to the central point. The upper limit to the bending moment at the central load point occurs when the two yielded zones meet at the neutral surface, and the stress distribution through the beam is as shown in Fig. 7.7b. The plastic moment M_{pl} for a beam that is w wide and d deep can be calculated from the internal moments on the cross-section as

$$M_{pl} = \frac{wd^2}{4} T \tag{7.8}$$

Necking will not occur on the tensile side because of the support of the compressive side. With metals the initial stages of yielding shown in Fig. 7.7a are real, and the beam will remain very slightly bent. With plastics there is no evidence that any permanent deformation occurs before the plastic hinge collapse, and the non-linearity in the force–deflection relationship of the bending test is due to non-linear viscoelasticity. Hence large surface strains must be imposed to cause yielding. The recommended short term surface strains that can be used without causing yielding are higher for semi-crystalline plastics (ranging from 4% for polyamides to 8% polyethylene) than for glassy plastics (1.8% for polystyrene to 4% for polycarbonate). To cause yielding the sheet must be bent to a small radius of curvature R, as the surface strain

$$e = \frac{d}{2R} \tag{7.9}$$

where d is the sheet thickness. For a 2 mm thick sheet of polyethylene the deflections involved in bending a sheet to a radius smaller than 12 mm would be far in excess of those acceptable for the function of any product.

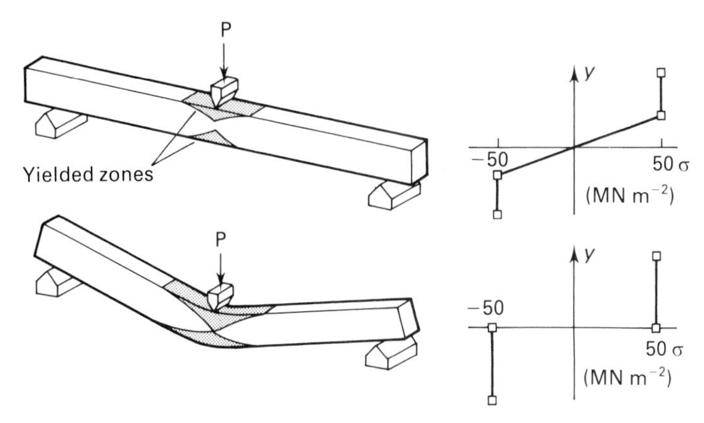

Fig. 7.7 Yielded zones in a 3 point bend test (a) while the centre of the beam remains elastic and (b) when a plastic hinge has formed. The stress distribution through the beam under the central load is shown on the right, for a polymer that does not orientation harden

7.2.3 Buckling and yielding in compression

Uniform compressive yielding is rare in plastics products, because they are thin walled and the plastic has a low elastic modulus. It is much more likely that regions under compressive forces, such as the vertical sides of a bottle crate in a stack of similar crates, will fail by buckling if they are overloaded. Compressive collapse can also be a mechanism in beam that is overloaded in bending. Fig. 7.8b shows the arm of a garden chair which is a beam of an L section to achieve high bending stiffness. The lower part of the L, which is in compression, has buckled plastically as a result of inadequate design. Hence a design change to improve one characteristic (bending stiffness) has changed the failure mode from excessive deflection to buckling.

Fig. 7.8a shows a strut of length L that is part of an injection moulding. Its ends are 'built-in' to the rest of the moulding; consequently they cannot rotate or move sideways under the influence of the compressive forces F. If the strut bends so that the sideways deflection is v at a point P, the bending moment at P is given by

$$M = -Fv + M_0$$

where M_0 is the unknown moment that acts at both ends. This equation can be substituted into the general relationship between the curvature of a beam, its Young's modulus E, and the second moment of area I of its cross section

$$M = EI\frac{d^2v}{dz^2}$$

to obtain the differential equation

$$EI\frac{d^2v}{dz^2} + Fv = M_0$$

Solutions of this equation, with the correct end conditions and indeterminate amplitude, are

$$v = \frac{M_0}{F}\left[\cos\left(\frac{2\pi z_n}{L}\right) - 1\right]$$

where n is 1, 2, 3, . . . and the buckling force F is given by

$$F = \left(\frac{2\pi n}{L}\right)^2 EI \tag{7.10}$$

The lowest buckling force occurs for the buckling mode shown in Fig. 7.8a with $n = 1$. We can use this equation for a polyethylene strut of rectangular cross section subjected to 3 months of compressive creep loading. If the strut has width w and depth d the value of $I = wd^3/12$. Making the pseudo-elastic approximation for a viscoelastic material we replace E in equation (7.10) by $1/J(t)$, so the compressive stress that will cause buckling is

$$\sigma_b = \frac{F}{wd} = \frac{1}{3J(t)}\left(\frac{\pi d}{L}\right)^2 \tag{7.11}$$

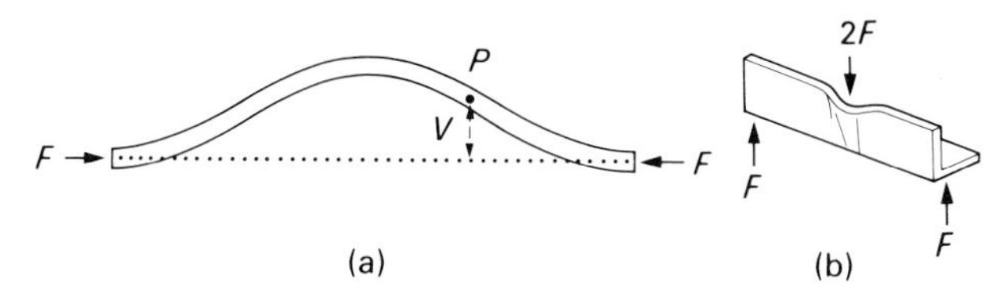

Fig. 7.8 (a) Elastic buckling of a strut with built-in ends due to axial compressive forces F. (b) Plastic buckling of a reinforcing rib in a beam, due to excessive bending moment

Substituting the values $L = 75$ mm, $d = 3$ mm and J (3 months) $= 0.3\ \mathrm{GN\,m^{-2}}$ gives a value of $\sigma_b = 1\ \mathrm{MN\,m^{-2}}$. This value is very low compared with the tensile creep rupture stress (see Section 7.3.1) of about $10\ \mathrm{MN\,m^{-2}}$. Consequently the strut should be redesigned with an L or U shaped cross section having a much higher bending stiffness; it might also be necessary to cross brace it to reduce the effective length L.

7.2.4 Localised yield in compression—hardness

Hardness tests are widely used as a non-destructive method of estimating the yield stress of metals, and hence checking whether heat treatments, or surface treatments like carburisation, have been carried out correctly. Although plastics have much lower hardnesses than metals the test is not widely used. This is partly because there are no heat treatment methods for modifying the yield stress, and partly because viscoelastic effects make the size of the indentation decrease with time. However plastics are tested for resistance to surface damage. We will deal with two extreme cases, when the depth of the indentation and the associated yielded region are either much smaller than the product thickness, or equal to it.

In conventional *hardness tests* the indentation size is much smaller than the product thickness, and the test can be carried out without noticeable damage. The mechanics-of-solids analysis of hardness tests assumes that the material has an infinite Young's modulus and that it yields at a constant shear yield stress k. For a strip indenter with face width w, one possible flow pattern in the material is shown in Fig. 7.9, with rigid prisms of material moving as shown. The indenter moves down with a displacement Δ; to avoid it overlapping with the triangular prism of material beneath it, the prism must have displacement $2\Delta/\sqrt{3}$. The other prisms move sideways and diagonally upwards by the same amount, so that the net effect is that the material under the indenter is extruded at each side of the indenter. The 10 dashed surfaces in Fig. 7.9 'slide' against a shear stress k, so the total energy dissipated at the surfaces is

$$W = 10 \times \frac{wL}{2} \times \frac{2\Delta}{\sqrt{3}} \times k = 5.8kwL\Delta$$

where L is the length of the indenter. When the above expression is equated with the work done by the indenter (area wL, times displacement Δ, times

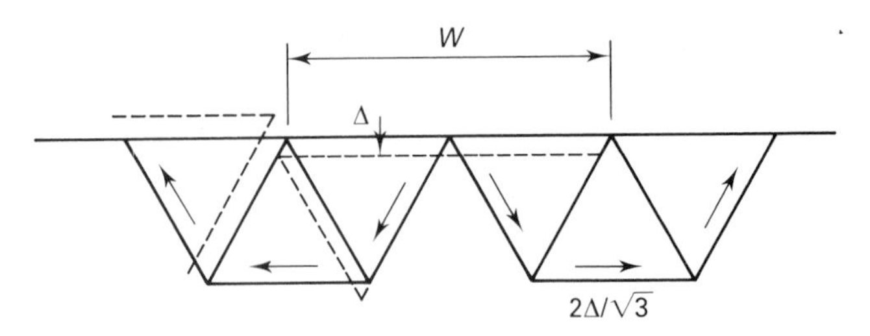

Fig. 7.9 Theoretical flow pattern in a solid beneath a line indenter of width w

pressure p) the identation pressure is predicted to be

$$p = 2.9 \times 2k$$

Hence the pressure needed on the face of the indenter is approximately three times the tensile yield stress $2k$. In the Vickers and Brinell hardness tests the flow is approximately radial from the indenter, rather than planar as in Fig. 7.9. Nevertheless the relationship

$$\text{hardness} \cong 3\sigma_y \tag{7.12}$$

is used widely as the basis of the analysing such hardness tests on metal products.

The pattern of yielding in plastics is different. Fig. 7.1 shows the shear band patterns when a long indenter with a flat base is pressed into a block of polystyrene. The geometry is chosen in preference to the conventional ball or pyramid indenters because plane strain yielding occurs (see Chapter 8). The indentation has formed by the plastic flowing towards the interior of the block. This is only possible because there are large strains in the surrounding elastic region. In spite of the different flow pattern the relationship between the indentation pressure and the tensile yield stress is still equation (7.12). This is a result of the boundary area of the yielded zone being much greater than the area of the indenter surface in both cases.

A stress analysis of the shear band pattern predicts that there are high tensile stresses parallel to the surface at the point where the yielded zone penetrates deepest into the block. The figure shows the crack that has formed because of the stresses. Therefore a compressive force applied on the surface of a material that is brittle in tension can cause fracture.

7.2.5 Localised yield—surface damage and wear

Most surface damage to plastics is caused by the contact of moving rather than stationary objects. Surface scratches, that occur in use where hard abrasive particles are dragged across a polymer surface (Fig. 7.10), have irregular shapes. The material that was displaced by the scratching particle has been displaced to the sides of the groves that have formed. Thus in this case the end result of the particle moving across the surface is similar to the two-dimensional indenter flow pattern in Fig. 7.9.

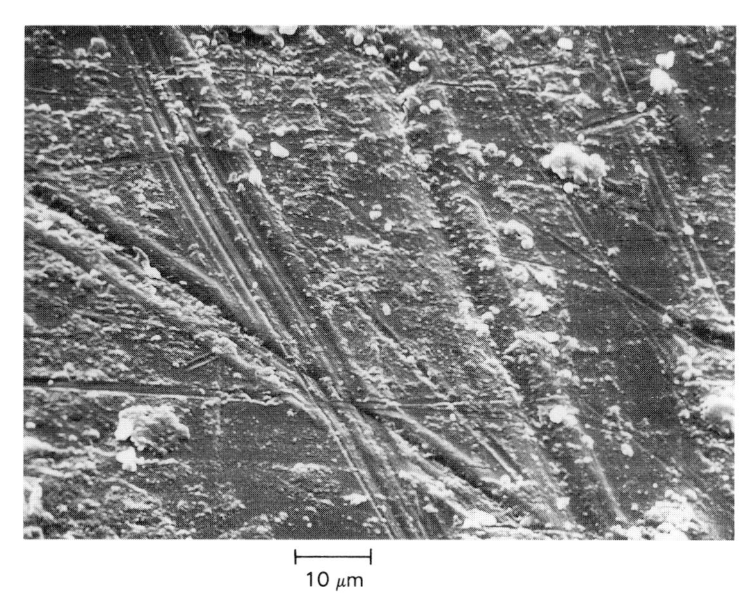

Fig. 7.10 Scanning electron micrograph of surface scratches on a polycarbonate visor, due to wiping away road dirt

Surface damage is classified as *wear* if particles of the plastic are lost from the surface. In the optical application above there has been roughening of the surface without any wear occurring. One test method for abrasive wear is to transverse a pin of polymer over a rough steel surface while maintaining a constant contact force. The data obtained (Fig. 7.11) show significant differences between plastics. The wear rate can increase rapidly as the surface roughness of the steel increases. The wear rate for several polymers on a steel of fixed roughness correlates with the quantity $1/\sigma^* e^*$, where σ^* is the tensile strength and e^* is the strain at failure in a tensile test. The best resistance to abrasive wear occurs with semi-crystalline polymers which have a moderate strength and a high failure strain. The situation is different where plastics are used as bearings, or applications where there is continual rubbing against the same metal surface. In these cases a transfer film of polymer can develop on the metal surface, and the resulting adhesive wear between polymer and transfer film ranks polymers in a different order to that for abrasive wear.

7.2.6 Localised yield—film or sheet penetration

When a pointed object (it could be a needle on a thin film, or a ball or conical spike on an injection moulding) penetrates a sheet of plastic the yielded region has a diameter that is larger than the sheet thickness. The former case is familiar to users of LDPE supermarket bags. The sheet is usually not supported by any underlying layer, so it can bend readily. Consequently the important stresses in the distorted sheet are in the radial r, hoop θ and thickness z directions (Fig. 7.12). As the three principal stresses differ we will

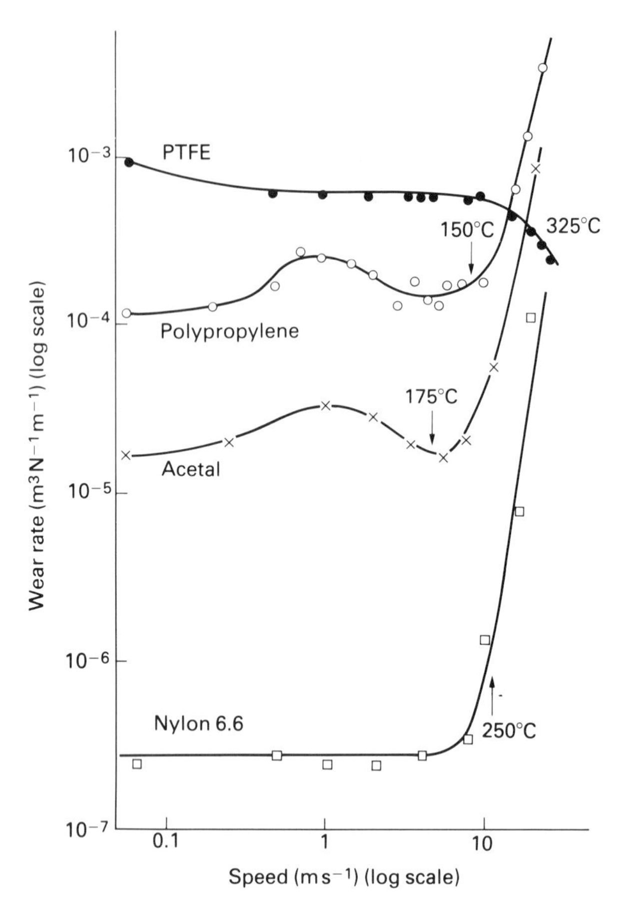

Fig. 7.11 Wear rate versus the speed of sliding on a steel surface. The surface temperature is indicated at the onset of rapid wear (from J. K. Lancaster, *Tribol. Int.*, 1971, **4**, 82, Butterworth Scientific)

need a *yield criterion* to be able to predict whether yielding will occur. These are empirical rationalisations of observed yield behaviour under combined stresses. *Tresca's* yield criterion states that the material will yield when the difference between the most tensile principal stress σ_1 and the most compressive σ_2 is given by

$$\sigma_1 - \sigma_2 = 2k \tag{7.13}$$

where the constant k is the yield stress under a shear stress alone. For a force F on the pointed indenter the radial membrane stress at a distance r from the indenter in a sheet of thickness t is given approximately by

$$\sigma_r = \frac{F}{2\pi r t} \tag{7.14}$$

If the plastic is isotropic, then as the hoop σ_θ and thickness σ_z stresses are

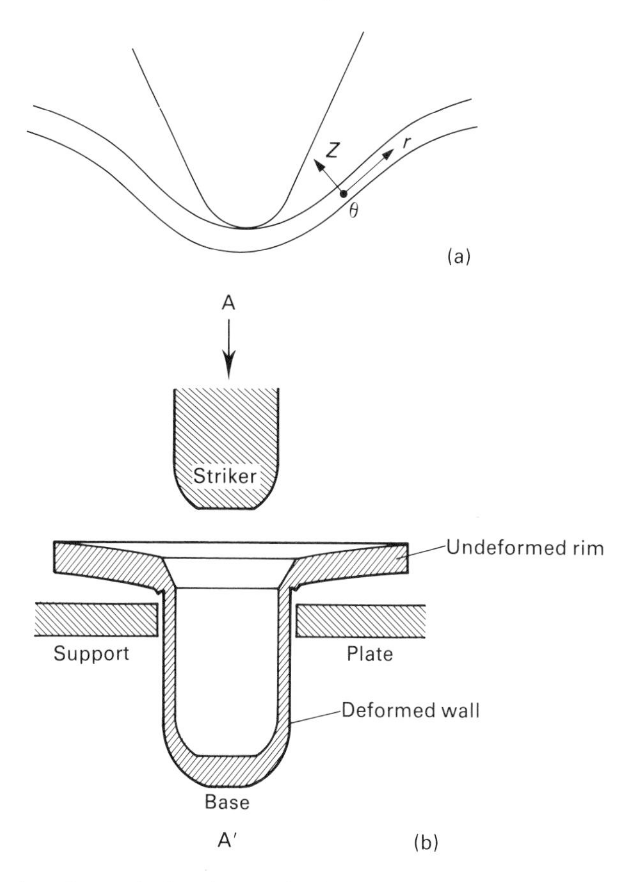

Fig. 7.12 (a) Penetration of a plastic sheet by a conical indenter, showing the polar coordinates used for stress analysis. (b) Sketch of the shape of a 3 mm thick LLDPE sheet after a 12 mm diameter nosed striker has impacted it at 5 m s^{-1}. There is axial symmetry about the axis AA′

relatively small, yielding will occur close to the top of the indenter when either

$$\sigma_r - \sigma_\theta = 2k \qquad \text{or} \qquad \sigma_r - \sigma_z = 2k \tag{7.15}$$

according to Tresca's yield criterion. Necking will occur followed by puncturing of the sheet. Fig. 7.12b shows that when a spherical nosed indenter is used the highly ductile LLDPE sheet can neck and envelope the indenter. The sheet resists thinning, and prefers to yield by stretching radially while reducing in length in the hoop direction. When the sheet bends to be in contact with the indenter sides, it can no longer compress in the hoop direction, so it is highly resistant to being punctured. For higher impact energies the yielded cylindrical region will fracture around its base.

Plastic films with biaxial orientation have a better resistance to puncturing than uniaxially oriented films. The films have anisotropic yielding behaviour, with the value of $\sigma_r - \sigma_z$ to cause yielding by thinning being larger than $\sigma_r - \sigma_\theta$ to cause constant thickness yielding.

7.3 YIELD ON DIFFERENT TIME SCALES

7.3.1 Creep rupture

If a plastic is subjected to a high enough creep stress for a long enough time failure will occur by yielding, crazing, crack growth or a combination of these. Creep rupture is important for products such as pressure pipes, where the shape is stable as creep proceeds, and less important in products that are bent or twisted where it is more likely that excessive deflections will occur before any rupture process starts.

Fig. 7.13 shows the equipment used for creep rupture testing of plastics used as pipes. Short lengths of pipe are fitted with mechanically screwed end fittings, then a fixed water pressure is applied internally, and the time recorded until a leak occurs. There are two reasons for testing the pipe itself rather than tensile specimens cut from it. Firstly the internal pressure p produces a biaxial stress state

$$\sigma_H = \frac{pr_m}{t} \qquad \sigma_L = \frac{pr_m}{2t} \tag{7.16}$$

where r_m is the mean radius and t the wall thickness of the pipe; the failure stress could be different from that when a uniaxial stress is applied. Secondly the pipes contain internal stresses from their fabrication (Fig. 5.16b), and this or other effects of processing may affect the creep rupture times.

The results of many tests are plotted as creep rupture time versus hoop stress in the pipe wall (Fig. 7.14) on logarithmic scales. The data fall on one or more lines, corresponding to different creep rupture mechanisms. The initial line of small negative slope is for ductile creep rupture, when the pipe wall balloons out and then necks, with the extension being mainly in the hoop direction. The molecular orientation of the necked material is in the hoop direction, thus this is also the easiest direction for crack growth, so the split

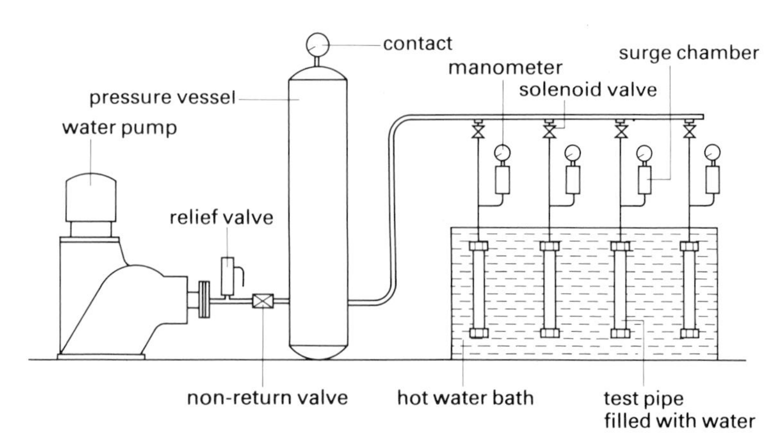

Fig. 7.13 Diagram of test rig with automatic pressure control for measuring the creep strength of plastic pipes (courtesy of Hoechst AG)

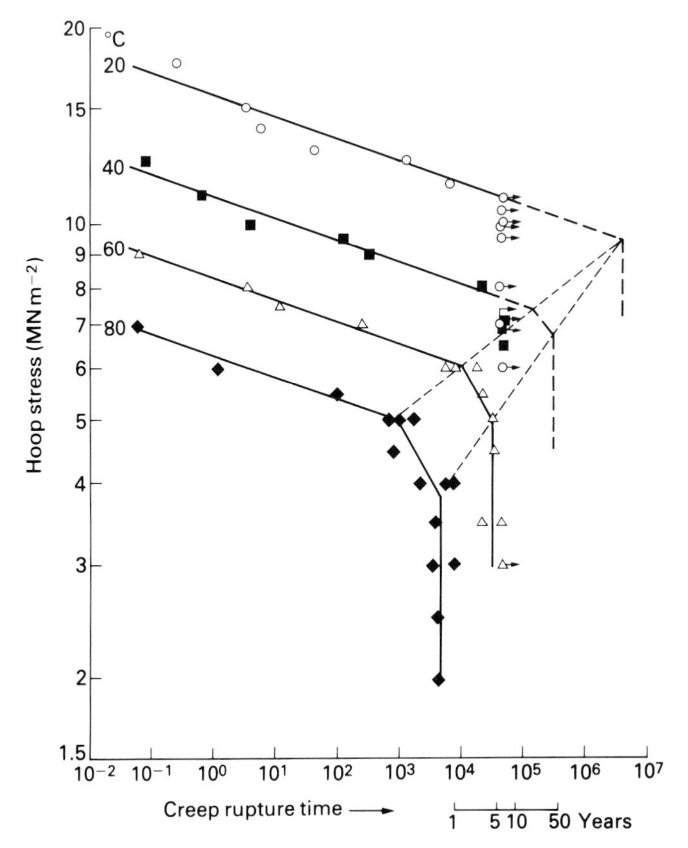

Fig. 7.14 Creep rupture data for polyethylene pipes made from Hoechst Hostalen GM 5010 T2 (courtesy of Hoechst AG)

occurs in that direction producing the characteristic 'parrot's beak' fracture (Fig. 7.15). Fractures which occur by 'brittle' or plane strain crack growth (see Chapter 8) fall on a steeper line which intersects the ductile creep rupture line. The intersection point changes if the molecular weight or chain branching characteristics of the polymer are changed.

Pressure pipes are usually designed for a 50 year life without creep rupture. Since the experimental data rarely extend to creep rupture times exceeding 10^4 hours it is necessary to make an extrapolation of the data. The simplest extrapolation is a straight line extension of the ductile rupture line. The British Gas specification for polyethylene pipe has a clause that the extrapolated 50 year creep rupture stress should be $\sigma_{50} > 10\ \mathrm{MN\ m^{-2}}$. The pipes will be designed with a safety factor S, so that the design hoop stress is given by

$$\sigma_H = \frac{\sigma_{50}}{S} = \frac{p}{2}(\mathrm{SDR} - 1) \tag{7.17}$$

The Standard Dimension Ratio (SDR) of the pipe is defined as the mean

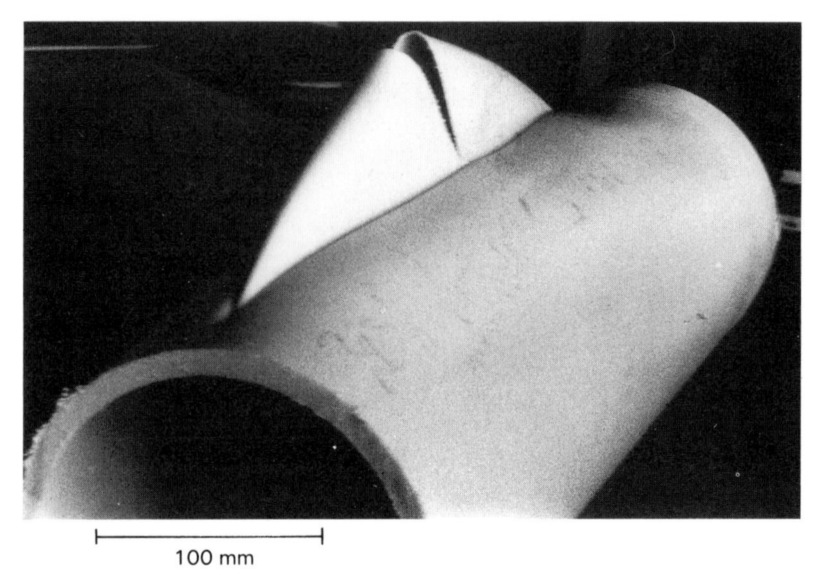

Fig. 7.15 A ductile 'parrot's beak' fracture in a pressurized MDPE pipe. The plastic strain in the neck is largely in the hop direction, so the split in the neck is parallel to the molecular orientation

diameter divided by the minimum allowed wall thickness. For gas pipe of SDR = 23 used at an internal pressure of 4 bar, equation (7.17) shows that the safety factor $S > 2.3$, assuming that the pipe material meets the minimum value of σ_{50}.

This linear extrapolation does not allow for the possibility of a brittle failure mode starting to operate at a time exceeding 10^4 hours. Section 9.5 deals with the environmental stress cracking that is the main cause of such a failure. It has been found that the brittle failure mode occurs at shorter times when the creep rupture testing is carried out at higher temperatures. For polyethylene the highest usable temperature is 80 °C; above this temperature the microstructure can change by the melting of rapidly cooled metastable crystals, and their slow crystallisation in a more stable form. Fig. 7.14 shows what happens for a particular HDPE, for which brittle fracture can be observed at 80 °C, 60 °C and 40 °C. It is then possible empirically to extrapolate the data to 20 °C by constructing an 'Arrhenius' plot of the logarithm of the creep rupture time versus the reciprocal of the absolute test temperature. This is usually found to be a straight line graph, which can be extrapolated to 1/293 K^{-1}.

As experience with creep rupture testing of polyolefins has accumulated, elevated temperature (80 °C) creep rupture tests have been used for quality control purposes, and standards have been set for specific applications, i.e. the creep rupture time for pipes for natural gas distribution must exceed 170 hours at a hoop stress of 3 MN m^{-2}. Care must, however, be exercised if a polyethylene made by a different process is introduced, because the real use temperature is closer to 10 °C buried in the ground, and the slope of the Arrhenius plot has been found to vary between different polyethylenes.

7.3.2 Strain rate dependence

There is a contributions to the tensile true yield stress from the true strain rate ε at which yielding occurs. The tensile yield stress is found to vary with the true strain rate according to

$$\sigma(\varepsilon) = A + B \log \dot{\varepsilon} \tag{7.18}$$

where the constants have the values $A = 44.7$ and $B = 4.67 \text{ MN m}^{-2}$ for HDPE at 20 °C. This means that the yield stress in an impact lasting 10 ms is twice as large as in a slow tensile test, where yielding occurs after 10 min.

7.4 STRAIN HARDENING

The term *strain hardening* implies a behaviour similar to that of metals, where a dislocation structure is developed as a result of plastic deformation, and even if the deformation direction is reversed (as when a copper wire is bent and unbent) the hardening continues. The effects of cyclic straining in polymers are not to harden the material, rather it softens both as a result of heat build-up and because the yielded material may have a higher fictive temperature than the original material. The following two cases of strain hardening of polymers are reasonably well understood.

7.4.1 Orientation hardening of solid polymers

There is a contribution to the true yield stress from the orientation hardening of the molecules as the true strain ε increases. When the strain rate in the neck of a tensile specimen was kept constant, by a feedback of the strain in the neck to the crosshead speed of the tensile testing machine, it was found that the orientation hardening contribution followed the true strain relationship predicted for crosslinked rubber networks in Chapter 2. Fig. 7.16 shows that the total stress follows a relationship of the form

$$\sigma = \sigma_0 + K[\lambda^2 - \lambda^{-1}] \tag{7.19}$$

The extension ratio λ has been written in terms of the true strain by inverting equation (7.1) to give $\lambda = \exp \varepsilon$. There is no constant term σ_0 in equation (2.28) for a rubber, but the second term in equation (7.19) confirms the role of entanglement networks in yielding. It is not understood how the empirical constant K is affected by the microstructure. There is no correlation with the entanglement molecular weight in the amorphous phase (Table 2.1), so it is possible that it is affected by the yield stress of the crystalline phase. It is impossible to continue to roll down plastic sheet in the same way as metal sheet, using intermediate annealing stages to remove dislocation structures. With plastic sheet any attempt at annealing would cause partial recovery of the original shape, but would not allow the molecules to flow back to an equilibrium state.

For semi-crystalline polymers, the orientation function P_2 for the crystal **c** axes can be calculated from X-ray diffraction measurements (see Chapter 2),

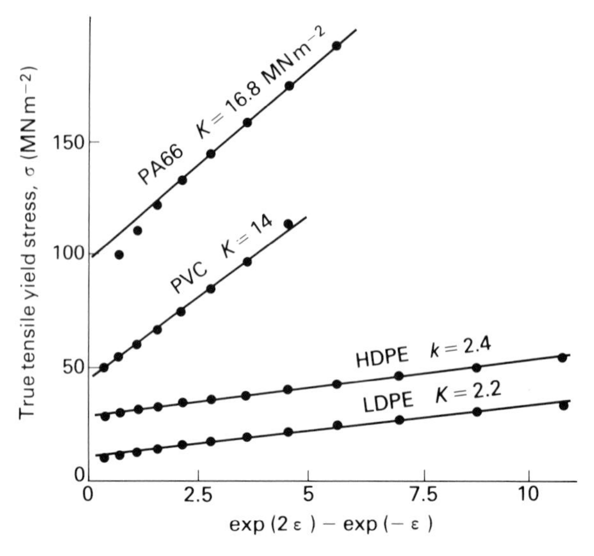

Fig. 7.16 Variation of the true tensile yield stress with the true strain hardening expression of equation (7.19). K is the slope of the lines in MN m^{-2} (from G'Sell and Jones, *J. Mater. Sci.*, 1981, **16**, 1966, Chapman and Hall)

and Fig 7.17 shows how it increases with the true strain, for different methods of fabricating polypropylene fibres and films. There is a linear increase while the spherulitic microstructure survives, but at a P_2 value of 0.9 the spherulites are destroyed and replaced by a microfibrillar structure. At this point there is a break in the P_2 versus true strain relationship. The figure indicates that it is impossible to achieve perfect **c** axis orientation by mechanically stretching a fibre or film, and that the length must increase exponentially in order to achieve a linear increase in P_2. The data in Fig. 7.17 can be compared with the pseudo-affine deformation model (Section 2.4.7). Using a true strain axis makes the initial increase in P_2 linear, and there is a steep upturn at a P_2 value of 0.8, close to the 'break' in the experimental data. Although the spherulitic microstructure is complex and there are deformation modes in the crystals, the simple pseudo-affine model is a good approximation to the behaviour.

The strong molecular and/or crystal orientation produced by cold drawing a polymer can increase the yield stress in the direction of drawing; Fig. 7.16 shows this for a semi-crystalline polyamide. However, the yield stress at right angles to the draw direction will hardly change or may even decrease. The resulting anisotropy of strength is usually acceptable in a fibre, but not in film products where biaxial orientation is more common. Fig. 7.18 shows how the yield stress of biaxially stretched amorphous PET increases with the draw ratio. The drawing process in PET causes crystallisation, and part of the figure shows how the crystallinity increases. The reference to heat set polymer is to a 10 s contact with a heated mould at 130 °C; this process is used to give dimensional stability to PET blow mouldings which will be subsequently filled with liquids at 90 °C. It can be seen that the biaxial orientation is the main

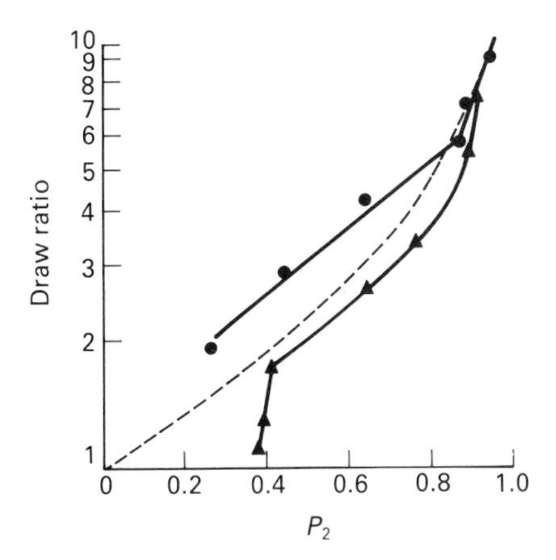

Fig. 7.17 The crystal orientation of polypropylene films as a function of the true strain in the deformation process. Draw temperatures: (●) 135 °C; (▲) 110 °C. The prediction of the pseudo-affine model is shown as a dashed line (from Samuels, *Structured Polymer Properties*, Wiley, 1974)

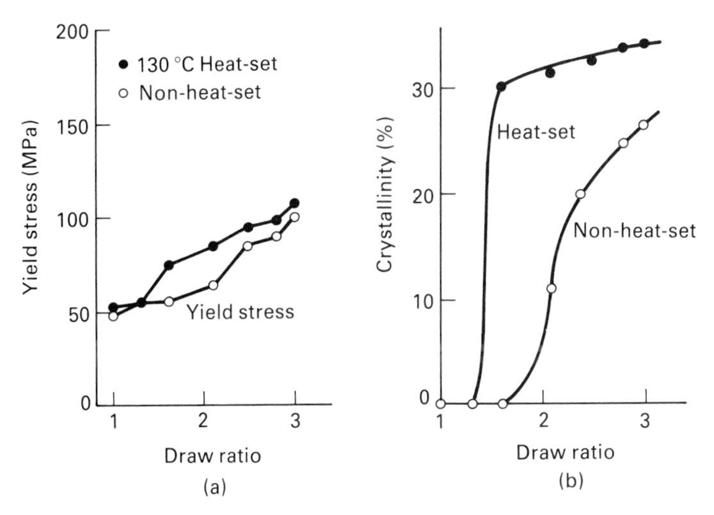

Fig. 7.18 Variation of (a) the yield stress and (b) the % crystallinity of biaxially stretched PET with the draw ratio (from Maruhashi, *Polym. Eng. Sci.*, 1992, **32**, 481)

cause of the increase of yield stress, rather than the increase in crystallinity. This is because the PET is below its T_g, and the main contribution to the increased yield stress is from the oriented amorphous phase.

7.4.2 Gas pressure hardening in closed cell foams

Chapter 3 described the microstructure of closed cell foams of low density. The strain hardening when the foam is compressed is due to the increase in the gas pressure in the cells. On the time scale of an impact there is no possibility of gas diffusing through the cell faces, so the response is that of a gas in a piston. The small cell size and the thinness of the cell faces means that there is some heat transfer from the gas, so the gas compression is somewhere between the adiabatic and the isothermal extremes. Gibson and Ashby (1988) have analysed the isothermal compression of the gas, assuming that the foam does not expand laterally. The gas obeys Boyle's law, and initially it is at a pressure $p_0 = 0.1\ \text{MN m}^{-2}$ and occupies a fractional volume $1 - R$ of the foam, where R is the relative density (equation 3.18). When the compressive strain is e the foam exerts a pressure

$$p = \frac{p_0 e}{1 - e - R} \tag{7.20}$$

on the loading surfaces. This contribution only explains the stress increase above the initial yield stress. Fig. 7.19 shows the compressive stress in a polypropylene foam plotted against the function on the right hand side of equation (7.20). A straight line can only be filled to the data if the p_0 value is $0.18\ \text{MN m}^{-2}$. This can be explained by saying that cell face tensions, induced by the buckling of some cell faces, are equivalent to the foam being filled by a gas at a pressure greater than atmospheric.

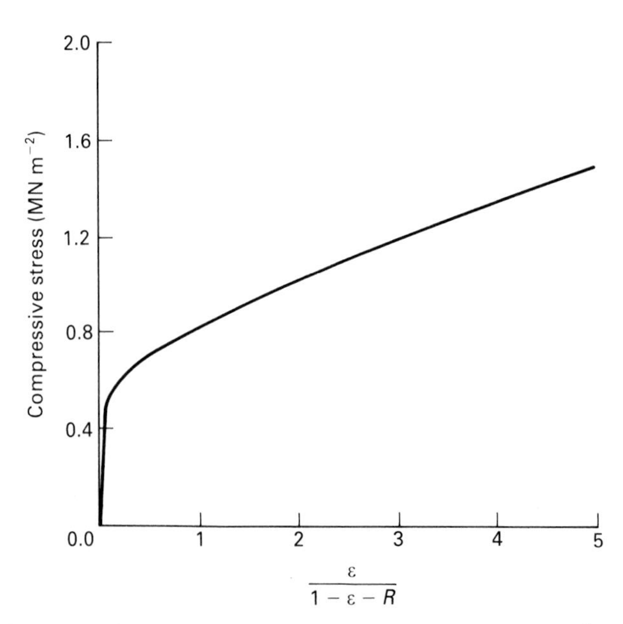

Fig. 7.19 The compressive stress in polypropylene foam of density 60 kg m^{-3} plotted against the gas pressure hardening factor of equation (7.20)

7.5 MICRO-YIELDING

7.5.1 Crazing

Crazing is most readily observed in transparent glassy plastics. For example if a strip of PMMA is bent and the surface moistened with ethanol, crazes form rapidly from the tensile surface. It also occurs in semi-crystalline plastics, but the opaque nature of the plastic makes the crazes more difficult to observe.

Crazes in glassy polymers reflect light, and look like small cracks that are <1 mm in length that have grown from the surface. It is not until thin sections across a craze are examined in a transmission electron microscope (Fig. 7.20a), or the fracture surface is examined in a scanning electron microscope (Fig. 7.20b) that the fibrils of oriented polymer bridging the craze, or the broken halves of these fibrils, are seen. The formation of fibrils from bulk polymer involves both yielding and the formation of voids (a craze typically has a 50% void content). Crazes form on planes that are perpendicular to the largest tensile principal stress; in this way they behave in the same way as cracks. There is a nearly uniform tensile stress across a loaded craze, the value being characteristic of the polymer, the environment, and the time scale. For polystyrene in air at 20 °C crazes appear in 30 s if a constant stress of 25 MN m^{-2} is applied, and in 24 hours for a stress of 10 MN m^{-2}. Instrumented impact tests (Section 8.5.1) suggest that the craze stress rises to 100 MN m^{-2} on a time scale of 5 ms. This is another incidence of the time-scale dependence of yielding processes in polymers.

The craze thickens by drawing in new material across the craze–bulk interface, this material then stretches to a characteristic extension ratio. In contrast with conventional yielding processes, such as the shear bands in Fig. 7.1, crazes can remain thin while allowing extension in the direction of the tensile stress, because the void content between the craze fibrils creates the necessary volume increase.

Crazes do not form unless the tensile strain exceeds a critical value. For example the tensile strain must exceed 2% before polycarbonate will craze in air, and 0.4% before polystyrene will craze. We shall see in Chapter 9 that these values are reduced when the polymer is exposed to certain liquids. If the applied strain barely exceeds the critical value the resulting crazes are widely spaced, and the craze spacing decreases as the applied strain is increased. This is further evidence for craze formation acting as a yielding process on the polymer surface.

The mechanism by which a craze tip advances into uncrazed polymer has been the subject of speculation. The stress across the craze is often much smaller than the yield stress of the bulk polymer (40 compared with 90 MN m^{-2} for polystyrene). Even though the stress can be locally higher at the craze tip the mechanism of craze advance cannot involve ultra-high stresses. For example if isolated spherical air bubbles have to grow by the whole of the surrounding volume yielding the required hydrostatic stress is nearly three times the tensile yield stress. However, the co-operative growth of parallel finger-like cracks is a widely observed phenomenon in the fracture of materials. Fig. 7.21a shows the fracture surface after liquid carbon

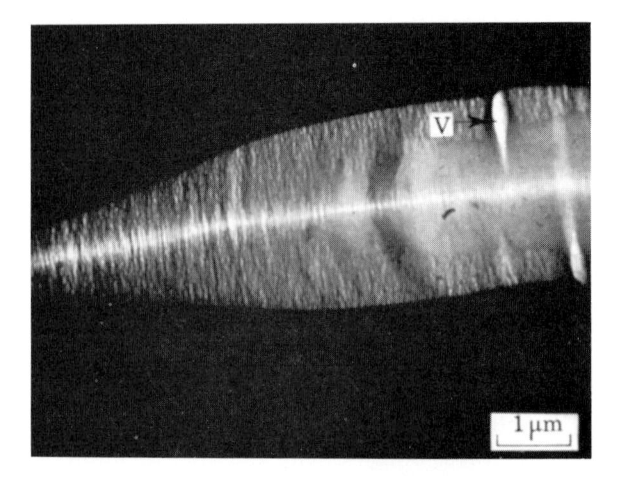

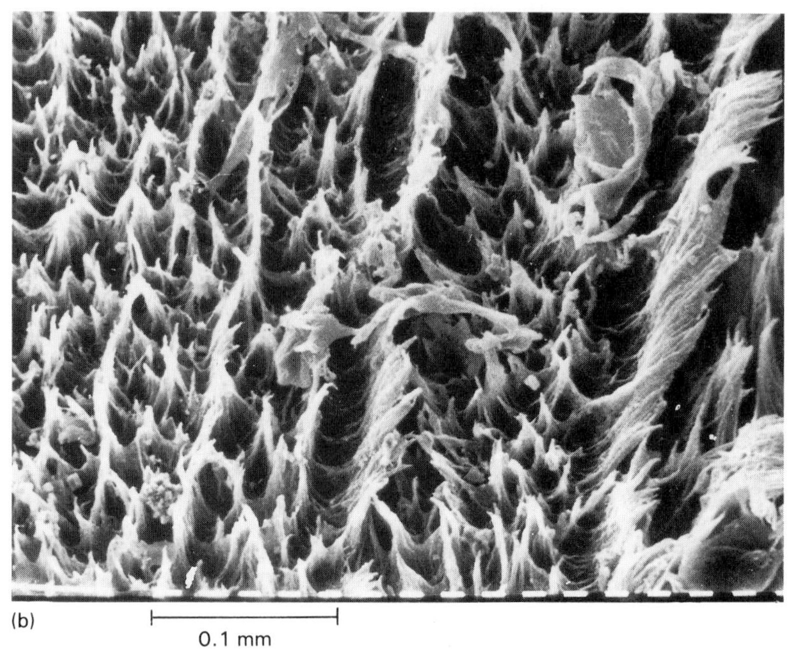

Fig. 7.20 (a) Craze in polystyrene seen in cross section in the transmission electron microscope (from Behan *et al.*, *Proc. Roy. Soc.*, 1975, **A343**, 530). (b) Craze remnants seen on the fracture surface of a polyethylene

tetrachloride has advanced through a craze in polycarbonate—the walls of the liquid channels fracture last and have a speckled appearance. It is reasonable to suppose that similar process could occur on a 1000 times smaller scale when air advances into a bulk polymer (Fig. 7.21b). This time the channel walls do not fracture, instead they fibrillate into strands of oriented polymer. In a high molecular mass polymer the fingers of air advance too rapidly for polymer molecules to disentangle and flow apart. Consequently there is considerable

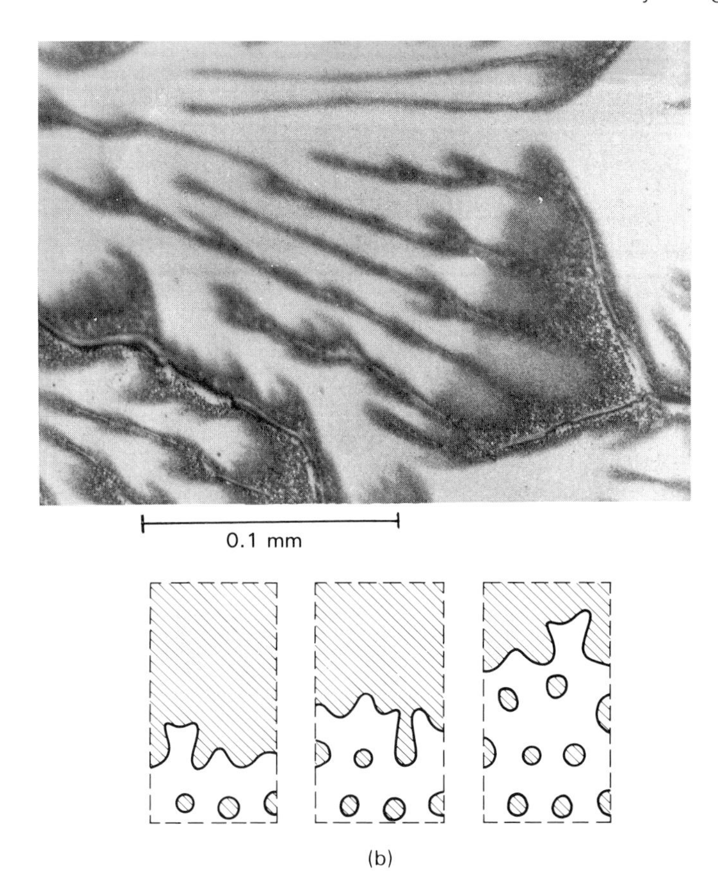

Fig. 7.21 Fingerlike advance of cracks lying in a plane. (a) The fracture surface of polycarbonate fractured in a carbon tetrachloride environment. The liquid has advanced down the channels. (b) Sketch of a proposed method of craze advance

chain scission as the craze advances. Crazes therefore are a locally weakened part of the microstructure; the large surface area of the fibrils also provide a driving force for the absorption of any liquids in the environment.

Crazes play an important part in fracture (Chapter 8) once they reach a critical size, so it is of interest to analyse the energies of craze growth. Fig. 7.22a shows single craze with a tensile stress σ_c across it, in a large sheet of material that has a tensile stress σ applied to the ends. This stress analysis problem can be decomposed into two parts—a uniform tensile stress σ_c in an uncracked sheet plus a sheet with a crack in it that has a stress $(\sigma - \sigma_c)$ applied to the ends. Therefore the energetics of craze growth is a modified form of the energetics of crack growth, so long as the applied stress exceeds σ_c. We shall assume that the ends of the sheet are held a fixed distance apart, so that craze growth can only depend on how the stored elastic energy in the sheet W changes with the craze area A. We assume that the growth velocity V is an increasing function $V(\partial W/\partial A)$. As the exact calculation of $\partial W/\partial A$ is rather complex we shall use the simplification that a craze of length a totally

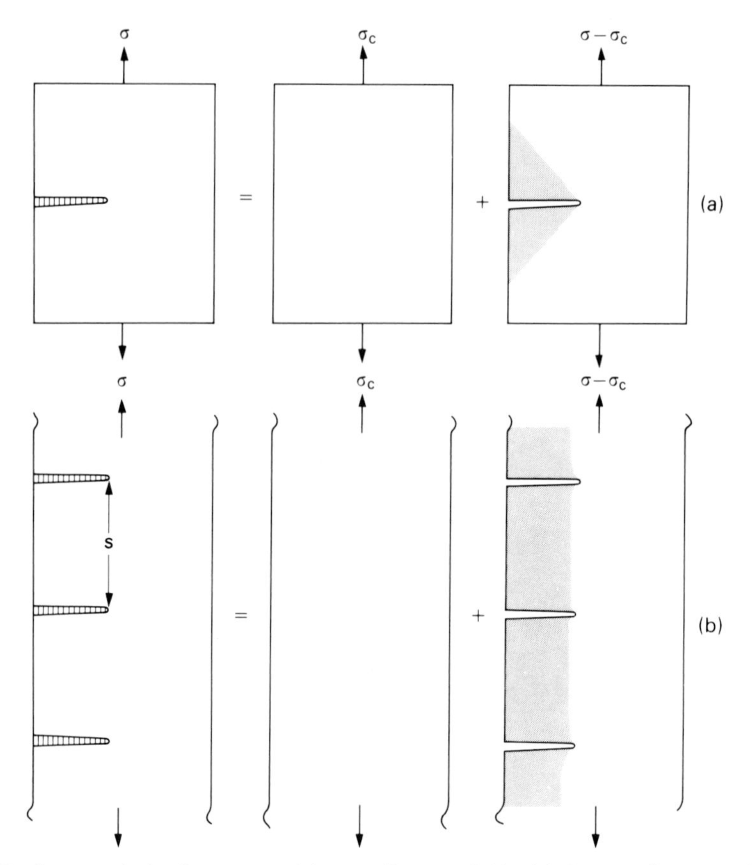

Fig. 7.22 Stress analysis of craze growth in a tensile stress field, with the stress free areas shown shaded. (a) For a single craze of a length a. (b) When there are equal length crazes separated by a distance s in the direction of the tensile stress

relieves the stress in two triangular areas with total area βa^2 where β is a constant. The elastic energy density in a uniform tensile stress field σ is $\sigma^2/2E$. Therefore the total stored elastic energy for the sheet of thickness t is

$$W = W_0 - \beta a^2 t \frac{(\sigma - \sigma_c)^2}{2E}$$

where W_0 is the stored energy in the sheet before the craze appeared. Differentiation gives

$$-\frac{\partial W}{\partial A} = \frac{\beta a}{E}(\sigma - \sigma_c)^2 \tag{7.21}$$

This means that the strain energy release rate is directly proportional to the craze length. Therefore if there are isolated crazes their growth rate will accelerate as their length increases, so long as the applied stress exceeds σ_c.

It is common for a large number of closely spaced crazes to form, the crazes nucleating on all surface defects above a certain size. In a simple two dimensional model of this (Fig. 7.22b) the parallel crazes are regularly spaced at separation s. The region ahead of the crazes is at the uniform stress σ, whereas behind the craze tips the stress has reduced to σ_c. When the problem is split into two parts as in Fig. 7.22a then the advance of each craze by a length da transfers a volume $st\,da$ from a stress of $(\sigma - \sigma_c)$ to a stress free state. Hence

$$-\frac{\partial W}{\partial A} = \frac{s}{2E}(\sigma - \sigma_c)^2 \qquad (7.22)$$

This shows that the strain energy release rate remains constant, and the parallel crazes can grow in a slow stable manner.

Equations (7.21) and (7.22) shows that the energetics of craze growth can change dramatically depending on whether the craze is isolated or interacts with its neighbours. This is but one example. We shall see in the next chapter that crazes can form at a crack tip, and that the ease of that crack growing depends on whether a single craze, or a bunch of crazes, forms. Therefore it is important to observe the number and geometry of the crazes involved in a fracture process.

7.5.2 Plastic collapse of closed cell foams

When polystyrene foams are compressed the cell faces buckle permanently (Fig. 7.23). The deformation mechanism is similar to that of the crumpling of thin sheet steel in the wing of a car after a frontal crash, in that there are *plastic hinges* at intervals across the faces. The plastic buckling of the 1 μm thick cell faces is in contrast to the crazing and fracture that would occur with 1 mm thick polystyrene. Part of the difference in behaviour may be due to some biaxial orientation of the cell faces. Crazes have been observed in very thin polystyrene films that have been stretched in transmission electron microscopes, but it is unlikely that thin films would craze when bent. In closed cell polyethylene foams this permanent deformation does not occur, rather there is a viscoelastic recovery to nearly zero strain, even after severe compressive impacts. Again this contrasts with the plastic hinges that form in severely bent 2 mm thick HDPE sheet. We can conclude that the very thin cell faces are more ductile than the corresponding bulk polymer.

Two microstructural features affect the cell deformation mechanisms.

(a) The gas pressure in the cells affects the mode of buckling of the faces. Any buckling mode that allows the volume of one cell to expand, while the neighbouring cells contract, is ruled out because a pressure differential would develop between cells. The likely buckling mode is one in which the vertical cell faces concertina as in a bellows; it allows Poisson's ratio to be zero. If the number of half waves in the buckled shape is even then the volume of each cell is equally compressed.

(b) A cell face is stabilised by the neighbouring faces; a similar effect occurs in a honeycomb structure. For a particular mode of buckling to occur in a face, there must be buckling in the neighbouring faces and some plastic deformation near the edges that join them to decouple the buckling modes. It could be

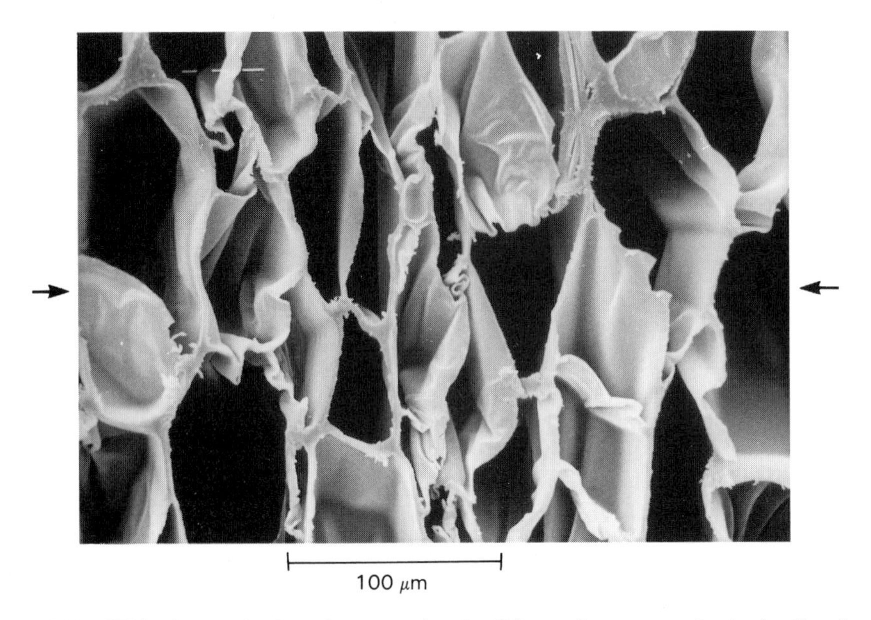

Fig. 7.23 SEM micrograph of a polystyrene closed cell foam after compression in the direction indicated to a strain of 80%

that this near-edge yielding is the controlling event for the stability of the closed cell shape, and hence the critical event for compressive yield of the foam. The axial compression of a honeycomb made from a ductile material has been analysed and the axial compressive yield stress σ_c predicted to be

$$\frac{\sigma_c}{\sigma_Y} = 5.6\left(\frac{t}{L}\right)^{5/3} \tag{7.23}$$

where t is the uniform wall thickness and L the wall length. For a closed cell foam, in which all the polymer is in the cell faces, the relative density R is proportional to the ratio t/L. Hence the honeycomb axial buckling model leads us to expect that the compressive yield stress of polystyrene foam would vary with the 1.7th power of the relative density.

Experimental plots of the yield stress versus the density for polystyrene and for high density polyethylene foam on logarithmic scales have slopes $\cong 1.5$. If the lines are extrapolated to the density of the solid polymer, it is found that the predicted yield stress is close to σ_0, that measured on the solid at static strain rates. Hence the yield stress of the foam σ_Y can be written as

$$\sigma_Y = \sigma_0 R^{1.5} \tag{7.24}$$

For high density polyethylene cell faces which behave in a non-linear viscoelastic manner, the stress distribution in a bent cell face (Fig. 6.9) is not very different to that in a plastic hinge (Fig. 7.7b), so it is not surprising that the exponent in the yield stress–relative density relationship is the same as for polystyrene. We shall return to these materials in the design studies in Chapter 13.

8

Fracture

8.1 INTRODUCTION

Plastics products should be designed so that the expected loads will not cause any type of fracture. There will always be some extreme impact that can initiate fracture, so in large products fast crack growth should be avoided. For example polyethylene pipes for natural gas distribution will be designed to avoid slow crack growth from defects at welded joints. It will always be possible to fracture the pipe by the careless use of a mechanical excavator, in which case the polymer should be selected so that rapid crack growth down the length of the pipe is impossible.

The chapter starts with fracture surface examination for clues to the causes of fracture. It then describes the mechanical and environmental causes of crack initiation. Enough fracture mechanics theory is included to explain the criteria for crack growth, and to show which parts of the theory are necessary for the varied fracture phenomena in polymers. In contrast with ceramics which are nearly always brittle, and with metals which do not display time dependent fracture phenomena at room temperature, polymer fracture is extremely varied in nature. The possibility of high anisotropy, and the widespread use of thin film products, makes a comprehensive treatment difficult. Finally it is necessary to evaluate the impact tests that are the most common type of fracture data.

8.2 FRACTURE SURFACES AND THEIR INTERPRETATION

Failure investigations involve a detailed examination of the relevant fracture surfaces. The geometry of the crack front at the various stages can be deduced, and this gives information about the type of loads that were acting. The basic principle is that the *crack plane is perpendicular to the most tensile principal stress*. Markings on the fracture surface can indicate the direction of crack growth and the approximate velocity of the crack. Arrest lines on the surface can show the sequence of crack front positions. These principles will be illustrated with a couple of examples.

Fig. 8.1 shows the appearance of the fracture surface from a tensile test carried out on a brittle plastic. The fracture surface is perpendicular to the length of the bar, which is the direction of the principal stress. As there is a uniform stress the crack could form at any site, but crack initiation is most likely at a surface where there could be scratches or contamination from finger grease. The flat mirror-like area on one edge is where a craze has nucleated. The crack has then propagated at a rapidly increasing speed, leaving parabolic markings (Section 8.4.4). The noses of the parabolas always point to the crack source, so the direction of crack growth is known to be

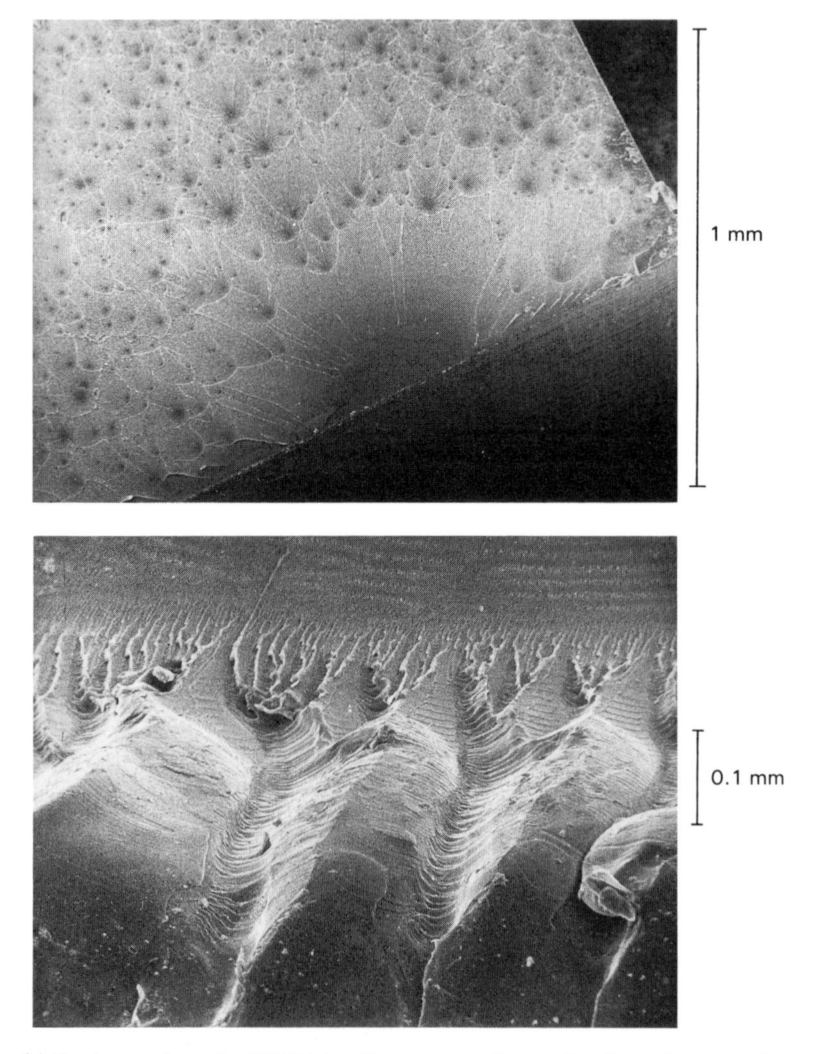

Fig. 8.1 (a) Fracture surface of a PMMA tensile specimen. A craze has formed on one edge, then failed, leaving a flat region. (b) Bending fracture surface in PC showing the splitting of the crack, as it moves from top to bottom of the picture

towards the top left of the figure. Eventually there will be a rough fracture surface as subsidiary cracks accompany the main one.

Most plastics mouldings are thin walled, and bending and twisting failures are far more common than tensile loading. Fig. 8.1b shows the fracture surface for such a case. Cracks have formed on the top side that was placed in tension by the bending moment; there are some markings parallel to the crack front positions that are remnants of craze failure. As the crack approached the lower side of the panel it has twisted and broken up into many parallel cracks. The formation of these can be explained by the combination of the tensile stress T from the bending moment, and a shear stress S from the torsional loading (Fig. 8.2a). The crack plane wants to rotate about the growth direction to remain perpendicular to the tensile principal stress σ. This adjustment can only occur in a short growth distance if the crack breaks up into an echelon of sub-cracks (Fig. 8.2b). The sides of the sub-cracks eventually join up (Fig. 8.2c) to make a stepped fracture surface as in Fig. 8.1b.

A common surface feature in the secondary stage of fracture is a series of parallel markings shaped like a quarter of an ellipse (Fig. 8.3a). The surface crack initiation has been due to a blunt object pressing on the surface of the product (Fig. 8.3b). As this crack spreads sideways the object penetrates the product and twists the two sides in opposite directions. This double torsion

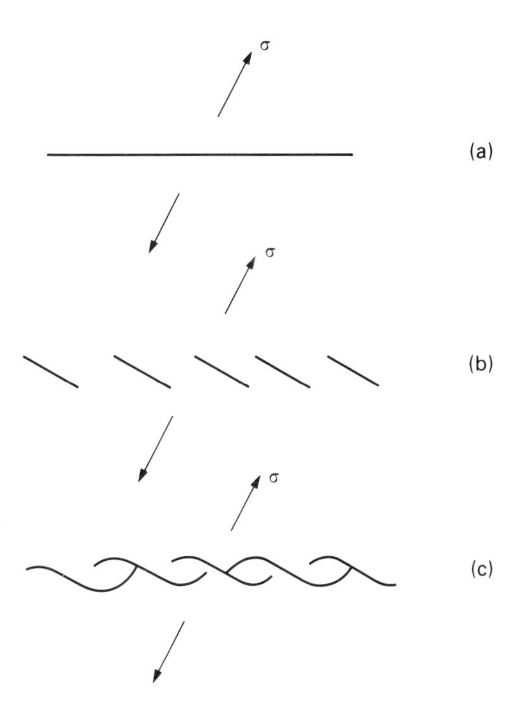

Fig. 8.2 (a) A crack growing into the paper with the crack plane at an angle to the principal tensile stress; (b) the crack splits into sub cracks that advance as fingers in echelon; (c) the cracks link up further back from the crack tips

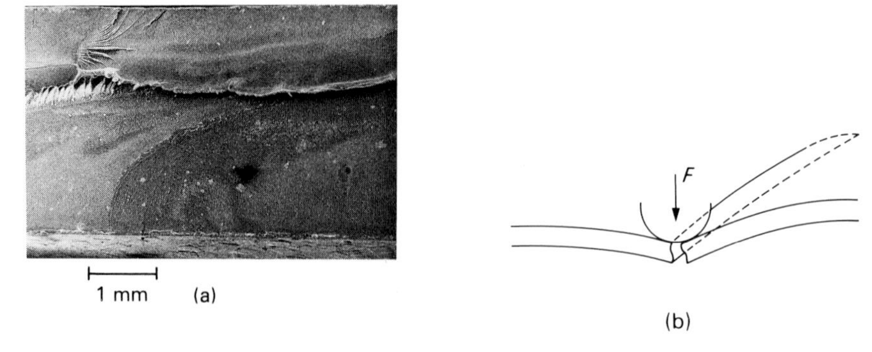

Fig. 8.3 (a) Elliptical markings due to momentary crack arrests on a polycarbonate fracture surface. (b) The double torsion loading that occurs when a crack is driven by a remote surface force F

loading causes the crack to advance more rapidly on the lower surface (the lower surface is in tension while the upper surface is in compression), and the characteristic markings are due to the crack momentarily hesitating. Detailed investigation shows that a craze forms at the crack tip while the crack hesitates, then the crack advances rapidly to the craze tip and the process repeats. The crack advance step often spreads sideways from one end of the craze to the other, like a dislocation running through a crystal.

8.3 CRACK INITIATION

Failure investigation covers a number of discipline areas, and it is important that one's background does not limit the range of possibilities considered. For example a mechanical engineer would be aware of the effects of stress concentrations in a product, whereas a polymer technologist might be more inclined to blame poor processing. A polymer chemist might consider the molecular weight distribution of the polymer, and whether it had been reduced by environmental degradation. Several of these factors may be involved in any particular case of crack initiation. The environmental changes will be analysed at greater length in Chapter 9, but their mechanical consequences will be examined here.

8.3.1 Elastic stress concentrations

Holes or sharp corners are the most familiar type of stress concentrating feature. The *stress concentration factor* q is defined for the most tensile principal stress as

$$q = \frac{\text{local maximum stress}}{\text{stress that would exist if the feature were absent}} \tag{8.1}$$

q has no dimensions, and it should not be confused with the stress intensity factor K introduced later, which has dimensions of N m$^{-1.5}$, and which allows

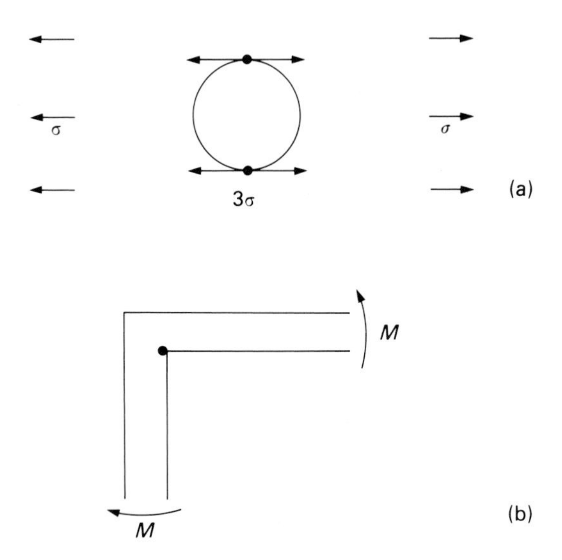

Fig. 8.4 (a) Stress concentration factor $q = 3$ at a cylindrical hole in a plate under tension and (b) $q > 10$ at an internal low radius corner in a box to which bending moment M is applied. The peak stresses are at the points ●

the stress at any point of the crack tip stress field to be calculated. For a cylindrical hole in a wide plate, subjected to a tensile stress σ, the maximum stress is 3σ at the side of the hole, so $q = 3$ (Fig. 8.4a). If a hole is needed in a product it can either be moulded-in, or drilled in a subsequent operation. The surfaces of drilled holes tend to be rough as a result of the localised heating and tearing of the plastic. If the hole is to be moulded-in then the melt must divide to pass on either side of a steel pin, and a weld line forms on the downstream side (Section 5.2.4), leaving a surface groove. Both these effects tend to increase the value of q.

The internal corners of any product should have a generous fillet radius r, but this may be omitted (Fig. 8.4b) to reduce the mould machining costs. The elastic stress analysis of such a part when it is subjected to bending is complex, but we expect that

$$q \propto r^{-0.5}$$

An ideally sharp internal corner with a value $r = 0$ would produce an infinite value of q, but with a machined steel mould the corner radius will exceed 5 μm. Surface scratches are also stress concentrating features.

The major limitation to the use of the stress concentration concept is that yielding will nearly always occur in polymers before crack initiation. However the knowledge of the location of the elastic stress maximum will indicate where yielding is likely to occur. It is impossible to use q with Charpy impact tests (Section 8.5.1) to calculate the stress at failure. Even the formation of craze(s) is a form of localised yielding, which may modify the stress distribution in the product. We shall see in Section 8.4.4 that the breakdown of a craze into a crack may be controlled more by the opening displacement of the craze rather than the stress across it. We shall now analyse the mechanism of localised yielding to see how cracks can be initiated.

8.3.2 Yield stress concentrations

The fracture behaviour of some of the tougher plastics can appear to be inconsistent. For example polycarbonate can sometimes fail by yielding, but on other occasions a crack can form and brittle fracture follows (Fig. 8.5). The explanation lies in the analysis of localised yielding (Section 7.2.4), which showed that the indentation pressure could be 3 times the tensile yield stress. A change of a similar magnitude can occur in the yield stress during tensile loading.

Careful examination of a notched region, after a slight impact load has started the yielding process, can reveal a pattern of shear bands (Fig. 8.6a) in some glassy plastics. The yield strain occurs inhomogeneously; the shear strain is about 1 within the shear bands, and zero between them. The overall pattern is remarkably similar to a particular *slip line field* pattern (Fig. 8.6b); this is a theoretical concept used in the analysis of metal working under conditions where all the non-zero permanent strains occur in one plane. The slip line field shown consists of 2 families of logarithmic spirals, which in polar coordinates r, θ have the equation

$$\ln r = \ln a \pm \theta - \theta_0 \tag{8.2}$$

where a is the notch radius and θ_0 is a different constant for each spiral. The family of α slip lines is orthogonal to the β family, and it is conventional that the more tensile of the principal stresses lies in the first quandrant of the $\alpha\beta$

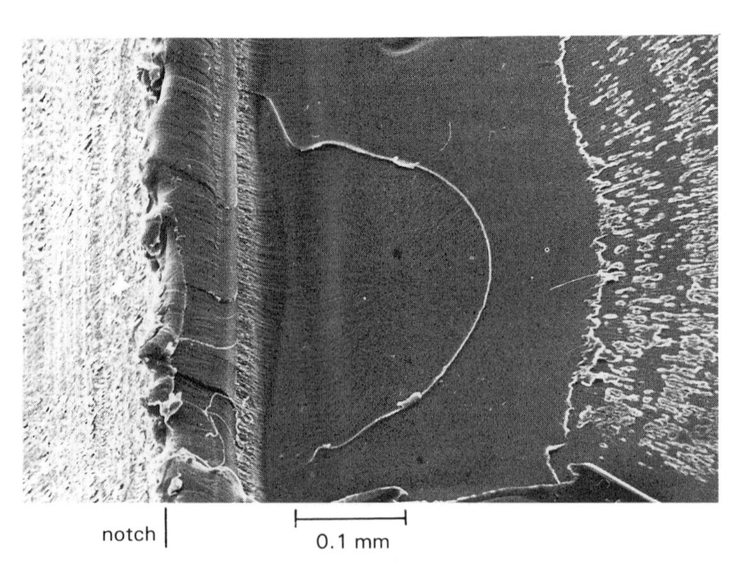

Fig. 8.5 Fracture surface of polycarbonate in which a crack has initiated at A, 0.55 mm below the root of a notch of radius 0.25 mm, in a sheet 6 mm thick

axes that occur at every intersection of the two families of lines. In the Mohr circle representation of the stress state, the points at the end of the vertical diameter of the circle represents the stress components in the $\alpha\beta$ axes (Fig. 8.6c). In these axes there are equal biaxial tensile stresses of magnitude $\bar{\sigma}$ (the coordinate of the centre of the circle), and the shear stresses have their maximum value k, according to Tresca's yield criterion. In order to find out how $\bar{\sigma}$ changes along an α slip line we consider a special case where only the α slip lines are curved, with a radius of curvature r. Fig. 8.6d shows the stress components, on the surface of the prism marked out by neighbouring α and β slip lines, that contribute to the moment about the point O. The prism is in static equilibrium so the moment of the forces on it about O is zero. For a unit length of prism in the direction perpendicular to the paper we therefore have

$$kr^2\,\mathrm{d}\phi - k(r+\mathrm{d}r)^2\,\mathrm{d}\phi + (\bar{\sigma}+\mathrm{d}\bar{\sigma})r\,\mathrm{d}r - \bar{\sigma}r\,\mathrm{d}r = 0$$

where $\mathrm{d}\phi$ is the angle by which the α line has rotated. Expanding this and ignoring the $\mathrm{d}r^2$ term gives

$$\mathrm{d}\bar{\sigma} - 2k\,\mathrm{d}\phi = 0$$

so on integration

$$\bar{\sigma} - 2k\phi = \text{constant} \tag{8.3}$$

along an α line. We shall use this equation to find how the stress state changes from the surface of the notch at A, to the top of the yielded zone at B in Fig. 8.6b. At A there is a tensile principal stress $2k$ parallel to the surface, and a zero stress perpendicular to the surface, so $\bar{\sigma} = k$, and the left hand circle in Fig. 8.6c represents the stress state. The angle of the slip line increases by θ between A and B (the slip line remains at 45° to the radius) so from equation (8.3)

$$\bar{\sigma}_A - 2k\phi_0 = \bar{\sigma}_B - 2k(\phi_0 + \theta)$$

so

$$\bar{\sigma}_B = \bar{\sigma}_A + 2k\theta \tag{8.4}$$

The stress state at the tip of the yielded zone is represented by the right hand circle in Fig. 8.6, and the most tensile principal stress

$$\sigma_{max} = 2k(1+\theta) \tag{8.5}$$

Equation (8.5) shows that the most severe notch has parallel sides, in which case the tensile stress at the tip of the yielded zone can rise to 2.57 times the unrestrained tensile yield stress. The yielded zone would be quite large at this stage; equation (8.2) shows that the point B would be $4.81a$ from the centre of the notch. In practice the yielded zone rarely becomes this large because either

(a) the tensile stress becomes high enough for a crack to form in which case no further yielding occurs, or
(b) the yielded zone becomes too long compared with the specimen thickness t. For all the yielding to occur in the plane of Fig. 8.6b the thickness t must be

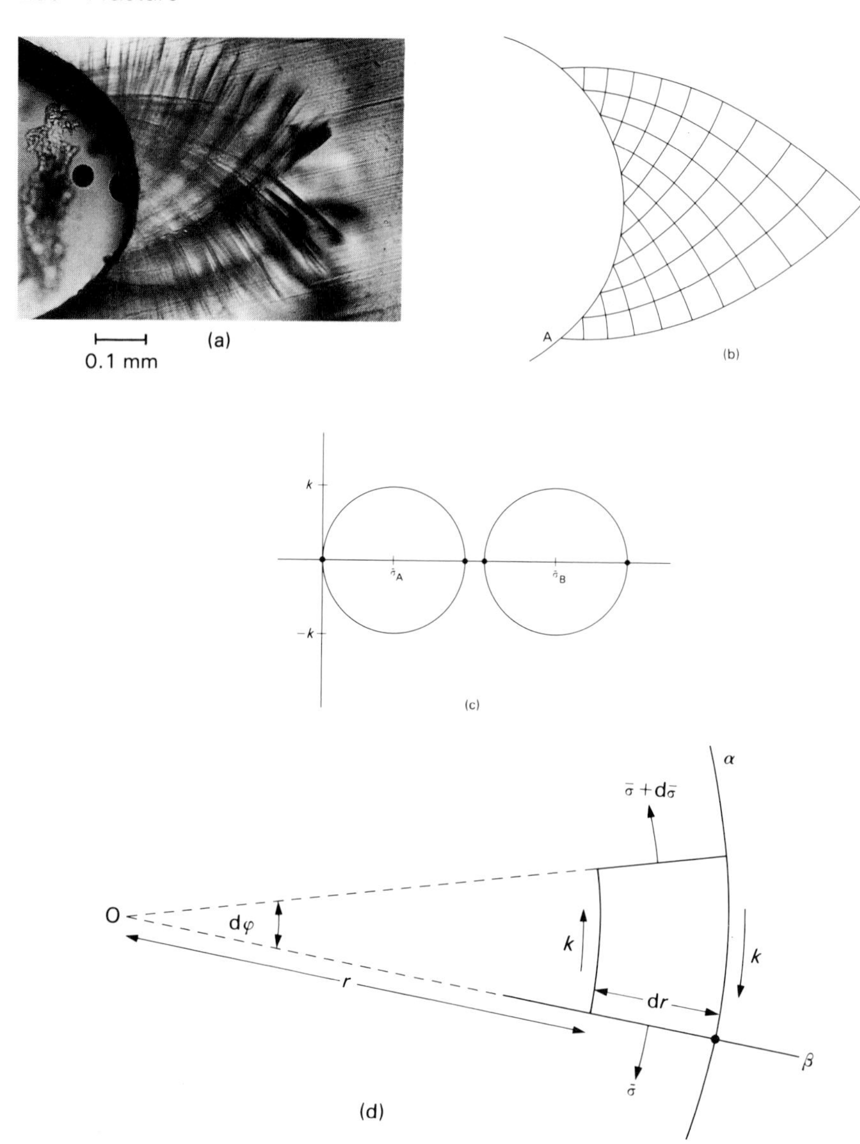

Fig. 8.6 Analysis of yielding at a notch. (a) Shear band patterns seen in a thin section cut from a polycarbonate specimen. (b) Slip line field pattern for yielding. (c) Mohr stress diagram for the state of stress at points A and B in (b). (d) Stress components on the surface of the prism marked out by neighbouring α and β slip lines

at least 5 times as great as the length of the yielded zone. For the common notched impact test dimensions of $a = 0.25$ mm, $t = 3$ mm; this occurs when $\theta = 1.22$ radian; however, the usual notch angle of 45° means that the largest possible θ value is 1.18 radian. We conclude from this analysis that crack initiation can occur in semi-tough plastics if any notch has a radius that is considerably smaller than the specimen thickness; the notch tip radius needs to be smaller than $t/10$ to maximise the effect (Section 8.5.1).

8.3.3 Cracks in brittle surface layers

In order to improve the surface gloss and appearance, layers of polystyrene or chromium have been applied to rubber toughened plastics like ABS. This is one example where a thin layer of a brittle material is deliberately applied to the surface of a relatively tough material. In-service deterioration (see Chapter 9) can also convert the surface layer of a plastic into a brittle state, for example by the strong surface absorption of the ultraviolet part of the sun's spectrum. The effects of these surface layers are the most marked when the product is bent, with the brittle layer being on the tensile side. The tensile failure strain of the surface layer will be smaller than that of the substrate so it will fail first. A series of sharp cracks will form perpendicular to the surface tensile stress; each will relieve the surface stress over a limited length. The cracks will start at the outer surface, and will travel rapidly when they have traversed the surface layer. What happens then depends on the substrate. If it is tough the cracks will stop, but if it has a low resistance to the fast propagation of cracks, the cracks will continue.

8.3.4 Residual stresses and environmental stress cracking

Chapter 5 mentioned the residual stresses that occur in a flat product when it is cooled rapidly from both sides. There are biaxial compressive stresses in the surface layers and there are biaxial tensile stresses in the interior. If a hole is drilled through such a product it cuts through the tensile stress region, and acts as a stress concentrating feature with a q value of 2. With the ingress of a stress cracking fluid a set of radial cracks may form from the bore of the hole, perpendicular to the residual circumferential stresses. These cracks will be at the mid-thickness of the product.

Another pattern of residual stresses can arise if a hole is drilled with a blunt bit in a product that is initially stress free. The drilling operation generates enough heat to melt an annulus of plastic surrounding the hole. When this cools down it will contract. The effect is the converse of shrink fitting a metal rim on a wheel, so a thin layer of plastic surrounding the hole will have a residual tensile circumferential stress. Consequently cracks may start in a radial direction, but they will turn to follow the boundary of the over-heated layer. These two examples show that the pattern of cracks that form are evidence of the type of residual stress field that existed in the product.

In summary the site of crack initiation is governed by several factors. It is likely to be on the surface for a number of reasons (bending or torsion loading, surface scratches causing stress concentrations, or degradation/ environmental factors) but may be internal (yield stress concentrations, or weak interfaces).

8.4 CRACK GROWTH

8.4.1 Fracture mechanics and the stress intensity factor

Once a sharp crack has formed it is possible to analyse whether it will grow using the concepts of fracture mechanics. The subject was developed for the failure of large metal structures, and the simplest theory is for a crack in an elastic material, hence the name *linear elastic fracture mechanics*. This is concerned with the stresses and strains in the region around the crack tip. In most loading situations the crack faces move directly apart (so-called mode I deformation) rather than sliding over each other in one of two directions (mode II or III). In Fig. 8.7a the crack front lies along the z axis and the crack faces move apart in the $\pm y$ direction.

When the specimen is loaded, by convention the crack tip is kept at the origin of the xy axes. All other points in the body will be displaced from their unloaded positions. There are two simple *boundary conditions* for the opening of a crack (Fig. 8.7a). Firstly the elastic displacement component u_y becomes two-valued along the crack; this means that there is a sudden jump in the value from $+\delta/2$ to $-\delta/2$ on crossing the crack, where δ is the crack opening displacement. Secondly both the shear stress on the crack plane and the tensile stress normal to it are zero; this is the case for all free surfaces. If the position in the stress field is described by the complex number $z = x + iy$, we seek suitable functions of z to describe the stress field and displacement field. $z^{0.5}$ (or $z^{1.5}$ or $z^{2.5}$ . . .) is two valued along the crack position so we can try

$$u_y = K^* Im(z^{0.5}) \tag{8.6}$$

for the discontinuous part of the displacement field, where K^* is a constant. Along the crack, where $z = -x$, the imaginary part of $z^{0.5}$ takes the values $\pm x^{0.5}$, so the crack tip has a parabolic shape as required. The strains are obtained from the displacement by differentiation, using equations such as

$$e_y = \frac{\partial u_y}{\partial y} = \frac{K^*}{2} Im\left(\frac{1}{\sqrt{z}}\right) \tag{8.7}$$

Therefore the strains vary as $z^{-0.5}$ (or $z^{0.5}$ or $z^{1.5}$ . . .). Close to the crack tip the $z^{-0.5}$ solution will dominate, so we use this term alone to describe the crack tip strain field. For an elastic material the stress components are linearly related to the strain components, so they too vary with $z^{-0.5}$. For convenience the results are re-expressed in terms of the polar $r\theta$ coordinates giving

$$\bar{\sigma} \equiv \frac{\sigma_{xx} + \sigma_{yy}}{2} = \frac{K_I}{\sqrt{2\pi r}} \cos \theta/2$$

$$\tau_{max} \equiv [0.25(\sigma_{yy} - \sigma_{xx})^2 + \sigma_{xy}^2]^{0.5} = \frac{K_I}{2\sqrt{2\pi r}} \sin \theta \tag{8}$$

The mean tensile stress $\bar{\sigma}$ and maximum shear stress τ_{max} (see Fig. 8.6c) are given in terms of the scaling constant K_I for the crack opening mode I

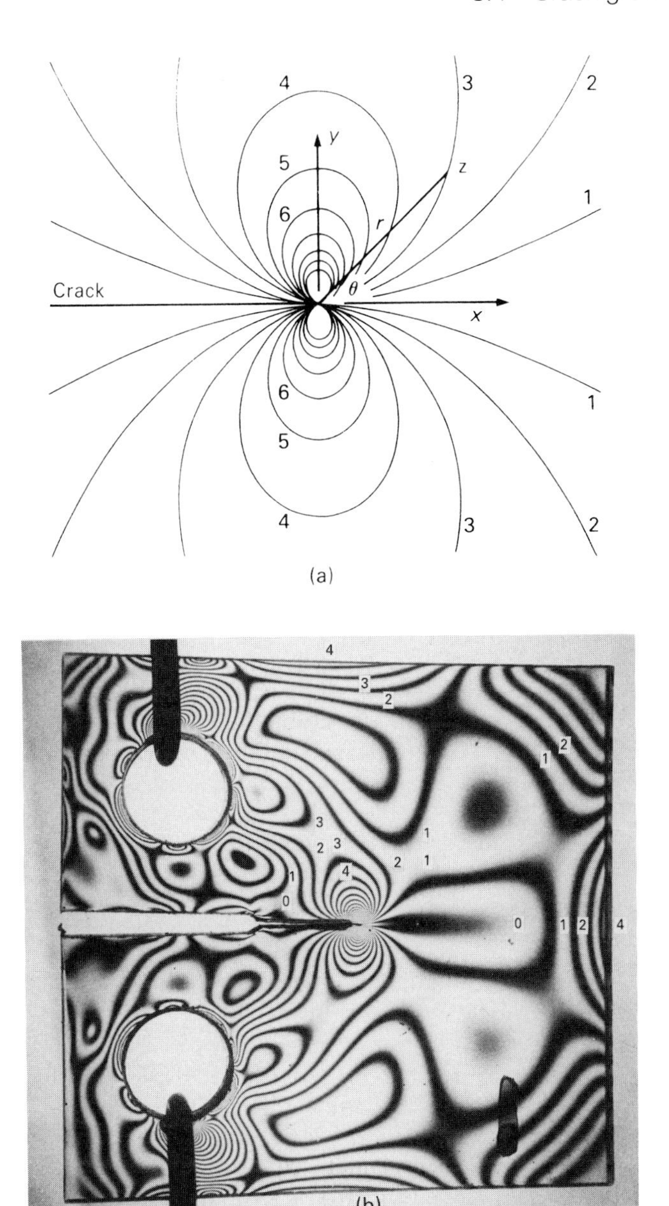

Fig. 8.7 (a) The theoretical isochromatic photoelastic pattern at a crack tip, with the contour levels of maximum shear stress. (b) Experimental isochromatic pattern of a loaded polycarbonate compact tension specimen

deformation. K_I is known as the *stress intensity factor*, and is proportional to K^* in equation (8.7). The inclusion of the $(2\pi)^{0.5}$ term is a result of the 1939 definition of K_I; logically it could be omitted. Note how both stress components in equation (8.8) are zero along the crack surface, as required by the boundary conditions. The stress intensity factor has the units of [stress × (length)$^{0.5}$] in contrast with the stress concentration factor in Section 8.3.1. There are parallels with other elastic stress fields, for instance that of an edge dislocation of Burger's vector **b**, where there is a constant displacement discontinuity of b across the dislocation plane, and the displacement field can be derived from the complex function $b \ln z$ in the equivalent of equation (8.6). A more familiar pattern is the magnetic field of a bar magnet which can be visualised by scattering iron filings on a sheet of white paper on top of the magnet. The shape of the pattern is always the same, but the magnetic field at any point is determined by the pole strength of the magnet, and the pole strength is the analogue of K_I.

Equations (8.8) state that the stress field close to the crack tip always has a particular shape. One way of revealing the stress field is to use the *photoelastic effect.* The refractive indices n_1 and n_2 in the directions of the principal stresses σ_1 and σ_2 are related by

$$n_1 - n_2 = C(\sigma_1 - \sigma_2) = 2C\tau_{max} \tag{8.9}$$

The magnitude of the stress optical coefficient C varies by an order of magnitude from polymer to polymer (Table 10.5, p. 270) and slightly with wavelength, so monochromatic light is used for photography (Fig. 8.7b). C is high for polycarbonate that contains benzene rings in the polymer backbone. Two dimensional models are cut from cast epoxy resin or carefully annealed PC sheet—the polymer will have some optical anisotropy unless it is in an equilibrium state. The *isochromatic fringes* are contour levels of the maximum shear stress τ_{max}. Each dark fringe can be assigned an integral fringe order f, with the first fringe to appear on loading having $f = 1$ etc. Since the maximum shear stress is proportional to the fringe order, then the radial distance r of any fringe from the crack tip is given by equation (8.8) as

$$r = A\left(\frac{\sin\theta}{f}\right)^2$$

where A is a constant. Consequently each fringe has a nearly elliptical shape with its major axis perpendicular to the crack (Fig. 8.7a).

The characteristic crack tip stress field can be recognised in the isochromatic pattern of any cracked body. Fig. 8.7b shows the *compact tension* specimen, often used to measure crack growth behaviour. The stress intensity factor K_I can be calculated from the isochromatic fringe positions close to the crack tip, by plotting the fringe number f versus $r_{max}^{-0.5}$, where r_{max} is the maximum excursion of the fringe from the crack tip. Equations (8.8) describe the elastic stress field close to the crack tip, where this dominates. The distant stress field is dominated by the point loading and the bending of the specimen as a C shaped beam.

8.4.2 Fracture toughness K_{IC}

The premise of fracture mechanics is that every *material* has a constant called the *critical stress intensity factor* or the *fracture toughness*, given the symbol K_{IC}. Crack growth will occur if and only if the stress intensity factor K_I satisfies

$$K_I > K_{IC} \tag{8.10}$$

Note that K_{IC} is a materials constant, whereas K_I is a mechanics parameter that changes with the applied load on the specimen. Measurements of the fracture toughness are useful for materials selection, and for safety calculations of cracked plastic structures under load. A set of fracture toughness values can be generated, using for example the compact tension specimens shown in Fig. 8.7(b). If the force applied to the loading pins is F, and the specimen thickness is B (this must be sufficiently large for the fracture to be plane strain—see Section 8.4.4) then K is calculated from the equation

$$K = \frac{F}{Bw}\sqrt{a}\left[29.6 - 185.5\left(\frac{a}{w}\right)^{0.5} + 655.7\left(\frac{a}{w}\right)^{1.5} - \ldots\right] \tag{8.11}$$

Although this and similar equations for other specimen geometries seem complex, the terms can be identified as follows: there is a tensile stress F/Bw multiplied by the square root of the crack length a. These terms must always occur to satisfy the units of K. The polynomial in brackets is the result of curve fitting a set of computer calculations of the K value for different ratios of the crack length a to the specimen width w (both are measured from an origin on the line joining the loading points), and it only is valid for a certain range of a/w. It is possible to calculate analytical solutions for certain simple geometries, for instance the polynomial in equation (8.11) is replaced by π for a central crack in an infinitely wide sheet under tension. However testing an infinitely wide sheet is wasteful of material!

The fracture toughness depends on the microstructure of the polymer, the polymer grade and the test temperature. The overall range of results in Table 8.1 is between 0.6 and 6 $\text{MN m}^{-1.5}$. This contrasts with the situation for metals where K_{IC} values can vary from 5 up to 100 $\text{MN m}^{-1.5}$.

Table 8.1 Range of K_{IC} values found for polymers at 20 °C

Polymer	K_{IC} ($\text{MN m}^{-1.5}$)
Polyester thermoset	0.6
Polystyrene	0.7 to 1.1
Polymethyl methacrylate	0.7 to 1.6
Polyvinyl chloride	2 to 4
Polycarbonate	2.2
Polyamide (nylon 6,6)	2.5 to 3
Polyethylene	1 to 6
Polypropylene	3 to 4.5
Polyoxymethylene	4

8.4.3 Yielding at a crack tip and the Dugdale model

For all plastics there will be yielding in the high stresses near the crack tip. This will make the propagation of the crack more difficult, and we wish to know whether the concept of K_{IC} is still valid. K_I describes the form and magnitude of the elastic stress field around the crack tip. In Fig. 8.7(b) there is undoubtably some yielding at the crack tip and yet the photoelastic pattern follows equation (8.8) for radial distances of 0.1 mm to 10 mm. Even if this elastic field surrounds the yielded region, the mechanism of crack growth occurs in the yielded region, and we need to be convinced that the critical parameter of this mechanism is directly related to K_I. If there is extensive yielding, or if the crack is growing into a highly anisotropic polymer, the stress field pattern implied by the use of K_I no longer exists. In this case other methods must be used to characterise fracture resistance.

Dugdale in 1960 proposed a particularly simple model of the yielded zone that gives useful quantitative predictions of the zone size. Fig. 8.8 shows that he assumed that the yielded zone is an extension of the crack, and that its thickness is much smaller than its length. The tensile stress σ_{yy} across it is assumed to be constant at σ_0, although the strain in the yielded zone can vary with the distance x from the crack tip. This means that the material is assumed not to work harden after yielding. The stress analysis problem for the surrounding elastic region could be solved because the shape of the yielded zone was predetermined, and the stresses at its boundary were known. Suppose that a stress intensity factor K_I acts at the crack tip before any yielding occurs (alternatively K_I describes the magnitude of the elastic stress field surrounding and slightly away from the small yielded zone). Consider a hypothetical crack that is R longer than the real crack, made by cutting around the yielded zone. When the zone is cut out closure forces must be applied to the boundary of the cut to prevent the shape of the elastic region changing. The loads on the hypothetical crack are the original ones, plus the pairs of closure forces acting on each length element dx between 0 and R, on the sheet of thickness t (Fig. 8.8). Dugdale assumed that, because of the yielding, the total stress intensity at the hypothetical crack tip is zero;

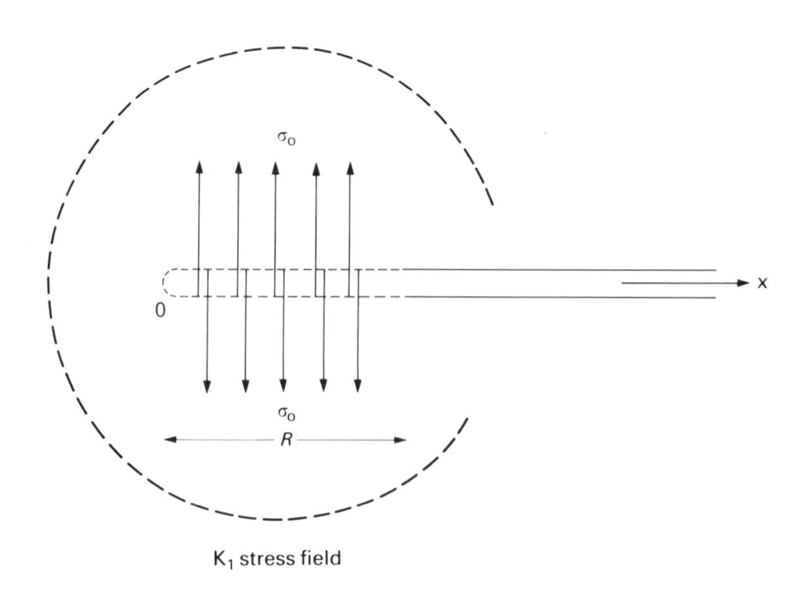

Fig. 8.8 Dugdale's model for crack tip yielding under plane stress conditions

otherwise there would be infinite stresses there according to equations (8.8), so

$$K_{\text{total}} = K_{\text{I}} + K_{\text{I}}^{\text{B}} = 0$$

where K_{I}^{B} is the stress intensity factor from the closure forces. The small increase R in crack length does not increase the original K_{I} value significantly. The stress intensity factor due to a pair of closure forces F acting a distance x from the tip of a crack is $-F/t(2/\pi x)^{0.5}$ so K_{I}^{B} is the sum of such values

$$\begin{aligned} K_{\text{I}}^{\text{B}} &= \int_0^R \frac{\sigma_0 t}{t}\left(\frac{2}{\pi x}\right)^{0.5} \mathrm{d}x \\ &= -\sigma_0\left(\frac{2}{\pi}\right)^{0.5} [2x^{0.5}]_0^R \\ &= -\sigma_0\left(\frac{8R}{\pi}\right)^{0.5} \end{aligned} \tag{8.12}$$

The integral is finite and increases as $R^{0.5}$, as the yielded zone grows in length. Therefore the yielded zone grows in length until the condition of zero K_{total} is obeyed, which on substituting equation (8.12) gives

$$R = \frac{\pi}{8}\left(\frac{K_{\text{I}}}{\sigma_0}\right)^2 \tag{8.13}$$

This result is only valid when the yielded zone length is small compared with the crack length, and when the zone has the shape assumed by Dugdale. The analysis can be continued to calculate the opening of the hypothetical longer crack; this then gives the amount by which the edges of the yielded zone move apart. At the real crack tip this quantity is called the crack tip opening displacement δ_0 and is given by

$$\delta_0 = \frac{K_{\text{I}}^2}{\sigma_0 E^*} \tag{8.14}$$

where E^* is equal to the Young's modulus E under plane stress conditions, or $E/(1-\nu^2)$ under plane strain conditions. Equation (8.14) is the basis of an alternative criterion to equation (8.8), that crack growth occurs when

$$\delta_0 > \delta_{0\text{C}} \tag{8.15}$$

where $\delta_{0\text{C}}$ is the critical crack tip opening displacement. If this criterion is obeyed for the failure in the yielded zone, then equation (8.14) shows that the alternative K_{IC} criterion can also be used, if the size of the yielded zone is small compared with the width of the specimen. Hence the linear elastic fracture mechanics criterion is still valid.

8.4.4 Plain strain fracture of a craze

Some of the jargon of fracture mechanics can be confusing because similar phrases are used more generally in mechanics. In the analysis of metal working *plane strain deformation*, such as the rolling of a sheet, means that the non-zero plastic strains e_{xx}, e_{yy} and e_{xy} all relate to a particular (xy) plane. In a *plane strain fracture* all the plastic strains in the yielded zone occur in the xy plane of Fig. 8.9a. In particular the strain $e_{zz} = 0$ so the sides of the specimen do not move inwards, and the fracture surface will appear macroscopically flat. If a crack grows continuously through a craze in a polymer then a plane strain facture will result. The voiding in the craze allows the craze to open while the strain e_{zz} in the craze remains zero.

The Dugdale model can be applied to a craze at a crack tip, because the craze is effectively an extension of the crack plane and there is believed to be a constant tensile stress across the craze (Chapter 7). If the craze length is measured while K_I is known then equation (8.13) can be used to calculate σ_0 and obtain a value of the craze stress. The measured thickness profile of a

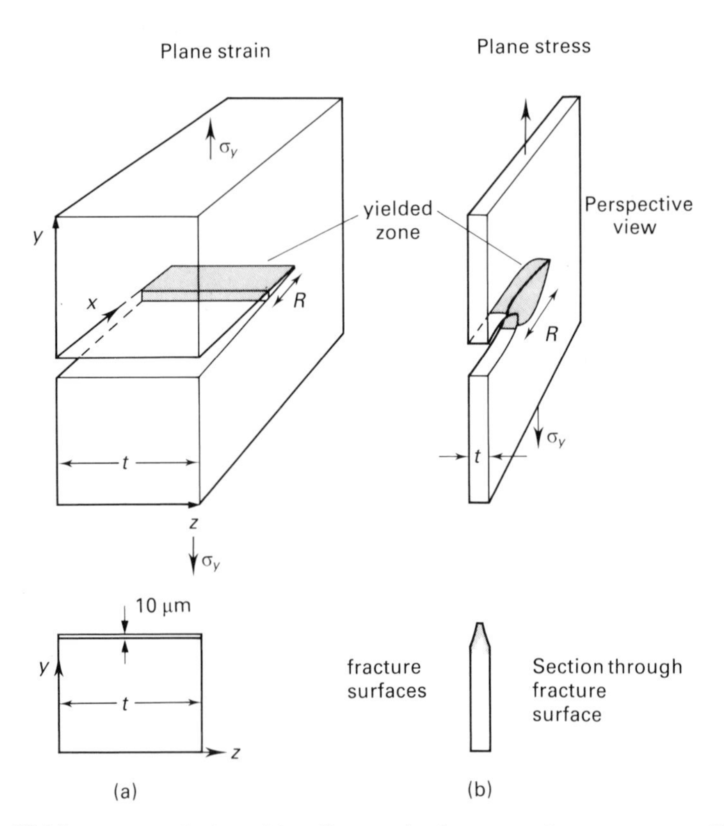

Fig. 8.9 Yielding at a crack tip and its effect on the fracture surface appearance. The two limiting cases are (a) plane strain fracture in thick specimens and (b) plane stress fracture in thin specimens

craze (from the interference fringe pattern seen when the craze is viewed normally using reflected monochromatic light) can be compared with the shape of the opening displacement profile calculated from the Dugdale model (Fig. 8.10). There is good agreement of the cusp like shapes of the two profiles.

The equivalent of condition (8.15) for crack growth is the condition that a craze fails when its opening displacement reaches a critical value. This does not provide an explanation of why the craze fails; it could be a failure of the entanglement network in the craze fibrils. As crazes in different polymers are observed to fail either at their mid-planes or at the bulk/craze interface there must be several mechanisms for craze failure. For viscoelastic materials in which both the craze stress and the Young's modulus vary with the strain rate, equation (8.14) predicts that the crack tip opening displacement and the stress intensity factor do not remain proportional to one another. Fig. 8.11 shows that the crack opening criterion (8.15) is more nearly obeyed for PMMA than the K_{IC} criterion.

The δ_{0C} criterion only applies when the fracture mechanism is unchanged; for the PMMA a crack is progressing down the midplane of a growing craze. When the crack velocity in PMMA exceeds 0.05 m s^{-1} it suddenly accelerates to over 100 m s^{-1} without an increase in K_I. The appearance of the fracture surface changes from a flat surface of a single advancing crack front to a surface with parabolic markings (Fig. 8.12). These latter are seen with many polymers and are ridges where there are changes of level of about 10 μm. The

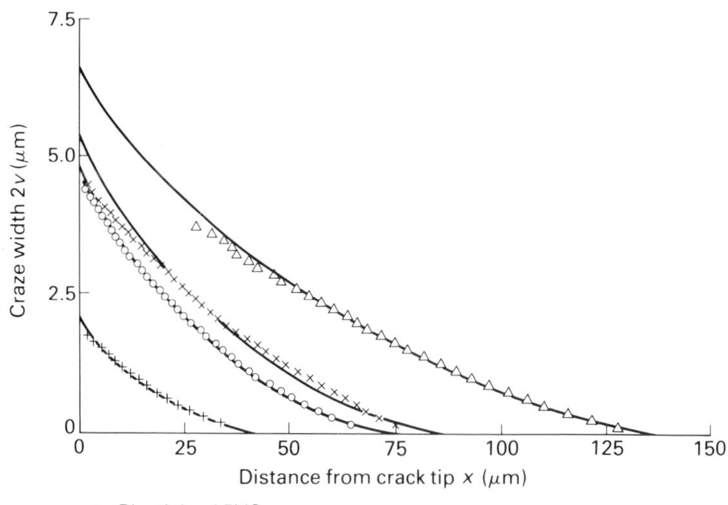

Fig. 8.10 The thickness profile of crazes in PES, PMMA, PC and plasticised PVC fitted by the prediction of Dugdale's model (from W. Döll, *Adv. Polym. Sci.*, 1983, **52/3**, 119, Springer Verlag)

fracture surface appearance can be explained as the result of a nucleation of penny shaped cracks ahead of the main crack front. These have only grown a small amount before they are overrun by the main crack front. However, the continued nucleation of cracks causes a 'leap frog' effect and a much higher crack speed than was possible with the growth of a single stable crack. A critical δ_{0C} criterion clearly cannot apply when there are many crazes present each in different stages of growth.

8.4.5 Plane stress fracture in thin sheet

The mechanics concept of a *state of plane stress* means that the internal forces, from which the stress components are defined, all act in a particular plane. If this is the *xy* plane then the tensile stresses σ_{xx} and σ_{yy} and the shear stress σ_{xy} are the only non-zero stress components. In a *plane stress fracture* there is a state of plane stress in the yielded zone ahead of the crack. The tensile stress σ_{zz} parallel to the crack front is zero, and the non-zero stress components are σ_{xx}, σ_{yy} and σ_{xy}. The stress component σ_{zz} will always tend to zero at the free surfaces of the specimen, so it is reasonable to suppose that plane stress fractures occur in thin specimens. The condition is more restrictive than this because straight slip lines must pass through the yielded zone from one free surface to the other to keep σ_{zz} equal to zero. Therefore plane stress fracture can only occur when the height of the yielded zone is at least as large as the specimen thickness (Fig. 8.9b). Slip on these lines will cause the yielded zone to contract parallel to the *z* axis, so there will be necking visible on the fracture surface.

Real fracture processes may contain some aspects of both plane stress and plane strain behaviour. The initial stages of crack growth will always be under

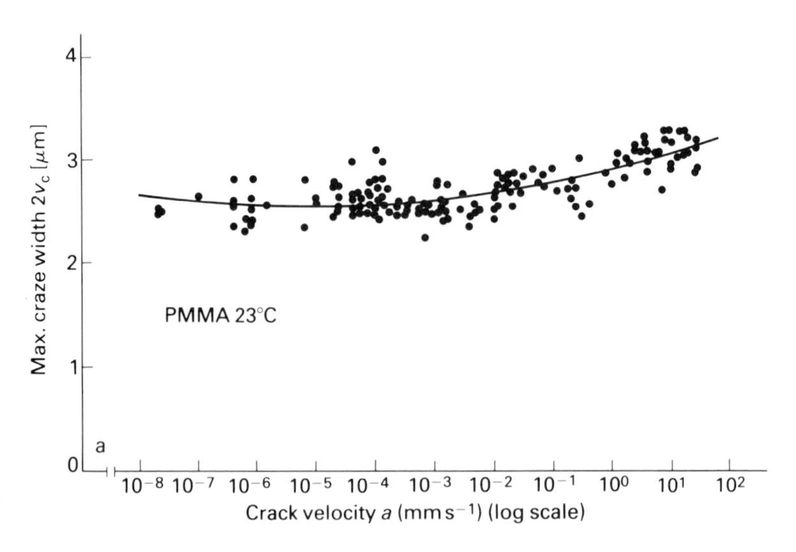

Fig. 8.11 Variation of crack tip opening with crack velocity in PMMA (from W. Döll, *Adv. Polym. Sci.*, 1983, **52/3**, 120, Springer, Verlag)

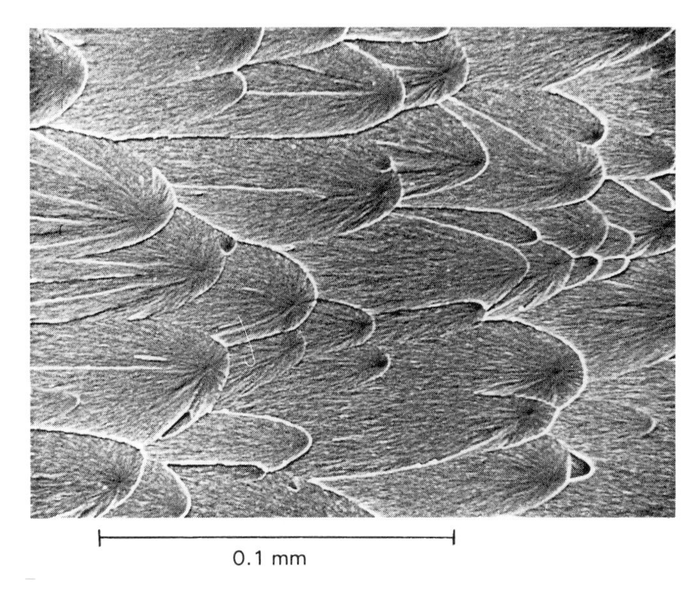

Fig. 8.12 Parabolic markings on a PMMA fracture surface caused by the nucleation of disc like cracks ahead of the main crack, which is moving to the left

plane strain conditions because the yielded zone must grow to a certain size before plane stress conditions can develop. It is common to observe plane stress 'shear lips' at the edges of a flat fracture surface.

When the Dugdale model is applied to plane stress yielding in a thin sheet of polymer, a distinction must be drawn between the opening displacement δ and the height of the yielded zone h. The opening displacement is the distance that the points on either side of the neck have moved apart as a result of the neck forming, and it is less than h. The yielded zone height h must be greater than or equal to the sheet thickness t; there is no upper limit as the neck can grow in height by drawing more material into itself. Fig. 8.13 shows the yielded zone in 0.2 mm thick polycarbonate, viewed by polarised monochromatic light so that the yielded region can be distinguished from the elastic material. Although the yielded zone shape is less elongated than Dugdale assumed, equation (8.13) predicts its length to within 20%. Polycarbonate happens to have a tensile stress strain curve with a yield drop, then insignificant orientation hardening until the strain exceeds 1 unit, so the constant stress condition of the Dugdale model is met. If the experiment is repeated with isotropic PVC sheet, or with biaxially stretched PET from a carbonated drink bottle, the yielded zone will be of a quite different shape. The biaxially stretched PET orientation hardens very rapidly after the initial yield point, and the crack plane tries to turn to be parallel to the sheet surface. The material prefers to 'delaminate' rather than to allow through-thickness crack growth.

The necking process that occurs in a plane stress yielded zone produces an oriented material, with anisotropic yield properties. There has been a high shear strain in the through-thickness yz plane but a lower shear strain in the

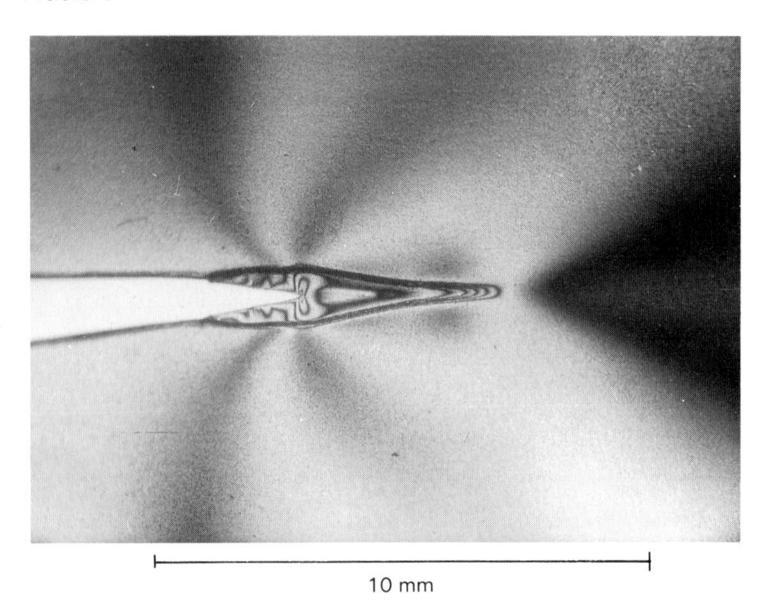

Fig. 8.13 Crack growth under plane stress yielding conditions in polycarbonate 0.2 mm thick, viewed with circularly polarised light

xy plane. Further shear can occur in the neck in the *xy* plane as the orientation hardening is less for this mode of deformation. This shear yielding allows the crack faces to open to an acute angle. For the crack to advance it must tear through stretched material in which the covalent carbon–carbon bonds are preferentially oriented across the crack path. Therefore the fracture toughness under plane stress conditions is several times higher than under plane strain conditions. Fig. 8.14 shows how K_{IC} varies with specimen thickness for polycarbonate.

It is useful to be able to estimate the thickness at which the level of K_{IC} falls rapidly. A semi-empirical limit has been derived from equation (8.13), which is used to calculate the longest possible yielded zone when $K = K_{IC}$. If we then use the previous observation that the thickness should exceed 5 times the yielded zone length to achieve plane strain conditions (Section 8.2.2) we find a transition thickness of

$$t_c = 2\left(\frac{K_{IC}}{\sigma_0}\right)^2 \tag{8.16}$$

If we substitute the values $K_{IC} = 2.2\ \mathrm{MN\ m^{-1.5}}$ and $\sigma_0 = 57\ \mathrm{MN\ m^{-2}}$ for polycarbonate to yield on a 1 min time scale at room temperature equation (8.16) predicts a transition at a thickness of 3 mm. For an impact loading time scale the transition thickness is only 25% of this.

These predicted transition thicknesses provide a useful selection criteria for polymers. The transition thickness varies from 0.6 mm for a PMMA with $K_{IC} = 1.6\ \mathrm{MN\ m^{-1.5}}$, $\sigma_0 = 90\ \mathrm{MN\ m^{-2}}$, to 195 mm for a MDPE with $K_{IC} = 5\ \mathrm{MN\ m^{-1.5}}$, $\sigma_0 = 16\ \mathrm{MN\ m^{-2}}$. Evidently the polyethylene is far more

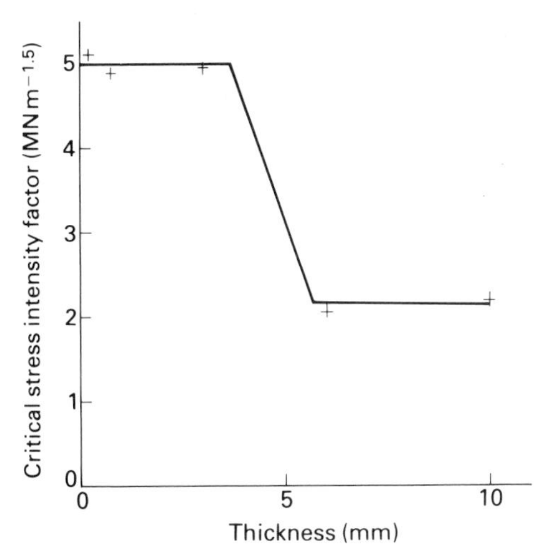

Fig. 8.14 Variation in the critical stress intensity factor with specimen thickness for slow fracture tests in polycarbonate

suitable than the PMMA for a thick walled gas pipe which must not be liable to high speed plane strain fracture.

8.4.6 Strain rate and crack velocity effects

Chapters 7 and 8 have shown that the moduli and yield stresses of plastics are time dependent; consequently we expect that the fracture properties will also be time or crack velocity dependent. Cracks can be made to propagate at a steady velocity V at stress intensity factors lower than K_{IC} in contradiction of the criterion (8.10). Fig. 8.15 shows how the crack velocity increases with increasing K_I values for a glassy polymer, using apparatus that imposes the double-torsion loading mentioned in Section 8.1. Above a certain velocity the crack can become unstable with the velocity jumping to half the speed of sound without any further increase in the applied K_I. We are forced to redefine the fracture toughness K_{IC} as the level of stress intensity factor that causes the crack to propagate at a speed exceeding 100 m s^{-1}. For slow crack growth the experimental function $K_I(V)$ must be determined to characterise the polymer. Both this function and the fracture mechanics function $K(F, a, w)$, that relates the stress intensity to the applied force F, the specimen dimensions w and crack length a, must be known if the life of a cracked product is to be calculated.

A further manifestation of the rate dependence of plastics is the effect of the rate of load application on a fracture toughness test specimen. If a cracked compact tension specimen of 5 mm thick polycarbonate is loaded in a tensile testing machine then there is time for a neck to develop from the crack tip and plane stress fracture to occur at a crack velocity of 5 m s^{-1}. If the load is applied in 1 ms by impact loading the fracture is plane strain, at a low K_{IC} value, and the crack velocity exceeds 200 m s^{-1}. It is necessary to develop

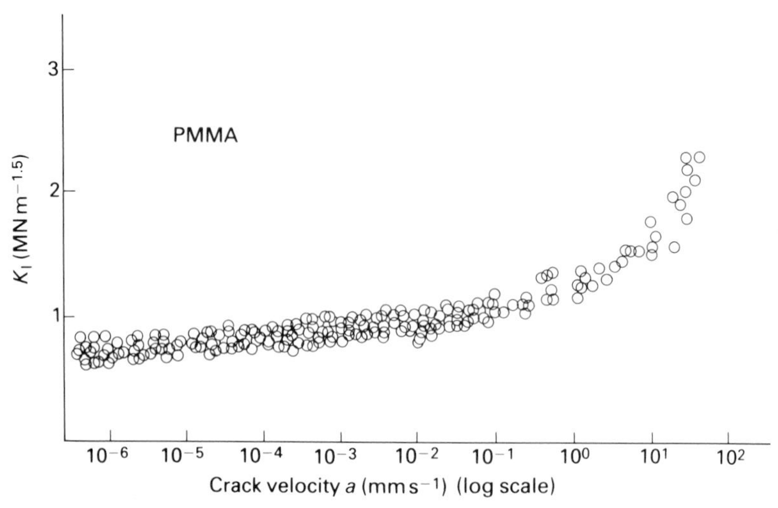

Fig. 8.15 Variation of stress intensity factor with crack velocity in PMMA (from W. Döll, *Adv. Polym. Sci.*, 1983, **52/3**, 119, Springer Verlag)

special methods of instrumention, with quartz crystal force gauges that can respond in 0.1 ms, and a grid of resistance lines on the surface to monitor the crack velocity.

8.5 IMPACT TESTS

Plastics products are most likely to break under impact conditions. This is partly a result of the higher yield stresses at high strain rates, and partly because large forces can be generated by low energy impacts on stiff structures. There are a variety of impact tests used for plastics. Usually a weight falls of the order of 1 m to hit the test specimen with a velocity of about 5 m s^{-1}. The high strain rates that occur are of the same magnitude as when a product is dropped 1 or 2 m, but they are lower than in a vehicle collision or a ballistic impact. It is more important to know the uses and limitations of the general categories of tests, rather than the details of particular tests. Three types of test are examined to bring out these points.

8.5.1 Izod and Charpy impact tests on bars

The Izod test is a variant of the notched bending test, where a standard size small bar is hit by a pendulum (Fig. 8.16) and the loss in the kinetic energy of the pendulum is measured in joules. The specimen has a width of 12.5 mm, and should be of a thickness representative of the use. The 45° notch is 2.5 mm deep and has a notch tip radius of 0.25 mm. Section 8.3.2 showed that the tensile stress in the yielded zone at the notch tip can be up to 118% higher than the yield stress in a tensile bar deformed at the same rate. It has become common to standardise on 3.2 mm thick specimens, which severely restricts

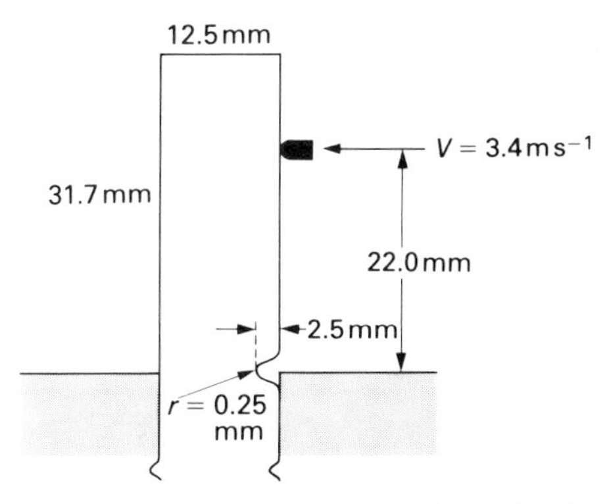

Fig. 8.16 Conditions used in the Izod impact test on a notched bar in bending

the value of the data. The half of the bar which projects above the vice is struck 22 mm above the notch by the pendulum. The results are quoted as the absorbed energy divided by the specimen thickness, in $J\ m^{-1}$.

The effects of notch tip radius on the impact energy for Charpy impact tests is shown in Fig. 8.17. The Charpy specimen is notched in the centre of the bar, then struck on the face opposite the notch, while supported in 3 point bending. The figure shows that ABS containing dispersed rubber particles is much less sensitive to the notch radius than is glassy polycarbonate. It also illustrates that the notch tip radius needs to be less than 100 μm for the low impact strength of 3 mm thick polycarbonate to be apparent, whereas the standard test uses a notch tip radius of 250 μm.

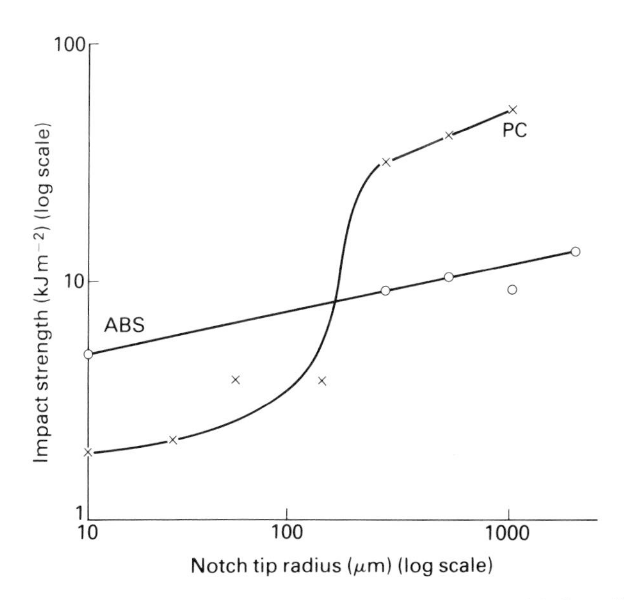

Fig. 8.17 Variation of the energy absorbed in a Charpy impact test with the radius of the notch tip, in 3 mm thick bars (from R. M. Ogorkiewicz (ed.), *Thermoplastics—Properties and Design*, Wiley, 1974, courtesy of ICI)

The Izod and Charpy impact strengths are quoted by materials manufacturers as this is a quick and easy test to perform. The values can be used to rank plastics in order of toughness for preliminary materials selection, but they are misleading or worse for design purposes. The analysis in Section 8.3.2 shows that the notch may initiate a crack after some yielding has occurred. Therefore the energy measured is a combination of crack initiation and propagation energies, in unknown proportions. The way in which the data are quoted misleads the reader into thinking that the result is independent of the specimen thickness. However there is a plane stress to plane strain fracture transition as the thickness increases, and the impact strength data vary in the same manner as for K_{IC} in Fig. 8.14. It is vital that the toughness is known for the same thickness as that in the proposed product. The Charpy test results in Fig. 8.17 are quoted in units of $\mathrm{J\,m^{-2}}$, the absorbed energy having been divided by the area of one fracture surface. This suggests that the crack uses a constant amount of energy per unit area of crack surface to propagate, whereas the energy consumption may be a strong function of the crack velocity. Consequently the test results are only suitable for quality control purposes.

The time to failure of 5 ms in impact tests is reduced by a factor of 2000 from the 10 s in slow bend tests. This will increase the value of the yield stress, without necessarily increasing the stress for crazing, so for some polymer/temperature combinations there may be a change-over from yielding to crazing and plane strain fracture. If the tests are performed with conventional Charpy or Izod impact machines this is all that will be noticed. If an instrumented impact tester is used it may also be possible to measure fracture mechanics parameters. In these the force on the striker is measured using a very stiff quartz piezoelectric force cell, and the data are digitized at time intervals of the order of 10 μs. Care is needed in both the experimental conditions and the interpretation of the data to avoid artefacts. The experimental factors in the initial impact are the velocity V of the striker or pendulum, and the contact stiffness k between the striker and the specimen, defined as the slope of the force deflection graph for the local elastic deformation of the plastic by the steel striker. Usually the contact stiffness is orders of magnitude greater than the bending stiffness of the specimen, in which case there is an initial force peak (Fig. 8.18) of magnitude

$$F_1 = V\sqrt{mk} \tag{8.17}$$

where m is the effective mass of the specimen. The effective mass is a point mass moving with the velocity of the contact point, which has the same momentum as that of the bending beam; for a cantilever beam struck at the end, m is 39% of the specimen mass.

This initial force peak from the elastic 'collision' between the striker and the specimen can excite a variety of flexural vibrations in the beam (Fig. 8.18). In the Izod test, where the specimen breaks near the clamped end, these oscillations are not transmitted to the failure region, and are an unwanted addition to the recorded force signal. They can be largely removed by reducing the contact stiffness, using a thin layer of high hysteresis polyurethane rubber on the striker face. For Charpy tests, where the failure

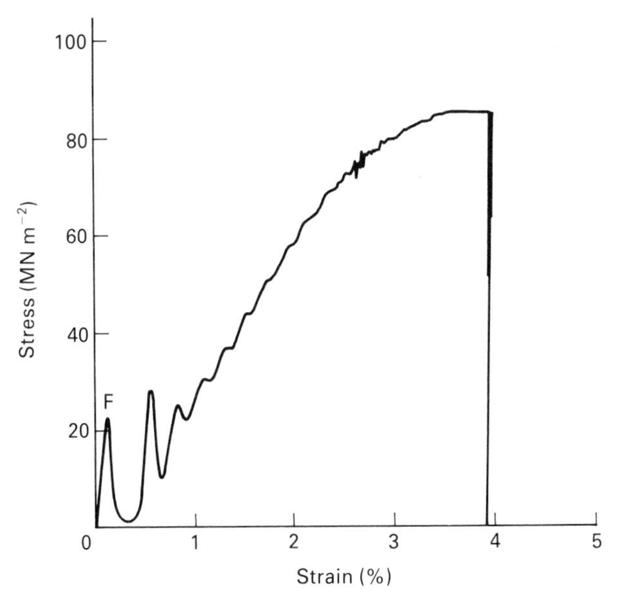

Fig. 8.18 The stress–strain graph from an instrumented Izod test on an un-notched polystyrene bar. The initial peak F is predicted by equation (8.17) and the subsequent force oscillations are damped by the polyurethane layer on the striker (see Mills and Zhang, *J. Mater. Sci.*, 1989, **24**, 2099)

site is at the centre of the bar opposite the striker, it is possible that the force oscillations could influence the failure process. It is not acceptable to remove the force oscillations by electronic filtering frequencies >200 Hz, as this will distort the shape of the force–time data. If un-notched bars are impacted it is possible to generate stress–strain curves for the maximum stress region on a 5 ms time scale. Fig. 8.18 shows that the fracture stress of polystyrene has risen from 40 MN m^{-2} in a conventional tensile test to 100 MN m^{-2} as a result of the time dependence of the craze stress.

8.5.2 Impacts on flat sheet

Injection moulded sheet, or a flat part of a product, or the curved surface of an extruded pipe is impacted by a falling mass that usually has a hemispherical end. The impact velocity can be modified from 1 m s^{-1} up to 100 m s^{-1} by dropping the projectile from different heights or by firing it from a gas-propelled gun. The difference from the Izod test is that no notch is introduced, and the axial symmetry of the test means that any cracks will form in the weakest direction of the sheet (Fig. 8.19). The test reveals molecular orientation in mouldings, and is sensitive to the presence of weak surface layers. It is carried out with sufficient impact energy that the sheet must either yield locally and stop the mass, or crack and allow the mass to pass through, or yield then fail by ductile tearing.

The stress analysis of the test is complex and it depends on the phenomena that occur during the test.

100 mm

Fig. 8.19 Crack along the flow direction due to a surface impact on a weathered ABS sheet

(a) While the disc remains elastic: The lower surface of the sheet is subjected to a balanced biaxial tensile stress, which has its largest value at the centre of the sheet. If a central force F is applied as a uniform pressure to a disc of contact radius a, on a sheet of thickness t supported at a radius R, the central biaxial tensile stress is

$$\sigma_\theta = \sigma_r = \frac{3F}{2\pi t^2}(1+\nu)\left(\ln\frac{R}{a} + \frac{a^2}{4R^2}\right) \tag{8.18}$$

Outside the contact radius the stresses decay as $\ln(a/r)$, so the high tensile stress region is limited to a small central region of the lower surface of the plate. If the plate fails due to crazing or crack propagation from some defect in the high stress region, the failure load could be small.

If the outside rim of the sheet is clamped, or the plate is a flat part of a larger product, large deflections cause in-plane stretching of the sheet and membrane stresses to be set up. These are constant in value through the thickness of the sheet in contrast with the bending stresses.

(b) When the central region of the plate yields: The yield geometry is of the type described in Section 7.2.6. The stresses are related to the yielding mechanism, and cracks may initiate at a critical level of plastic strain. The force on the striker, for the experiments shown in Fig. 7.12b, goes through a maximum value given by

$$F_{max} \cong \pi D t \sigma_y$$

where D is the diameter of the cylindrical portion of the punch and σ_y is the tensile yield stress. If the plate deforms to this degree it is not in danger of brittle fracture at low impact energy levels. Hence the real screening power of the test is for injection mouldings that fail at forces well below F_{max} due to a weak surface structure.

8.5.3 Product testing

In many cases the failure mechanisms that occur in small laboratory specimens bear no relationship to service failures. Hence product tests have been developed by the British Standards Institution and other national standards organisations. The tests simulate the use or abuse of the particular product, using simple apparatus that is reliable to use. This can be illustrated from BS 6658 : 1985 'Protective helmets for vehicle users'. Among the various test requirements is one for the energy absorbing properties.

Crack initiation is encouraged by wiping the outer surface of the helmet with a toluene/isooctane mixture. This will rapidly produce small crazes or cracks at highly stressed places in polycarbonate helmet shells (see Section 9.5). One location is where a webbing chin strap is riveted to the shell. In real life such cracks can develop over a period of years of use (Fig. 8.20) and the environmental agent may be less severe.

Rapid crack propagation is encouraged by the high energy impact. The helmet is cooled to $-20\,°C$ before falling at $7\ m\ s^{-1}$ on to a hemispherical steel anvil with a 50 mm radius. The kinetic energy of the headform and helmet is of the order of 150 J. The test temperature is at the lower limit of weather conditions, but this compensates for the low impact speed and the very blunt nature of the anvil. The test is much simpler than accident reconstruction using vehicles and dummies. Thermoplastic helmet shells should buckle inwards without yielding, and the polystyrene foam liner underneath should crush. If any small cracks propagate rapidly the shell could split into two or more sections, and the load-spreading function of the shell would be lost. The shell normally absorbs about 30% of the impact energy, so if it fractured the peak acceleration of the headform would exceed the 300 g test limit. The large size of the product gives space for cracks to accelerate, and the high impact energy provides the driving force for such high speed crack growth.

Many product standards contain *performance tests* rather than specifying which materials may be used, what fracture toughness they must have, or the detailed design of the product. It is then up to a manufacturer to reach the required performance level in whichever way he wishes. It is possible to design a helmet shell without rivet holes, or change to a plastic that has a higher fracture toughness. It is very difficult to design a laboratory fracture test in which the crack geometry is the same as in Fig. 8.20. The molecular orientation, the residual stresses and local details of the product shape are different in a laboratory specimen than in a helmet shell moulding, so the correlation between laboratory fracture tests and helmet performance is poor. Consequently product testing, under conditions that simulate in-service failures, is essential.

There are a few products of simple shape where it is possible to carry out a fracture mechanics analysis of the failure mode. One is the polyethylene *pipe for the local distribution of gas*. To avoid the risk of an accidental breach of the pipe, by a careless digging operation, propagating at high speed down the length of the welded network, the K_{IC} value must be adequate. In Chapter 7 the stress analysis of the growth of parallel crazes at a separation S was made. Equation 7.22 gave the differential of the stored elastic energy W with the

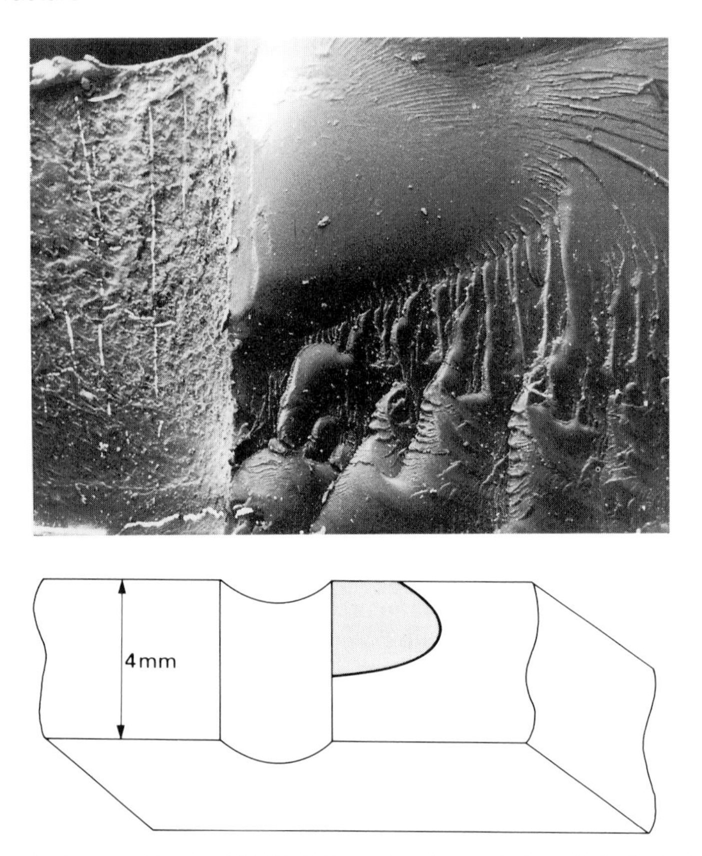

Fig. 8.20 Crack at the side of a drilled hole in a polycarbonate helmet shell. The hole has been loaded by accident forces on the chin strap rivet that passed through it

crack area A. This analysis can be applied to the pipe on the assumption that the crack is travelling so fast that the gas has no time to flow along the pipe to escape. The crack separation S is replaced by the circumference of the pipe πD_m, where D_m is the mean diameter, so the equation becomes

$$\frac{\partial W}{\partial A} = -\frac{\pi D_m}{2E}\sigma_H^2 \tag{8.19}$$

and the hoop stress is given by equation (7.15). In order to express this result in terms of the stress intensity factor, a general relationship from fracture mechanics is used

$$K_I^2 = -E\frac{\partial W}{\partial A} \tag{8.20}$$

This can be derived by a consideration of the energy release when a crack grows by a small amount. Using equation (8.19) gives the stress intensity of the cracked pipe as

$$K_{\mathrm{I}} = \sigma_{\mathrm{H}} \sqrt{\frac{\pi D_{\mathrm{m}}}{2}} \tag{8.21}$$

This shows that large diameter pipes are more at risk, as the design hoop stress is constant. By using MDPEs with a low melt flow index and a K_{IC} value of 6.0 MN m$^{-1.5}$, and keeping the hoop stress low, the risk of high speed brittle crack propagation down the length of the pipe is removed. This was not the case with some thick-walled rigid PVC pipes used as water mains; fortunately the high speed brittle fracture was limited to a 12 m length, as mechanical seals were used between lengths of pipe.

9

Environmental effects

9.1 INTRODUCTION

The properties of polymers deteriorate because of environmental effects of many kinds. The chemical structure of the polymer plays a far more important role in this than it does in the phenomena described in previous chapters. Details of the chemical reactions of particular polymers are given in specialised texts. Examples will be given of the range of phenomena that occur and the methods used to combat deterioration. There are many ways of classifying environmental agents, none of which is ideal. Here we shall start with the effects of melt processing; additives may have been used to make melt processing possible, and the chemical after-effects of processing can affect subsequent degradation. The effects of heat and oxidation indoors are described before the extra effects of outdoor weathering. Finally, some environmental stress cracking mechanisms are analysed, where failure is due to a combination of stress (Chapter 8) and the environment.

An approximate ranking of the stability of polymers in a vacuum can be found by heating them at a constant rate until half of the initial mass has been lost. Fig. 9.1 shows that this temperature T_h correlates reasonably well with the estimated bond dissociation energy of the weakest bonds in the polymer. Not only does this indicate the relatively low decomposition temperatures relative to most other man-made materials, but also the relative stability of the fully fluorinated structure PTFE.

9.2 DEGRADATION DURING PROCESSING

The conditions during the short period of melt processing are extreme compared with those during the subsequent life of a product. The melt temperatures are high, the melt has a high diffusion coefficient for oxygen compared with the solid state, and there can be high mechanical stresses applied. Yet it is expected that the chemical structure and the molecular weight of the polymer should not be significantly changed by processing. The chemical additives that permit such processing may be taken for granted, but

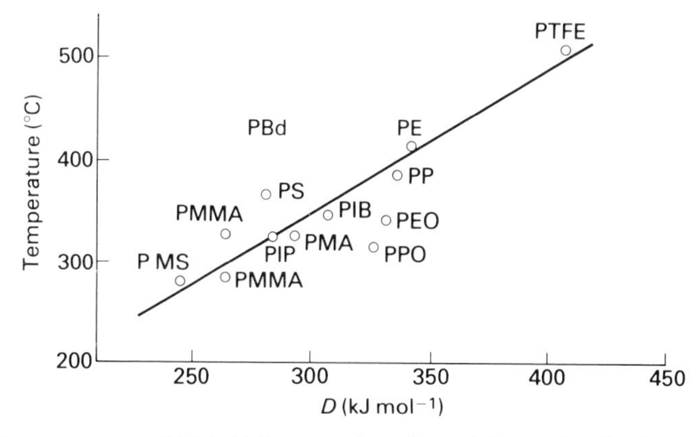

Fig. 9.1 Temperature at which half the mass of a polymer is lost on heating in a vacuum versus the bond dissociation energy (from T. Kelen, *Polymer Degradation*, Van Nostrand Rheinhold, 1983)

they cannot prevent degradation if the polymer is heated for too long, or recycled too many times.

Of the polyolefins, polypropylene is more susceptible to melt degradation than is polyethylene, because of the presence of the tertiary carbon group (where the carbon atom is bonded to three other C atoms). At the temperature of up to 270 °C used for injection moulding polypropylene, some alkyl free radicals $R^\bullet$ can be generated thermally. If oxygen is present there is a rapid reaction ($R^\bullet + O_2 \rightarrow RO_2^\bullet$) to produce a peroxide radical, which then reacts further to form hydroperoxides ($RO_2^\bullet + RH \rightarrow RO_2H + R^\bullet$). If the dissolved oxygen has been used up there is a greater chance of the chain scission reaction occurring.

$$-\overset{\overset{\displaystyle CH_3}{|}}{C^\bullet}-CH_2-\underset{\underset{\displaystyle H}{|}}{\overset{\overset{\displaystyle CH_3}{|}}{C}}- \longrightarrow -\overset{\overset{\displaystyle CH_3}{|}}{C}=CH_2+\overset{\overset{\displaystyle CH_3}{|}}{\dot{C}H}-$$

Hindered phenols are one type of stabiliser that is highly reactive towards alkyl radicals and scavenges them, preventing chain scission. They are used in concentrations of less than 1%, and can increase the period of melt stability by an order of magnitude.

Melt degradation of polypropylene has been used commercially to narrow the molecular weight distribution, to improve fibre production and blow moulding processes. In principle the random chain scission that should occur at any of the tertiary C—H bonds will eventually produce a 'most probable' molecular weight distribution with $M_W/M_N = 2$. In practice the distribution, produced by deliberately degrading the polypropylene with added peroxides, does not become as narrow as this, but the higher molecular weight molecules are preferentially degraded. Fig. 9.2 shows the melt viscosity versus shear stress curves for polypropylenes that have been degraded from different initial melt flow indices to an MFI of 200. As the molecular weight

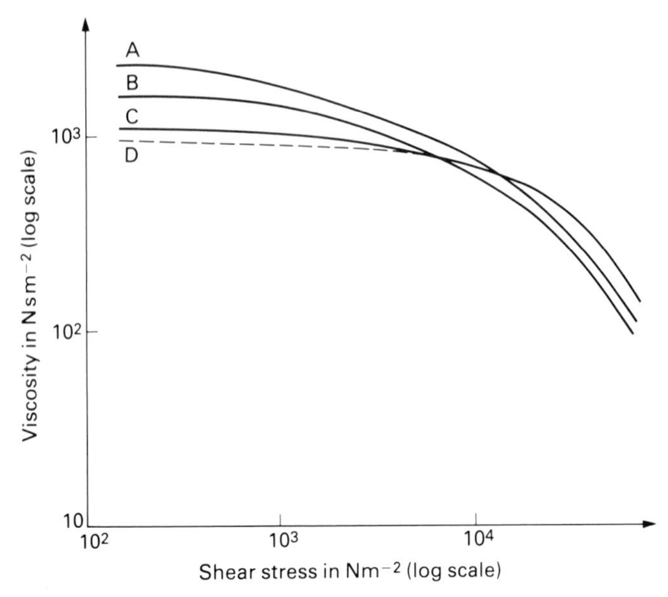

Fig. 9.2 Apparent shear viscosity versus shear stress for polypropylenes all of MFI = 200, produced by increasing amounts of oxidative degradation. The M_W/M_N ratios are (A) 12.5, (B) 9.5, (C) 5.0, (D) 4.5 (from N. S. Allen (ed.), *Degradation and Stabilisation of Polyolefins*, Elsevier Applied Science, 1983)

distribution narrows the viscosity curve becomes less non-Newtonian. If an already high MFI grade is being used for injection moulding, degradation that increases the MFI must be prevented, because the resulting lower molecular weight product will have inadequate toughness and other undesirable physical properties.

There will have been sufficient oxidation in the melt processing stage to produce a significant number of carbonyl groups C=O and hydroperoxide groups —COOH in any polyolefin. The significance of these is that they will dramatically speed up the start of the photooxidation process in outdoor weathering. Also a proportion of the stabiliser will have been used up. Consequently any recycling of faulty mouldings, by regrinding them in a granulator and mixing the regrind with virgin polymer, must be limited to a low percentage if the product quality is not to suffer.

In contrast to polypropylene, the molecular weight of PVC increases during the typical melt degradation process. PVC has the particular problem that it is not stable at temperatures of 220 to 230 °C at which the crystalline phase melts. As the crystallinity of commercial PVC is only of the order of 10%, processing in the semi-solid state is not an insuperable problem, but the apparent viscosity is much higher than for most other polymer melts. There are chain branches in PVC and consequently the tertiary chlorine atoms (the chlorine atom is bonded to a carbon atom that is bonded to 3 other carbon atoms) at these are weak points. The elimination of a hydrogen chloride molecule occurs with an activation energy, estimated for the decomposition of model compounds, which is 21 kJ mol^{-1} less than for a secondary chlorine atom (bonded to a carbon atom which is bonded to two other carbon atoms).

$$-CH_2-\underset{Cl}{\overset{|}{\underset{|}{C}}}- \longrightarrow -CH=\overset{|}{C}- + HCl$$

Once the double bond is formed, it is itself a weak site and further HCl loss occurs from neighbouring units. The resulting conjugated double bond polyene structure

$$=\underset{H}{C}-\underset{H}{C}=\underset{H}{C}-\underset{H}{C}=\underset{H}{C}-$$

is on average 5 to 6 carbon atoms long. The polyenes formed are highly coloured so the PVC rapidly goes brown and then black when it degrades. The polyenes are highly reactive and participate in secondary reactions such as crosslinking. This causes the average molecular weight to rise and eventually a gel is formed. The resulting increase in the already high melt viscosity is hardly welcome, and without careful design degraded PVC will build up in slow flow regions of the processing equipment. Since much PVC is compounded using large internal mixers, laboratory scale stability testing is carried out with a small internal mixer in which the torque is measured as a function of time. Fig. 9.3 shows how the molecular weight average M_W increases with the mixing time at 190 °C until it is no longer measurable. The torque falls from its initial peak after fusion occurs, but rises again later; G indicates the gel point after which an insoluble fraction occurs. If oxygen was

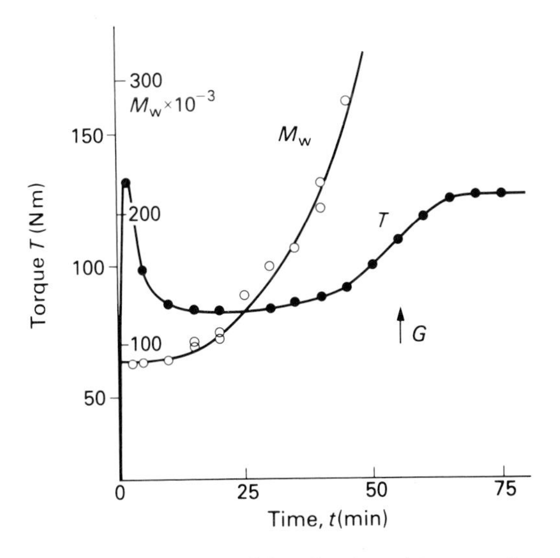

Fig. 9.3 Degradation with time in the mixing chamber of a torque rheometer at 190 °C. The torque T and molecular weight M_W are shown, with the gel point G being shown by an arrow (from T. Kelen, *Polymer Degradation*, Van Nostrand Rheinhold, 1983)

allowed free access to the mixer the degradation would be many times faster, as the polyenes readily oxidise into peroxides, which in turn decompose and liberate further HCl.

There are many additives used for stabilising PVC; two of the most common types are mixed metal salts of fatty acids, and organotin compounds. For example 2 or 3% of barium stearate/cadmium stearate can be used. The metal stearate, written for conciseness as MSt_2, reacts with free hydrogen chloride, and by removing it slows down the degradation of PVC.

$$MSt_2 + HCl \rightarrow MStCl + HSt$$

$$HCl + MSCl \rightarrow MCl_2 + HSt$$

The reason for using mixed metal stearates is that chlorides such as $CdCl_2$ act as a catalyst for dehydrochlorination; the following reaction regenerates some cadmium stearate

$$CdCl_2 + BaSt_2 \rightarrow CdClSt + BaClSt$$

Synergistic effects of this kind are widely used in stabilising PVC. The more expensive organotin compounds replace the tertiary chlorine atoms in the PVC by more stable groups, as well as reacting with HCl. Therefore they can prevent dehydrochlorination and the formation of polyenes.

Many of the engineering polymers are condensation polymers, and their melt stability is strongly influenced by water absorbed prior to processing. For example polyethylene terephthalate (PET) has an equilibrium water content of about 0.3% at 25 °C and 50% relative humidity. The high molecular weight grade used for injection moulding of bottle preforms must be dried to a water content of about 0.003% (30 ppm), and even this content causes the intrinsic viscosity (a solution viscosity measure of molecular weight) to drop from 0.73 to 0.71 dl g^{-1}. The hydrolysis is quantitative in this case so one molecule of water (relative molecular weight 18) per PET molecule ($M_N = 24\,000$) will halve the number average molecular weight and produce a brittle, useless product. In order to dry the PET, a source of dry air (10 ppm water or less) is required. The slowest step in the drying process is the diffusion of water through the solid PET granules. The diffusion coefficient increases with increasing temperature, but above 150 °C hydrolysis starts at a slow rate. Consequently a compromise drying schedule of 4 hours at 170 °C is used, and the PET is kept blanketed in dry air until it enters the injection moulding machine.

9.3 DEGRADATION AT ELEVATED TEMPERATURES

9.3.1 Oxidation

In most polymer applications oxygen is available to assist in the degradation process. Therefore studies of polymer stability in a vacuum, or blanketed by an inert gas, are of little relevance and give optimistic measures of the stability. The diffusion of gases into polymers is dealt with in Section 10.1.3, where equation (10.9) gives the distance that a gas will diffuse in a transient flow. According to this equation it takes 6 hours for the oxygen concentration at a depth of 1 mm to reach 50% of the surface concentration, if the polymer (LDPE) was initially oxygen free.

Many oxidation studies have been carried out using films a few μm thick, in which there is a uniform oxygen concentration. The oxidation of polyolefins is auto-catalytic; once the reaction starts a chain reaction develops and the rate of reaction accelerates. Consequently there is an induction period followed by a sharply rising curve (Fig. 9.4). When antioxidants are present the induction period is longer, and it represents the period until the antioxidant is used up. Polyolefins begin to show signs of embrittlement at a time equal to the induction time.

Once the induction period is over, the oxygen concentration C is determined by the competition between oxygen use and diffusion from the surface of the sheet of polymer. The x axis is taken to be normal to the sheet surface and the rate of change of oxygen concentration

$$\frac{dC}{dt} = D\frac{d^2C}{dx^2} - kC \qquad (9.1)$$

is where D is the diffusion constant for oxygen, and k is an oxidation rate constant. There is a steady state solution of equation (9.1), with the oxidation process balancing the diffusion process. In this the oxygen concentration falling exponentially with distance. The concentration falls by a factor e in a distance y given by

$$y = \sqrt{\frac{D}{k}} \qquad (9.2)$$

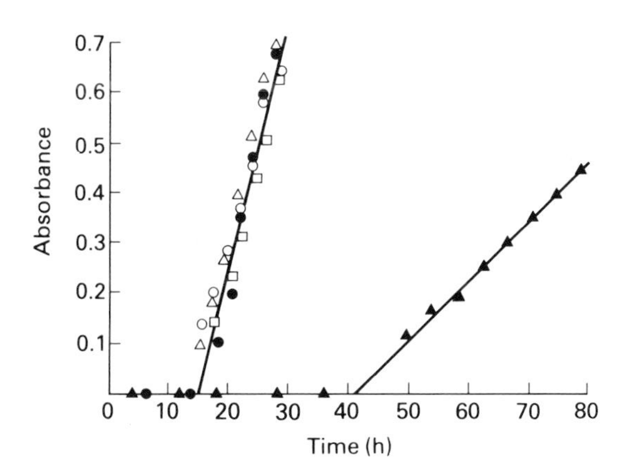

Fig. 9.4 The carbonyl absorption of polypropylenes versus the thermal oxidation time at 100 °C (from T. Kelen, *Polymer Degradation*, Van Nostrand Rheinhold, 1983)

Experimental results for polypropylene at 130 °C suggest that $y = 0.1$ mm. At the higher temperatures in melt processing y will decrease, because the activation energy for oxidation of about 100 kJ mol^{-1} is higher than that for diffusion. Consequently the inside wall of a polyolefin pipe, which is exposed to air during the cooling process, will only oxidise to a depth of about 10 μm if there is insufficient antioxidant present.

The hindered phenol antioxidants commonly used with polyolefins, such as BHT (butylated hydroxytoluene) can themselves diffuse

(Structure: HO– benzene ring bearing two t–Bu groups ortho to the OH, and –R para to it)

HO–C_6H_2(t–Bu)$_2$–R where t–Bu = $H_3C-C(CH_3)_2-$

and R = CH_3

In order to reduce the diffusion coefficient the molecular weight of R is increased. Nevertheless the loss of antioxidant from film or fibre, as it diffuses out and is removed from the surface by liquids, reduces the long term stabilisation.

There is still some dispute about the details of the embrittlement mechanism in oxidised polyolefins. The most obvious changes include a reduction in the elongation at break in a tensile test (Fig. 9.5). The molecular weight of the surface layers is found to have halved; however, as all of the oxidation is believed to occur in the amorphous regions the relative changes in these must be higher. It is not clear if oxidation occurs uniformly in the amorphous regions, or selectively in regions where there are residual tensile stresses caused by the volume contractions during spherulite growth. Nevertheless the

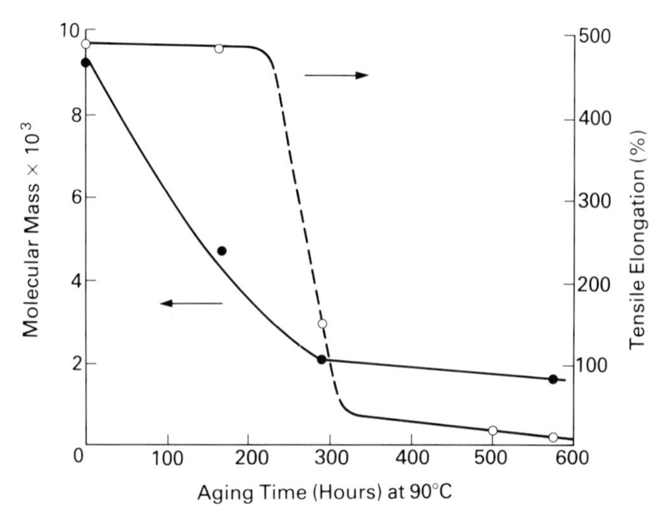

Fig. 9.5 Tensile elongation at break and number average molecular weight of polypropylene versus the exposure time to air at 90 °C (from N. S. Allen (ed.), *Degradation and Stabilisation of Polyolefins*, Elsevier Applied Science, 1983)

surface layer of the necked region in a tensile test is covered with cracks at 90° to the tensile stress, and it is the ductile growth of one of these cracks that terminates the necking process.

9.3.2 Hydrolysis

Hydrolysis can occur with solid polymers as well as with melts, one example being polycarbonate. The effect of prolonged storage under hot damp conditions is to cause random chain scission, and a reduction in the average molecular weight. Polycarbonate has a high melt viscosity, and in order to facilitate melt processing the average molecular weight is not taken much above that necessary to establish an entanglement network. Consequently a modest drop in molecular weight by hydrolysis reduces M_W to the level at which brittle failure occurs in a tensile test.

The effect of reduced molecular weight on the strength of glassy polymers can be seen in Fig. 9.6 where the properties of narrow molecular weight distribution polysulphone and polystyrenes are illustrated. The strength reaches a plateau value when M_N exceeds the entanglement molecular weight (Table 2.1) by a factor of about 4. Hence the effect of hydrolysis, or other degradation processes, is marked when the molecular weight is reduced to the level where there is no longer stress transfer between the molecules by means of an entanglement network.

Hydrolysis can be used to illustrate a common method of predicting lifetimes of plastics products. The assumption made is that there is a single degradation process, that is thermally activated. The degradation rate is proportional to $\exp(-E/RT)$ where E is the activation energy, R the gas

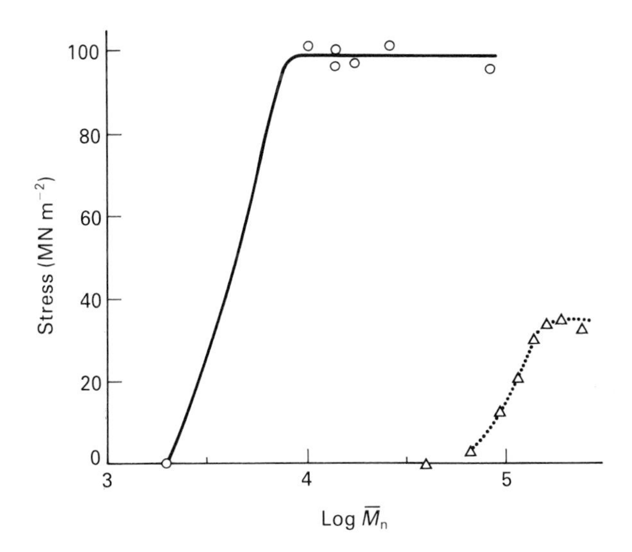

Fig. 9.6 The tensile strength of polystyrene △ and the compressive yield stress of polysulphone ○ versus M_N, for narrow MWD samples (from N. J. Mills, *Rheol. Acta*, 1974, **13**, 185)

constant and T the absolute temperature. If failure always occurs when the oxidation or hydrolysis has proceeded to a critical level, then the failure time t is given by

$$t_r = A \exp\left(\frac{E}{RT}\right) \tag{9.3}$$

where A is a constant. Consequently a graph of log (failure time) versus $1/T$ should be a straight line. Fig. 9.7 shows that this appears to be the case for the hydrolysis of polycarbonate. Extrapolating the line with activation energy 8.5 kJ mol^{-1} leads to a prediction of a 5 year life at 38 °C at 100% relative humidity. This extrapolation is by more than a factor of 10 in time, so if the degradation process changes its kinetics in the intervening temperature range, the prediction may be inaccurate.

9.3.3 The maximum use temperature

The only way to be certain of the maximum use temperature is to evaluate the product lifetime at a range of temperatures. At the materials selection stage it is necessary to estimate the maximum use temperature, by extrapolating from

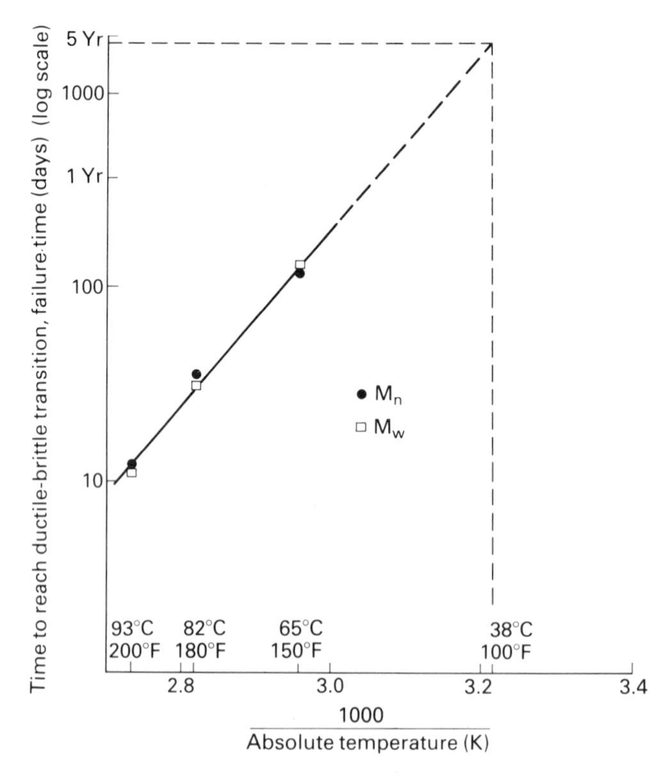

Fig. 9.7 The exposure time at 100% R. H. necessary to make polycarbonate brittle in a tensile test, versus the reciprocal absolute test temperature (from R. J. Gardner and J. R. Martin, *J. Appl. Polym. Sci.*, 1979, **24**, 1275, Wiley)

accelerated ageing tests. A guide to the maximum use temperatures has been contructed by Underwriters' Laboratories Inc., an American organisation. It relies on field experience with well established polymers. The maximum use temperature is for a life of about 10 years, and is based on a range of properties (tensile strength, flexural strength, impact strength, dielectric strength and other electrical properties); whichever of these deteriorates first to 50% of its initial value determines the lifetime. This implies that the design has a minimum safety factor of 2, but it recognises that one design aspect will be critical in determining the lifetime.

Any new polymer is compared with a reference polymer of a similar type that has an established upper use temperature. Samples of both, of the same thickness, are exposed in circulating air ovens at 4 temperatures chosen to give lives of roughly 1, 3, 6 and 12 months. Specimens must be removed at intervals for the relevant test to be performed. The test results are then plotted as log (half-strength time) versus reciprocal absolute test temperature, in a similar way to Fig. 9.7, and the best straight lines fitted. The line for the reference material will pass through its maximum use temperature at a particular time, say 60 000 hours. The maximum use temperature of the new material is then the coordinate of its line at 60 000 hours. Table 9.1 gives some maximum use temperatures generated in this manner. Some polymer manufacturers' brochures will quote much higher values than these, possibly because a restricted number of tests have been carried out.

Table 9.1 Maximum use temperatures in dry air

Plastic	Maximum use temperature (°C)
ABS	67
POM	87
PPO/PS	96
Nylon 6,6	96
PBTP	116
PC	120
PSO	145
PPS	165

The limitations of a single maximum use temperature should be obvious. It cannot apply to all products; in particular film and fibre products may oxidise faster. It does not allow for outdoor use, or the effects of chemicals in the environment, topics that will be dealt with later in this chapter. However, it shows that degradation occurs well below the glass transition or melting temperature of the polymer.

9.3.4 Fire and flammability

The temperature in a fire can easily reach 1000 °C within 10 min of the fire starting. The thermal radiation level from the red-hot material of up to 150 kW m^{-2} can cause such high surface temperatures on neighbouring polymer (remember the low thermal diffusivity) that combustible gases are evolved. The fire spreads rapidly when these gases spontaneous ignite.

Plastics provide particular hazards in fires. The major toxic hazard is carbon monoxide, which is produced from all organic materials if combustion is not complete. The nitrogen containing polymers (polyamides, polyurethanes, polyacrylonitrile, etc.) produce hydrogen cyanide when they burn in air. This and the dense smoke that is often produced have given plastics a bad name as a fire hazard. A further concern with electrical equipment (e.g. telephone exchanges) is the formation of corrosive gases such as hydrogen chloride.

Some basic information about flammability can be obtained from laboratory scale tests. The *limiting oxygen index* (LOI) test measures the percentage by volume of oxygen in an oxygen/nitrogen mixture that will just support the combustion of a particular polymer. Fig. 9.8 shows the apparatus used; the strip of plastic approximately 10 mm wide and 100 mm high is ignited at the top, and the gas mixture that will just allow the flame to burn for 3 min, or down by 50 mm, is found. Table 9.2 gives some values. It would appear at first that if the LOI exceeds the value for normal air of 20.8% then the polymer will not burn continuously. However, if the temperature of the polymer is raised the LOI falls, and at high temperatures in fires very few polymers will not sustain burning. An LOI of at least 27% is required to achieve fire retardation. The LOI figure correlates with the proportion of halogen atoms to hydrogen atoms in the polymer. It also increases linearly with the char yield when the polymer is carbonised in a nitrogen atmosphere. These two factors can be related to the processes of burning. Solid char can seal the surface of the polymer and prevent oxygen diffusing into the polymer from the flame, or the gaseous decomposition products from the polymer passing as fuel to the flame. The halogen content of polymers is liberated into the flame as hydrogen halide where it interferes with the free radical chain reactions in the combustion process

$$HX + OH^{\bullet} \rightarrow H_2O + X^{\bullet}$$

The halide radical then reacts with more fuel to regenerate the hydrogen halide

$$X^{\bullet} + RH \rightarrow R^{\bullet} + HX$$

The heats of combustion of polymers vary over a wide range (Table 9.2), and this indicates the fuel load that they can contribute to a fire. The correlation between the heat of combustion and the LOI is poor.

There are many different standard tests for the spread of fire. The easily performed laboratory tests on 150 mm long strips of plastic can be criticised because they lack the thermal radiation factor that accelerates the spread of flames. On the other hand a full scale simulation in which a 340 kg wood crib burns in a corner of a 'warehouse' that is 2.5 m high by 3.7 m by 5 m, with the walls lined with the test material, is realistic but very expensive to carry out. A compromise scale is used in British Standard BS 476 Part 7 in which a test panel 900 mm wide and 225 mm tall is exposed edge-ways on to a radiant heat source so that the radiation intensity decreases from 40 to 8 kW m^{-2} across the width of the panel. The materials are classified by the rate at which the flame spreads; for class 1 this is less than 165 mm in 10 min, whereas for class

Table 9.2 Data related to the fire performance of polymers

Material	LOI (%)	Char yield (%)	Heat of combustion (MJ kg^{-1})	Smoke density rating
Polyoxymethylene	15–16	0		0
Polyurethane foam	16.5			
Polymethyl methacrylate	17.3	0		
Polyethylene	17.4		46.5	15
Polypropylene	17.4			32
Polystyrene	17.8	0		94
ABS	18.8			97
PET	20			84
Polycarbonate	22–28			
Nylon 6,6	24			5
Neoprene rubber	26		29	
Polyphenylene oxide	31	40		86
Polysulphone	31	48		61
Phenolic thermoset	35	60		
PVC (rigid)	45–49	24		97
Polyvinylidene chloride	60		23	46
Polychlorotrifluoroethylene	95			
PTFE	95		4.5	

4 it is greater than 710 mm in 10 min. PMMA is rated as class 3 whereas PVC and polycarbonate are class 1. In an application such as glazing for public buildings, more stringent fire regulations have meant that PMMA, which has a better weathering resistance than the other materials, can no longer be used.

The data given so far have been for polymers which have not had fire retardants added to them. Since a multitude of compounds have been used in an empirical attempt to achieve fire retardation only a few of the major systems will be mentioned.

Polyolefins (polyalkenes) decompose by random chain scission to liberate alkanes and alkenes as fuel for the flame. Consequently they do not leave any char or generate much smoke (Table 9.2). A common flame retardant is a mixture of chlorinated alkane (a compatible substance that provides HCl to the flame) and antimony oxide. These two have a synergistic effect, as antimony trichloride is formed in the flame.

Flexible PVC, used widely for the insulation of electrical wiring, has a number of problems. Firstly the plasticiser, since it does not contain chlorine, reduces the LOI. For example a 60 parts per hundred addition of dioctyl phthalate reduces the LOI to 22. Consequently a flame retardant such as antimony oxide must be used. Alternatively phosphates such as tricresyl phosphate can be used but they generate large amounts of smoke. Since the smoke density rating by ASTM D 2843 is already high (Table 9.2) this does not help matters. Some countries have banned the use of halogen containing polymers for certain wire insulation applications because of the corrosive effects of HCl. There are few alternative polymers that are flexible enough for

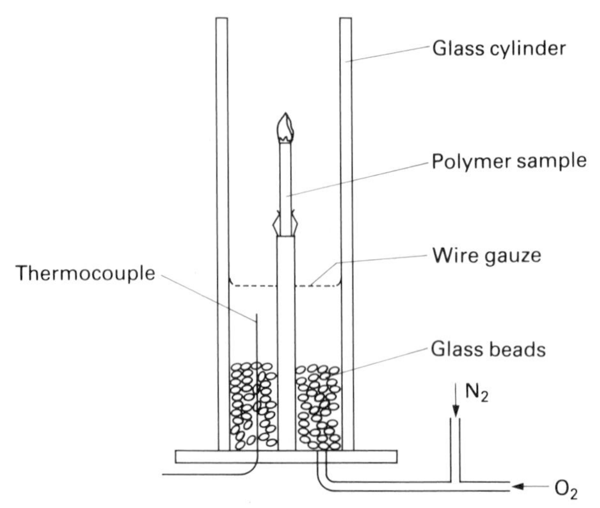

Fig. 9.8 Apparatus to determine the limiting oxygen index

the application. Polyethylene, crosslinked to prevent it dripping from the cable in a fire, has a high heat of combustion. The fluorinated hydrocarbons have a good fire resistance and good dielectric properties but they are very expensive. Consequently there is no ideal solution to the problem.

Polyurethane foams, once widely used as flexible foam in furniture and as rigid foam in insulated building panels, have a low LOI and burn very rapidly because of the high surface to volume ratio. The use of halogen and/or phosphorus additives to reduce flammability is expensive, and it adds to the release of toxic gases on combustion. A change in one of the constituents to form polyisocyanurates (PIR) increases the char yields to about 50% and improves the fire retardation. However, the rigid PIR foams are friable and do not bond as well to the surfaces of laminated building panels. Chemical changes of this kind, together with the addition of glass fibres to prevent the chair from cracking, have greatly improved the fire rating of wall and roof linings. The fire performance of flexible polyurethane foams in furniture has similarly been improved by using a fire retarding cotton interlayer between the foam and the fabric covering.

9.4 WEATHERING

When plastics are used outdoors they are exposed to solar radiation. Fig. 9.9 shows the short wavelength end of the solar radiation spectrum. Absorption alone is insufficient to damage the polymer. Unless the wavelength is specific to raise one of the bonds to an excited state it will merely heat the polymer. The energy E of a photon of light of wavelength λ is given by

$$E = hc/\lambda = h\nu \qquad (9.4)$$

where h is Planck's constant, ν the frequency and c the velocity of the light.

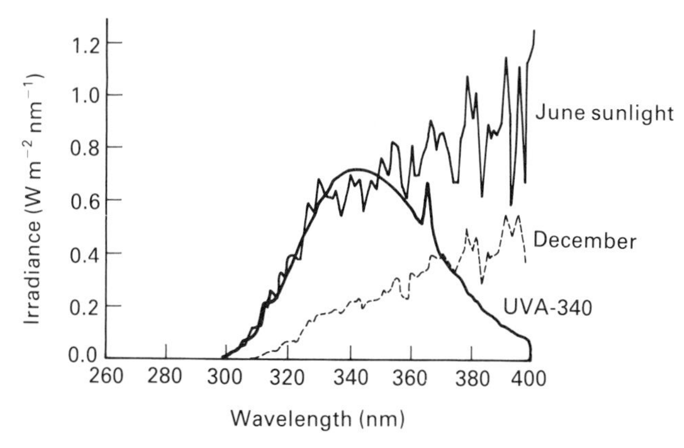

Fig. 9.9 UV spectrum from sunlight at noon in Cleveland, USA, in June and December, compared with the output of a UVA fluorescent tube (from the Q Panel Company, Cleveland)

Consequently the ultraviolet (UV) end of the solar spectrum contains the most energetic photons. If there were to be permanent damage to the ozone layer in the stratosphere the amount of UV light would increase drastically. Table 9.3 lists the UV wavelengths which cause damage to specific polymers. The absorbed photons raise electrons to an excited state and cause bond dissociation reactions, the process being referred to as photodecomposition. A typical example is the photodecomposition of hydroperoxide groups that were introduced during melt processing.

$$\mathrm{ROOH} + h\nu \rightarrow \mathrm{RO}^{\bullet} + {}^{\bullet}\mathrm{OH}$$

Table 9.3 Ultraviolet wavelengths which cause photodegradation

Polymer	Wavelength (nm)	Energy (kJ mol^{-1})
Polyethylene	300	400
Polypropylene	310	384
Polystyrene	318	376
PVC	320	372
Polycarbonate	293,345	405,347
SAN copolymer	290,325	414,368

The UV intensity in the polymer is attenuated by absorption and, in semi-crystalline polymers by scattering at the crystal surfaces (Section 10.4.2). A high attenuation absorption coefficient means that the light of this wavelength will penetrate the polymer to a limited distance. The intensity will fall exponentially with penetration distance x, according to

$$I = I_0 \exp\left(-\frac{x}{L}\right) \tag{9.5}$$

For unpigmented HDPE irradiated at a wavelength of 310 nm the constant L is equal to 1.25 mm.

Carbonyl groups are efficient absorbers of UV light. If the free radicals produced subsequently react with oxygen the process is then referred to as *photo-oxidation*. This is a much faster process than the thermal oxidation discussed in Section 9.3.1. Fig. 9.10 compares the oxygen uptake by the two mechanisms for polyethylene, the thermal oxidation taking place at a higher temperature. An oxygen uptake of 50 $cm^3 g^{-1}$ greatly exceeds the solubility of oxygen in polyethylene (~0.04 $cm^3 g^{-1}$) so the oxygen must diffuse in from the surface once the reaction is well under way. This, more than attenuation of the UV radiation, means that photo-oxidation is confined to a thin surface layer. For unpigmented ABS exposed outdoors for 3 years all the butadiene is oxidised in the surface 125 μm layer, then the butadiene deficit decreases exponentially further into the moulding. Assuming that it requires a certain amount of absorbed UV radiation per unit volume of polymer to oxidise the butadiene, the depleted surface layer thickness will increase with the logarithm of the exposure time. With the butadiene rubber converted to a glassy material, the toughening mechanism for the glassy SAN matrix has been lost, and the consequences are shown in Fig. 8.19.

The methods of protecting polymers against photo-oxidation can be classified according to where they act in the sequence of photo-oxidation events. Protective layers can be used on the outside of the polymer—however, the paint film itself may become brittle and cause problems. The UV radiation can be absorbed in various ways; the cheapest and most

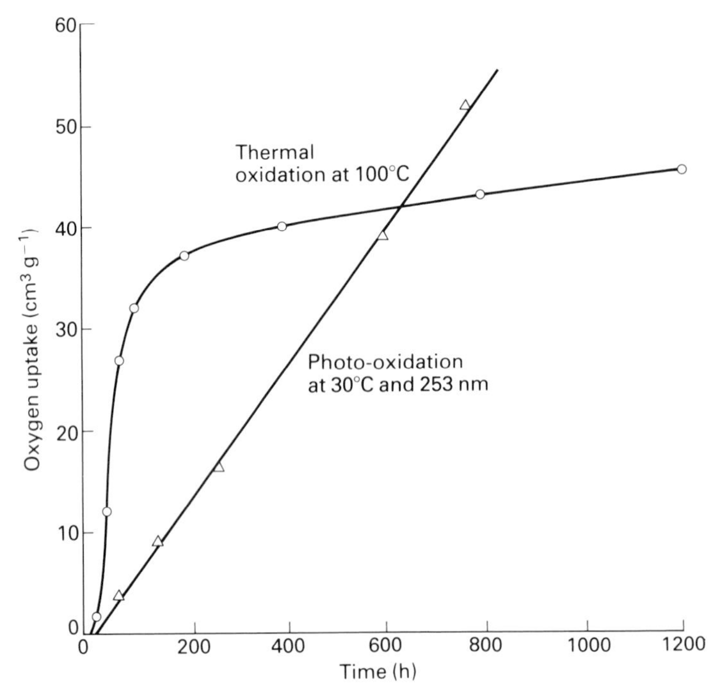

Fig. 9.10 Uptake of oxygen by a polyethylene film sample during thermal oxidation at 100 °C and photo-oxidation at 30 °C (from T. Kelen, *Polymer Degradation*, Van Nostrand Rheinhold, 1983)

effective way is to use well dispersed carbon black pigment. White pigments such as TiO_2 and ZnO are effective at scattering and hence reflecting UV radiation, but they do not absorb it to a significant extent. Fig. 9.11 shows the absorption increase as a result of adding a red organic pigment to HDPE. The attenuation in the natural HDPE is mainly due to Rayleigh scattering (p. 275) from small crystals. The red colour is achieved by absorbing the violet end of the visible spectrum; as the absorption peak is particularly broad it increases the absorption in the UV region. This pigment is itself photodegraded so the protection is limited. Both HDPE traffic cones and industrial helmets lose their red colour in the surface layers after a couple of years exposure; this could be an indicator that they need replacement!

There are organic additives that absorb UV radiation and convert it into heat. An example is hydroxy benzophenone which can transform between the keto and enol forms.

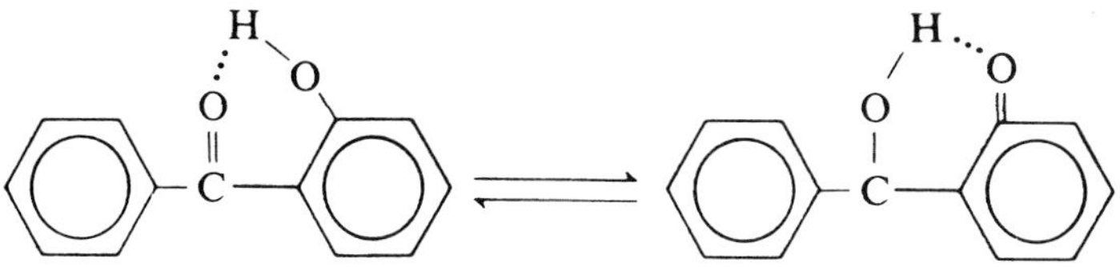

If the UV radiation is allowed to be absorbed by the polymer then free radicals will be formed in it. These can be scavenged by other forms of photostabiliser. A very effective type is the hindered amine light stabiliser (HALS).

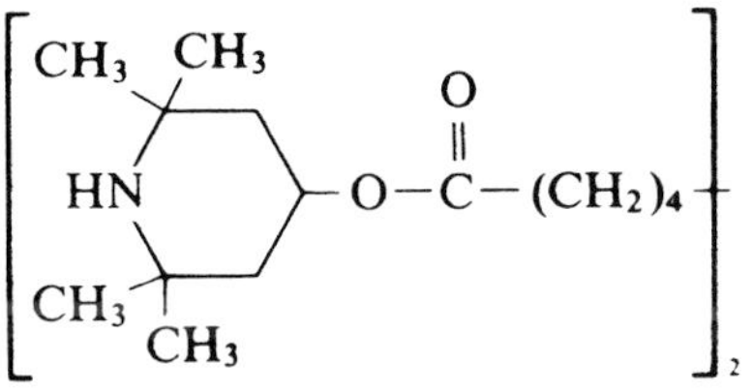

This is converted by oxidation into a nitrosyl radical ($NO^{\bullet}$) which is capable of scavenging polymer radicals and then regenerating itself. Although the HALS is highly effective at a 1% level in stabilising ABS, there is a synergistic effect if it is used in combination with a UV absorber. Accelerated exposure tests with a xenon light source showed that the impact strength of an unstabilised ABS fell to a specified level after 150 hours exposure. With 1% HALS added the lifetime was increased to 1000 hours, but with 0.5% HALS and 0.5% of a UV absorber it was increased to 1650 hours (1% of the UV absorber on its own only gave a 350 hour life).

Accelerated UV exposure tests are used to predict the performance of outdoor exposed polymers, as the natural ageing process takes several years. It is not possible to increase the intensity of sunlight as the polymer would get too hot and thermal degradation would dominate. If high pressure xenon sunlamps are used the majority of the infrared part of the spectrum must be filtered out to prevent the plastic overheating. Fluorescent tubes with a

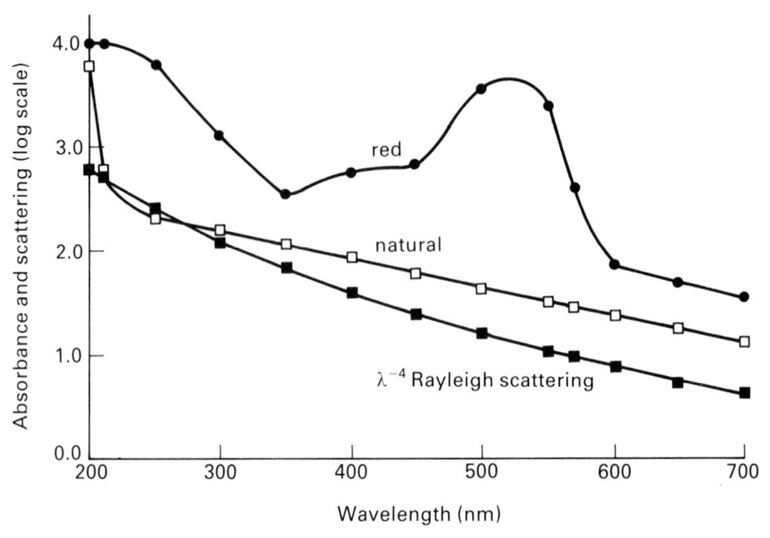

Fig. 9.11 Light attenuation of 0.39 mm of natural and red pigmented HDPE in the UV region (from Hulme, Birmingham University, unpublished)

suitable phosphor can produce an intensity versus wavelength relation in the UVB region, (280 to 320 nm) or UVA region (320 to 400 nm) that is similar to that of the solar spectrum (Fig. 9.9). Commercial exposure chambers using such tubes expose the plastic to levels of UVB of about 2 mW cm^{-2}. This level is measured with a total energy detector with zero attenuation at 310 nm, and with a filter width nearly identical with that of the tube spectrum. The maximum level of outdoor UVB in England in summer is 2 to 3 mW cm^{-2}, so the accelerated test achieves its ends not by having a high radiation level but by having continuous output. If the outdoor UVB levels are averaged over a year (Davis and Sims, 1983) the acceleration factor of the UVB chamber is found to be about 80. Fig. 9.12 shows the results of exposing natural HDPE of MFI = 5 in a UVB chamber. The bars were subjected to impact bend tests and the energy absorbed in the tests plotted against the exposure time. The type of failure changed from plastic hinge formation to crazing at an exposure time of 120 hours. At higher exposures the toughness increases again, as a result of the formation of more crazes in the high stress region. The estimated outdoor life of this unpigmented and non UV stabilised HDPE is 120 × 80 hours = 1.1 years.

Accelerated exposure predictions must be treated with caution as the effects (no visible cracks before impact testing) are not the same as those of outdoor exposure. Fig. 9.13 shows the pattern of surface cracks on a polyethylene garden chair after several years of use. Some of the cracks may be due to the low-cycle fatigue loads when the chair was used, but others are due to transient stresses when the surface temperature suddenly falls. The residual stresses that can occur due to the cooling stage of polymer processing are explained in Chapter 5. When a uniformly hot product, that has been in the sun, has its surface suddenly cooled by rain there will be a transient tensile

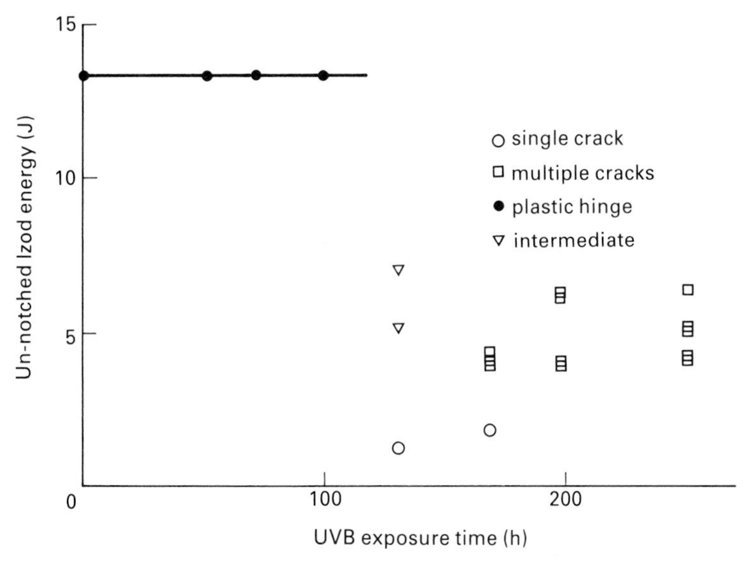

Fig. 9.12 The impact strength of un-notched HDPE bars as a function of the time of exposure to UVB radiation in an accelerated ageing chamber (Mills, unpublished)

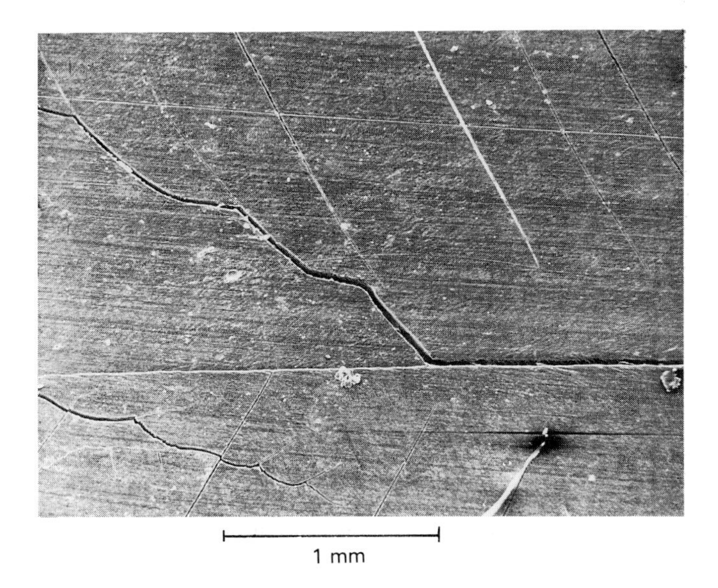

Fig. 9.13 SEM of a polyethylene garden chair surface showing cracks due to weathering and fatigue loads

stress on the surface. Similarly sudden reductions in the relative humidity can cause the surface layer to release water and go into tension. Once cracks have formed in the surface atmospheric pollutants can get into them. Dirt particles can wedge cracks open when the surface heats up and expands, and liquid pollutants can diffuse more readily into the polymer. Hence accelerated UVB exposure can only be a guide to one aspect of weathering.

Weathering is critical with film products, such as the 250 μm thick LDPE film coverings of agricultural greenhouses. It is accepted that the film will only last a couple of seasons, but the overall cost is lower than that for glass and the structure to support it.

9.5 ENVIRONMENTAL STRESS CRACKING

Environmental stress cracking (ESC) is the name used for the phenomenon called stress corrosion cracking by metallurgists. Crack initiation and growth occur by a combination of a tensile stress and an environmental liquid or gas. Fig. 9.14 shows a section through a welded joint in a polyethylene water pipe that has failed by this mechanism (in an accelerated laboratory test). The failures are worrying because they occur by slow crack growth at stresses that are less than the conventional design stress. Attempts to understand the phenomenon have concentrated separately on the crack initiation and crack propagation stages. Once the mechanisms have been found and a fracture mechanics analysis made of the crack growth, it is possible to predict the lifetime of a product in a particular environment.

9.5.1 Crack and craze initiation

Simple laboratory tests can be used for an initial screening of the susceptibility of a plastic to a range of chemicals. Strips of polymer are bent on to an elliptical former and the surface in tension exposed to various liquids. The elliptical shape means that the surface strain varies by a factor of about 10 typically from 0.3 to 3%. After exposure for a fixed time the surface can be examined with a microscope to find the minimum strain at which crazes or

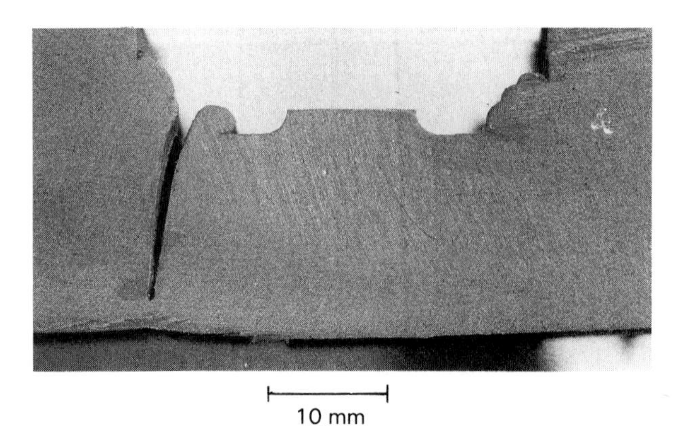

Fig. 9.14 Section through a welded polyethylene joint that has failed during a creep rupture test in water at 80 °C

cracks form. As stress relaxation occurs while the specimens are under test, it is impossible to tell whether a critical stress or a critical strain criterion is more accurate. The simplicity of the test has ensured its wide use and Fig. 9.15 shows the results for two glassy polymers exposed to petrol (gasoline) with different aromatic contents. Polycarbonate is clearly very susceptible to petrol with a high aromatic content, and this is the reason for wiping motorcycle helmets with a 1:1 isooctane–toluene mixture prior to impact testing.

The cracks that appeared in service in polycarbonate motorcycle helmets, manufactured prior to 1981, are shown in Fig. 8.20. They radiate from a hole through which a rivet passes to secure a webbing chin strap, and are often hidden from inspection until the helmet is dismantled. The residual stresses as the result of riveting, plus some environmental fluid, have caused the cracks to form. The webbing acts as a wick to the high stress region, and any liquid trapped near the rivet cannot easily evaporate. The problem was cured by using a rubber toughened shell material that was less susceptible to ESC, and by terminating the webbing strap on a steel hanger plate, which is riveted to the shell.

Semi-crystalline polymers such as polyethylene are less affected by organic liquids, but nevertheless, the amorphous phase is susceptible to attack. Both alcohols and surface active agents can eventually lead to crack formation. More severe conditions are used for laboratory quality control tests of the

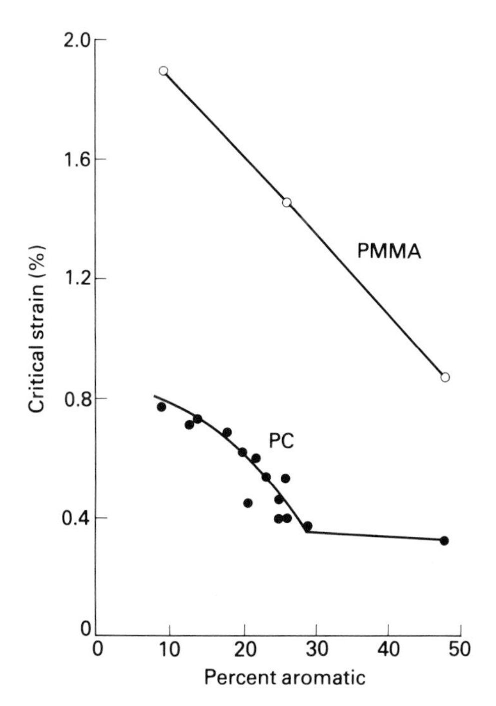

Fig. 9.15 Critical tensile strains for the crazing of polycarbonate and PMMA exposed to petrol with different aromatic content (from Wysgoski and Jacques, *Polym. Eng. Sci.*, 1977, **17**, 858)

ESC resistance because the resistance is better than for glassy polymers. In the Bell Telephone test a rectangular specimen 38 × 13 × 3 mm is bent into a U shape, and the time until cracks appear is measured. The maximum surface tensile strain is about 12% and the specimen is placed in a concentrated surface active agent at 50 °C. A razor cut is made parallel to the length of the specimen before it is bent, and the cracks form at the intersection of this cut with the outer surface. The cut does not create a stress concentration because it is parallel to the bending stresses; however cracks prefer to nucleate at corner positions. This may be because the two free surfaces assist the crack opening process.

Attempts have been made to rationalise the susceptibility of glassy polymers to liquid attack. There is a correlation between the critical strain and the equilibrium solubility of the solvent in polycarbonate (Fig. 9.16). The line extrapolates to a zero critical strain at a solubility of 0.19, a value that would be sufficient to reduce the T_g of the swollen polymer to 20 °C. It is not easy to predict the equilibrium solubility of a range of non-polar, polar and hydrogen-bonded liquids from tabulated thermodynamic data. Lists exist of the effect of common liquids on the major polymers. The critical strain is small if the solubility parameter of the polymer is similar to that of the liquid. The sensitivity to ESC is affected by the molecular weight of the polymer, and by molecular orientation at the surface of injection mouldings, particularly near the gate.

Table 9.4 Craze parameters for PMMA for equilibrium swelling

Environment	Volume fraction of liquid	T_g (°C)	Craze stress (MN m^{-2})
Air	0	115	100
Methanol	0.23	36	7.0
Ethanol	0.24	30	5.2
n-Propanol	0.24	32	5.3

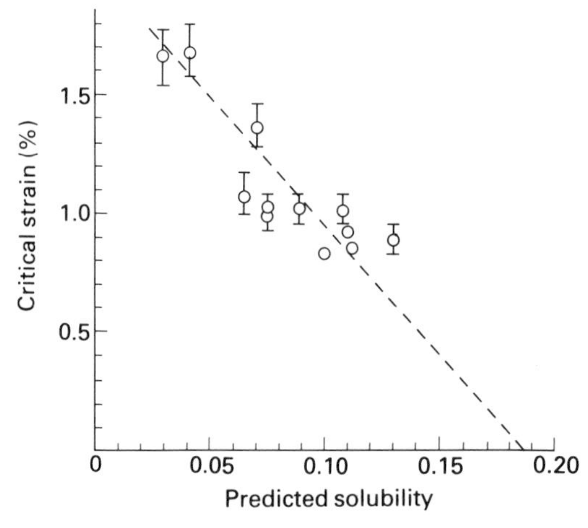

Fig. 9.16 Critical strain for the crazing of polycarbonate versus the solubility of straight chain hydrocarbons in polycarbonate (from C. H. M. Jaques and M. G. Wyzgoski, *J. Appl. Polym. Sci.*, 1977, **23**, 1159, Wiley)

As most polymers craze during ESC, the liquids must affect the mechanics of crazing. The main effect is to plasticise the craze fibrils, lowering their glass transition temperatures, and the tensile stress across the craze. Table 9.4 shows the effects of various alcohols on crazes in high molecular weight PMMA. The values in the second and third columns are for the equilibrium swelling of solid PMMA. Methanol swells the bulk PMMA, and reduces the glass transition temperature markedly. The diffusion coefficient in the swollen surface layer is orders of magnitude larger than with the glassy core. The position of the swollen front advances linearly with time, taking about 5 days to penetrate 1 mm. This is non-Fickian diffusion because of the moving sharp boundary between the swollen and unswollen material. In a short laboratory test methanol acts as an ESC agent for PMMA by reducing the crazing stress, without affecting the bulk of the material. Continuous immersion in methanol would eventually change the mechanical properties of the polymer, and lead to a completely different result.

Methanol absorption is fast for the porous craze, but there is a viscous resistance to the flow of any liquid through a craze, and this limits the growth kinetics of the craze. Fig. 9.17 shows a craze that is full of liquid to a length L. The pressure p_1 of the liquid is atmospheric at the beginning of the craze at $x = 0$, and p_2 at $x = L$ where the surface tension forces of the liquid keep the pressure constant (the value of p_2 is uncertain because the dimension of the channel that is filled with liquid may be the gap between the craze fibrils, or the craze height). If the liquid in the craze is moving with the same velocity V as the craze advances, and the pore area of the craze cross section has the constant value A, then D'Arcy's law for the flow of a liquid of viscosity η through a porous medium

$$V = -\frac{A}{12\eta}\frac{dp}{dx} \tag{9.6}$$

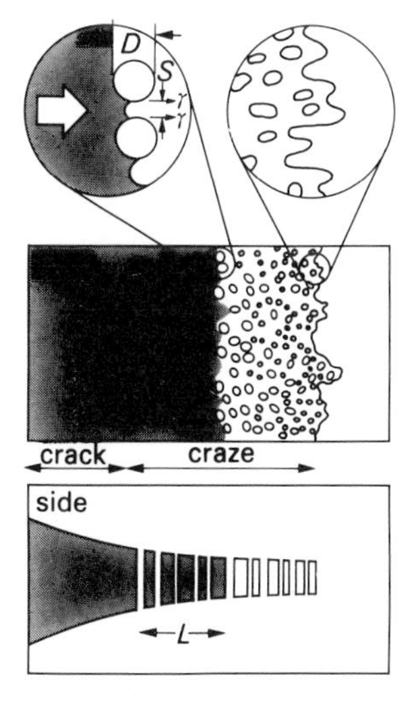

Fig. 9.17 Flow of a liquid into a craze, with a liquid/air boundary near the tip of the craze

can be used. The pressure gradient dp/dx can be put equal to $(p_2 - p_1)/L$, and the velocity to dL/dt and the equation integrated to give

$$L^2 = \frac{A}{6\eta}(p_1 - p_2)t \tag{9.7}$$

Experimental data on the growth of methanol filled crazes in PMMA confirm equation (9.7). The length increases with $\sqrt{t}$ until the equilibrium length according to the Dugdale model (equation 8.13) is reached.

9.5.2 Crack growth

Just because a liquid promotes the growth of crazes it does not mean that crack growth becomes easier. The effect of alcohols on PMMA is to encourage the formation of multiple crazes at a crack tip. This is more effective than a single craze at blunting the crack tip, and the K_{IC} value is increased above the value in air. For crack growth to become easier it is necessary for the crack to be preceded by a single craze. The presence of a liquid modifies the relationship between the stress intensity factor K_I and the crack velocity V. Fig. 9.18 shows that a liquid can allow slow crack growth at very much lower K values than in air. Two features of such graphs are noteworthy. The threshold value K_{th} at which the crack velocity is 10^{-9} m s^{-1} is relevant for the design of structures that have lives measured in years. A crack that can grow 1 mm in 12 days will rapidly become of a size that can cause fracture.

At higher velocities than those in Fig. 9.18 there can be a horizontal section of the graph where the crack velocity is independent of the K value. The

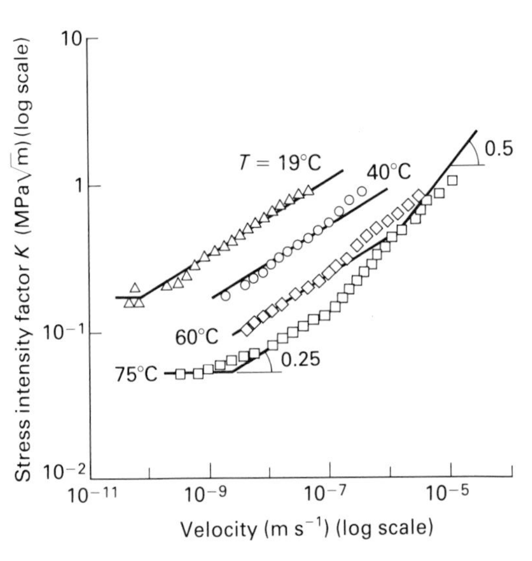

Fig. 9.18 Variation of the crack speed with the stress intensity factor for slow crack growth in a HDPE in a detergent environment (from J. C. Williams, *Fracture Mechanics of Polymers*, Ellis Horwood, 1984)

reason for this is that the flow of the liquid through the craze cannot keep up with the crack tip, when its velocity exceeds a certain value. This part of the graph represents the changeover, from a region where the craze is fully softened by the liquid, to one where the dry craze fractures and the liquid plays no part.

9.5.3 The complete failure process

The creep rupture tests of plastic pressure pipe in Chapter 7 are typical of whole-lifetime testing. It is unknown at which stage in the test cracks nucleate, but it is often assumed that there are small cracks present from the start of the service life. It is difficult to manufacture plastics products without incorporating foreign particles of a size about 0.1 mm; these can be undispersed pigment or stabiliser, or metal wear fragments from extruder screws. The adhesion between such particles and the polymer is so poor that there can be considered to be a crack at the interface.

It is possible to calculate the lifetime of a structure that has a constant load or stress applied to it if we know (i) the initial crack size and location, (ii) the K_I versus crack length calibration and (iii) the variation in the crack velocity V with K_I. This can be illustrated for an elliptical shaped surface crack that forms at the bore of a HDPE pipe under internal pressure p. The crack is normal to the hoop stress σ_H, and has its minor axis a perpendicular to bore and its major axis b parallel to the bore. The stress intensity factor of such a crack is given by Ewalds and Wanhill (1984) as

$$K_I = C_1 \sigma_H \sqrt{\pi a} + C_2 p \sqrt{\pi a} \qquad (9.8)$$

where the constants C_1 and C_2 depend on the ellipticity of the crack and the

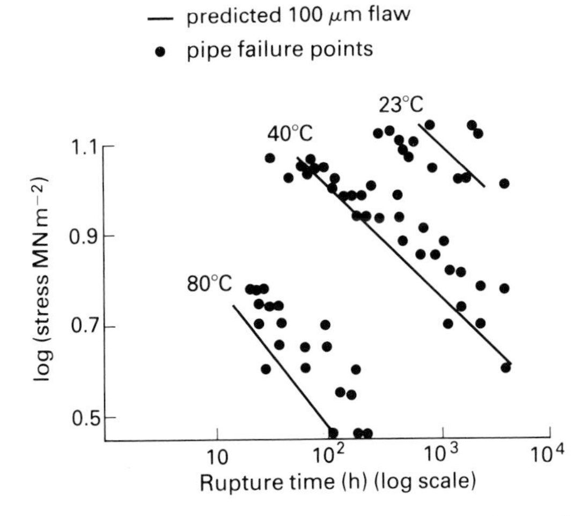

Fig. 9.19 Predicted creep rupture lives of an HDPE of MFI = 0.6 assuming an initial flaw size of 100 mm, compared with experimental points (from A. Gray *et al.*, *Plastics and Rubber Processing and Applications*, 1981, **1**, 53, Plastics and Rubber Institute)

Fig. 9.20 Fracture surface of an MDPE with good ESC resistance, after crack growth from left to right at 80 °C

size of the crack relative to the wall thickness. The crack velocity V versus stress intensity factor graphs for certain HDPEs have the form

$$V = AK_{\mathrm{I}}^{4.0} \tag{9.9}$$

where A is a constant, for velocities between 10^{-6} and 10^{-2} mm s^{-1}. The lifetime of a pipe can be calculated as the sum of the times $\Delta t = V/\Delta a$ required for the crack to grow a distance Δa. After each increment, equation (9.8) allows the new K_{I} value and equation (9.9) the new velocity to be evaluated. Fig. 9.19 shows that the predicted lives are slightly conservative when compared with experimental creep rupture data.

Tougher medium density polyethylenes used for gas pipes no longer have K–V relationships that can be expressed by equation (9.9), so the prediction of the lives is more difficult. An examination of the fracture surfaces shows crack arrest markings running vertically (Fig. 9.20). These show that the crack stops while a craze grows ahead of it; when the craze fails a rapid increment of crack growth occurs and the process repeats itself. The many coarse fibrils on the fracture surface represent the final products of breakdown. However, with the third generation MDPEs (Böcker et al., 1992) failure by slow crack growth no longer reduces the 50 year creep rupture design stress (see p. 320).

10

Transport properties

When substances travel through polymers the intention may be to restrict the transport to a minimum level, as with water transport through a LDPE damp-proof membrane, or to maximise the transport, as with the use of pipelines. There will be intermediate cases where selective transport is required—separating gases, or separating salt from water in a desalination plant. The subject of the design and use of pipes is covered in Chapter 13, so the main topic here is film products that transport limited amounts of substances. Solid transfer through polymer sheet is impossible, but woven polymers or polymer grids allow solid transfer on different scales. The optical properties of films are important in their commercial applications, and the transmission of light via fibre optics is explored because of its revolution of telecommunication links. The transmission of heat through solid polymers uses many of the same concepts as gaseous diffusion, so it is convenient to treat it here. The transmission of electrons and electromagnetic radiation is dealt with separately in Chapter 11 because of the size of the topic.

10.1 GASES

10.1.1 Solubility

The transport rate of gases through a polymer film depends both on the solubility of the gas in the polymer and on its diffusion coefficient. The solubility of the gas is affected by the strength of the intermolecular forces between pairs of gas molecules. In Chapter 1 the nature of the van der Waals bond was described. The strength of the bond can be characterised by the depth E_0 of the potential energy well shown in Fig. 1.1. Table 10.1 gives values of this constant for some common gases. E_0 measures the propensity of a gas to condense inside a polymer.

For the gases down to oxygen in Table 10.1 it is usually found that the gas concentration C is related to the gas pressure p by Henry's law

$$C = Sp \tag{10.1}$$

where the constant S is the solubility constant for the gas. Gas concentration

Table 10.1 van der Waals bond energies and solubilities in the amorphous phase in polyethylene

Gas	E_0 (10^{-23} J)	S^* (10^{-6} mol m^{-3} Pa^{-1})
He	14	5.4
H_2	52	—
N_2	131	18.4
O_2	163	34.3
CH_4	204	90.6
CO_2	261	201

can be expressed in various units; in S.I. units it is in mol m^{-3}, but it is more common to use m^3 of gas at STP per m^3 of polymer. Conversion between the units is made using the molar volume (at 0 °C, 1 bar) of 22.4×10^{-3} m^3. Solubility is expressed in S.I. units as mol m^{-3} Pa^{-1} (where 1 Pa = 1 N m^{-2} = 10^{-5} bar); a reduction to the base units of kg, m, s is not made because it confuses calculations.

In semi-crystalline polymers such as polyethylene and polypropylene that are above their glass transition temperatures the solubility constant is found to be proportional to the volume fraction V_{am} of the amorphous phase

$$S = V_{am} S^* \tag{10.2}$$

The solubility constant for 100% amorphous material S^* increases with the van der Waals bond energy.

For the more soluble gases such as methane and carbon dioxide the solubility behaviour is more complex, particularly for glassy polymers, for which the difference between the volume expansion coefficients of the liquid and glassy states is large. The values of $\alpha_L - \alpha_G$ for PET, PC and PMMA are 8.0, 4.3 and 1.3×10^{-4} C^{-1} respectively, and only the first two of these glassy polymers show this anomalous effect. Some of the gas is physically adsorbed on the surface of submicroscopic holes in the polymer. Fig. 10.1 shows how the concentration of CO_2 increases with pressure p in PET, a polymer that is used in carbonated drinks containers. The solubility equation is modified to

$$C = Sp + \frac{abp}{1 + bp} \tag{10.3}$$

The second term on the right is the *Langmuir adsorption isotherm* which describes the equilibrium concentration of gas molecules on a surface. The gas molecules bombard the surface and some of them stick; this process is in equilibrium with the rate of desorption from the surface. The area of the internal holes per unit volume of polymer is the disposable constant a, known as the hole saturation constant [7.9 cm^3 (STP)/cm^3] and b is the ratio of the adsorption to desorption rates (0.35 bar^{-1}) for CO_2 in PET. This Langmuir term dominates the solubility at pressures below 5 bar. The gas that is adsorbed on internal surfaces plays little part in gas transport through the polymer, and it can be ignored in calculations of gas permeation.

Water vapour is a gas with an anomalous solubility characteristic, this time because of the strong hydrogen bonding between the molecules. Fig. 10.2

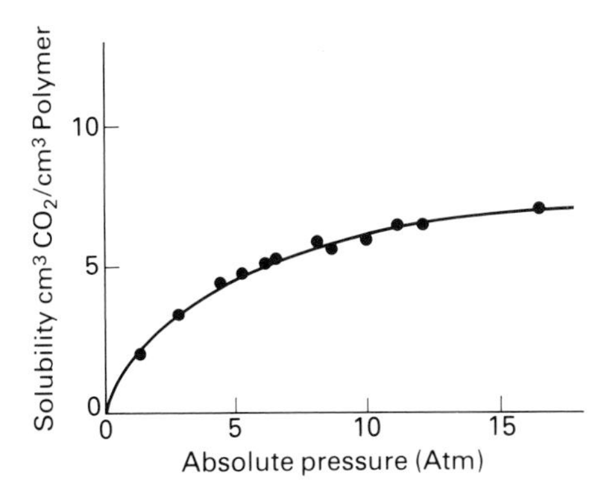

Fig. 10.1 Solubility of CO_2 at 25 °C in polyethylene terephthalate versus the gas pressure (from Hopfenberg (ed.), *Permeability of Plastic Films and Coatings*, Plenum Press, 1974)

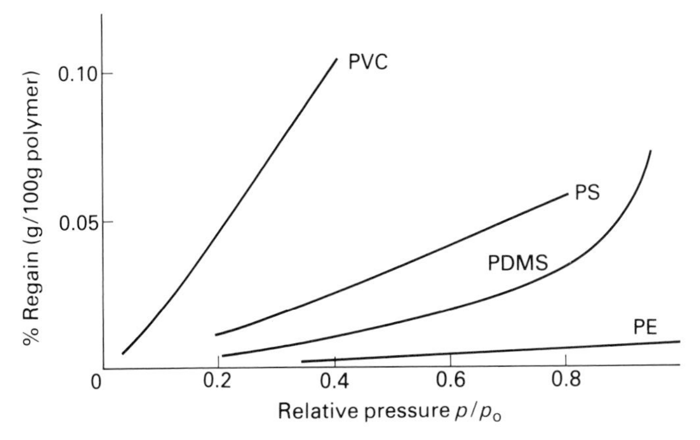

Fig. 10.2 Solubility of water vapour in various polymers versus the relative pressure (= vapour pressure/saturation vapour pressure) at 25 °C (PDMS at 35 °C) (from Crank and Park (eds.), *Diffusion of Polymers*, Academic Press, 1968)

shows that the sorption isotherms can curve steeply upwards as the relative pressure approaches 1. However, hydrophobic polymers such as polyolefins still obey Henry's law.

10.1.2 Steady state diffusion

The mathematics of diffusion of a gas in a polymer is exactly the same as for the diffusion of heat considered in Chapter 4. Thus, although the permeation constant is defined by steady state conditions, there may be significant transient effects. In considering the steady state transfer of a gas at a pressure p_1 on one side of a polymer film of thickness L, to a pressure p_2 on the other side (Fig. 10.3) there are two ways of defining the transfer constant. The gas concentration will be constant at C_1 and C_2 respectively in the polymer at the

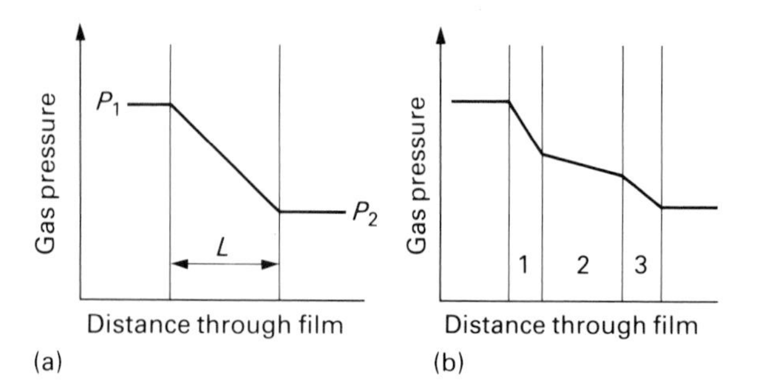

Fig. 10.3 Variation of gas pressure through a single polymer film, and a multilayer film in which the polymers 1, 2 and 3 have different permeabilities, when there is a steady state gas flow

two surfaces. The flow rate through an area A of film is then given either by

$$Q = DA\left(\frac{C_1 - C_2}{L}\right) \tag{10.4}$$

or

$$Q = PA\left(\frac{p_1 - p_2}{L}\right) \tag{10.5}$$

where D is the *diffusivity* and P the *permeability*. There are many different units in which these constants are quoted, because the apparatus used to determine them may measure gas volumes or mass changes, and the pressure units and time units can vary. The diffusion constant has the simple units m^2 s^{-1} so long as both Q and C use the same units for the amounts of gas. Table 10.2 gives some values for pure polymers in the unoriented state.

The diffusivity is related to the permeability by

$$P = DS \tag{10.6}$$

so the S.I. units of permeability are mol m^{-1} Pa^{-1} s^{-1}. As 1 mol of gas at standard temperature and pressure (STP) occupies 22.4 litres, it is also possible to use the units m^3 (STP) m^{-3} bar^{-1} s^{-1}. American permeability data are usually quoted in Imperial units using 'mil' (0.001 inch) for thickness, and a standard test area of 100 in^2. Older European data can be in cgs units, with pressure measured in cmHg. Hence the following conversion factors for permeability may be useful.

$$\frac{\text{mol}}{\text{m Pa s}} = 4.91 \times 10^{17} \frac{\text{cc mil}}{100 \text{ in}^2 \text{ atm day}}$$

$$= 2.95 \times 10^5 \frac{\text{cm}^3 \text{ (STP)}}{\text{cm cmHg s}}$$

$$= 2.95 \times 10^{15} \text{ barrer}$$

where

$$1 \text{ barrer} = 10^{-10} \text{ cm}^3 \text{ (STP) cm}^{-2} \text{ cmHg}^{-1} \text{ s}^{-1}$$

Table 10.2 Permeability data and diffusion constants at 25 °C

	Permeability ($mol\ m^{-1}\ Pa^{-1}\ s^{-1}$)			Diffusion constant ($m^2\ s^{-1}$)	
Polymer	O_2 $\times 10^{-18}$	H_2O $\times 10^{-15}$	CO_2 $\times 10^{-18}$	O_2 $\times 10^{-12}$	CO_2 $\times 10^{-12}$
Dry EVAL (33%E)	0.02				
PVDC	1.3	0.7	7		0.001
PET	14	60	30	0.36	0.054
PVC rigid	23	40	98		
Nylon 6	30	135	200		
Polyether sulphone	340		1900		2.0
HDPE	400	4	1000	17	12
LDPE	1100	30	5700	46	37
PP	400	17	1000		
PC	500	470	2900		5.3
PS	580	330	4000	12	1.3
Polyphenylene oxide	780		3000		
Butyl rubber	370		1500	80	5.8
Natural rubber	7000	770	37 000	158	110
Silicone rubber	205 000	14 500	1 095 000	1700	

For semi-crystalline polymers above T_g with a variable crystallinity the permeability is no longer proportional to the volume fraction amorphous phase V_{am}. This is because the gas must diffuse through the channels between the lamellar crystals, and the detailed morphology depends on the polymerisation, catalyst type, the thermal history and whether there is any orientation in the film. The permeability is found to be proportional to $V_{am}{}^n$ where the exponent n lies between 1.2 and 2 for different polymerisation routes. The effect on the permeability of the molecular weight M of the gas is approximately described by $P \propto M^{-0.5}$. This is not followed rigorously and it can be seen from Table 10.2 that the permeability of CO_2 in glassy polymers is higher in proportion to the oxygen permeability than it is for semi-crystalline polymers.

10.1.3 Transient effects in gaseous diffusion

There are transient effects when a plastic container is filled with gas. Fick's second law, derived in Appendix A, applies so long as the diffusion coefficient is independent of the gas concentration C. The differential equation for the gas concentration C, for one-dimensional diffusion along the x axis, is

$$\frac{dC}{dt} = D\frac{d^2C}{dx^2} \tag{10.7}$$

where t is time and D is the diffusion coefficient. The variation of D with temperature is described by

$$D = D_0 \exp\left(-\frac{E_D}{RT}\right) \tag{10.8}$$

where E_D, the activation energy for diffusion, is a measure of the rate at which D increases with increasing temperature, R is the gas constant and T the absolute temperature. The diffusion coefficients will be highest in semi-crystalline polymers if the amorphous phase is above its glass transition temperature T_g. Table 10.2 gives diffusion data for polyethylenes at 25 °C. The reason for the higher diffusion coefficient in LDPE is that the crystalline phase has a negligible diffusion coefficient, and LDPE has a higher percentage amorphous content. Equation (A.17) shows that after a time t the oxygen concentration will exceed 50% of its surface level in a layer x thick where

$$x = 0.94\sqrt{Dt} \tag{10.9}$$

When Fick's second law is obeyed the analytical methods in Appendix A can be used to predict the total flow of the gas through the film. If the film is initially free from the diffusing gas, and a constant concentration C is applied at time $t = 0$ at one surface, the total amount V that passes through unit area of sheet thickness L is given by

$$\frac{V}{LC} = \frac{Dt}{L^2} - \frac{1}{6} - \frac{2}{\pi^2} \sum_{n=1}^{\infty} \frac{(-1)^n}{n^2} \exp\left(-\frac{Dn^2\pi^2 t}{L^2}\right) \tag{10.10}$$

Fig. 10.4 shows the amount that diffuses through the sheet as a function of the elapsed time. The graph settles down to a steady permeation rate, and if this straight line is extrapolated back it cuts the time axis at a *time lag* t_L given by

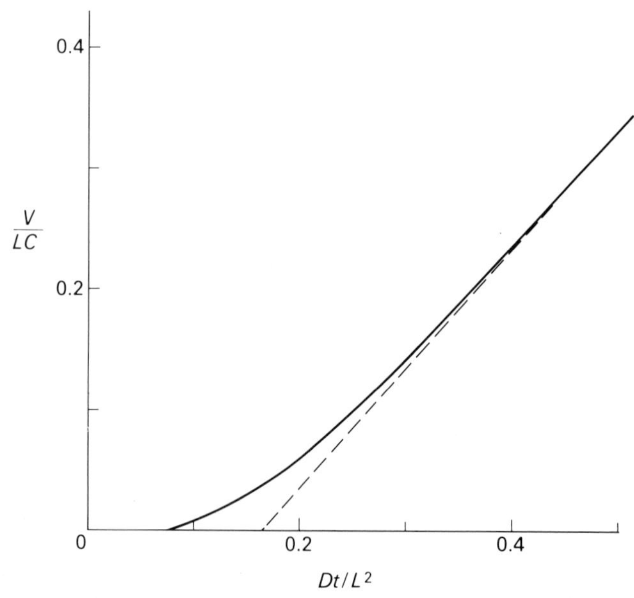

Fig. 10.4 Volume of gas passing through 1 m^2 of sheet of thickness L versus time according to equation (10.10)

$$t_L = \frac{L^2}{6D} \tag{10.11}$$

For natural gas (methane) passing through a 3 mm thick MDPE pipe wall, the time lag is 10.4 days, as $D = 1.0 \times 10^{-11}\ \mathrm{m^2\,s^{-1}}$. Consequently a steady state permeation test with this pipe is a long term experiment. If the diffusion constant is independent of concentration and Henry's law holds, then the diffusion constant can be calculated from the permeability.

Where the diffusion coefficient D increases with the gas concentration, equation (10.7) is replaced by

$$\frac{\partial C}{\partial t} = \frac{\partial}{\partial x}\left(D\frac{\partial C}{\partial x}\right) \tag{10.12}$$

This equation is best solved by finite difference methods on a computer. This case occurs with organic vapours diffusing through rubbers, and can occur with high concentrations of gases in glassy polymers (Fig. 10.1) where the gas swells the glass and can alter the T_g value.

10.1.4 Packaging applications

Biaxial orientation is often used to improve the in-plane tensile strength and toughness of polymer films. This also decreases the permeability, the main reason being the increase in crystallinity on stretching—see Fig. 7.18 for PET. It is useful to have data for the barrier properties of particular thickness films, to allow materials selection and package design. For multilayer films made by coextrusion or coating techniques the total permeability P_{TOT} is related to the total thickness L by

$$\frac{P_{TOT}}{L} = \frac{P_1}{L_1} + \frac{P_2}{L_2} + \ldots \tag{10.13}$$

It is often more convenient to be able to add directly the *resistance to the gas transfer* for each layer, defined as

$$R_i \equiv \frac{L_i}{P_i} \tag{10.14}$$

where L_i is the layer thickness, and then to calculate the steady state gas flow from

$$Q = (p_1 - p_2)/(R_1 + R_2 + R_3 + \ldots) \tag{10.15}$$

Table 10.3 gives some film resistances for commercially important films. Low density polyethylene, because of its domination of the film market, acts as a reference material by which others can be compared. Its good water vapour resistance cannot be improved upon by a large factor, but its oxygen resistance is relatively poor. Even the oxygen resistance of PET in the biaxially oriented form is not good enough for the preservation of oxygen sensitive foodstuffs such as beer. Consequently PET bottles can have an outer coating of a PVDC copolymer to meet this requirement. The oxygen resistance of each 4 μm thick PVDC coating can be calculated from the data in Table 10.3 as 3150 GN s mol^{-1}. The PVDC coating is sprayed on to the outside of completed bottles, and it is not capable of being recycled. The initial pressure loss in a carbonated drink bottle when it is filled is a result of transient diffusion and the Langmuir adsorption of the CO_2 in the PET.

Table 10.3 Strength and gas resistance data

Polymer	Film thickness (μm)	Tensile strength (kN m^{-1})	Oxygen resistance (GN s mol^{-1})	Water vapour resistance (GN s mol^{-1})
LDPE	25	0.35	23	0.8
Chill cast PP	32	0.9	45	1.4
Biaxially oriented PP	14	3.5	45	1.2
Biaxially oriented PET	12	2.0	1700	0.32
Biaxially oriented PET + 4 μm coat of PVDC	20	2.5	8000	2.5

Extruded ethylene vinyl alcohol copolymer (EVAL) is used as an oxygen-barrier packaging material. When EVAL is dry it has an extremely low permeability to oxygen, but the vinyl alcohol part of the copolymer is highly hydrophilic, and in the swollen wet state the permeability increases (Fig. 10.5) to be higher than that of PVDC. The construction of a multilayer squeezable bottle, for foods such as ketchup that are sensitive to oxygen yet

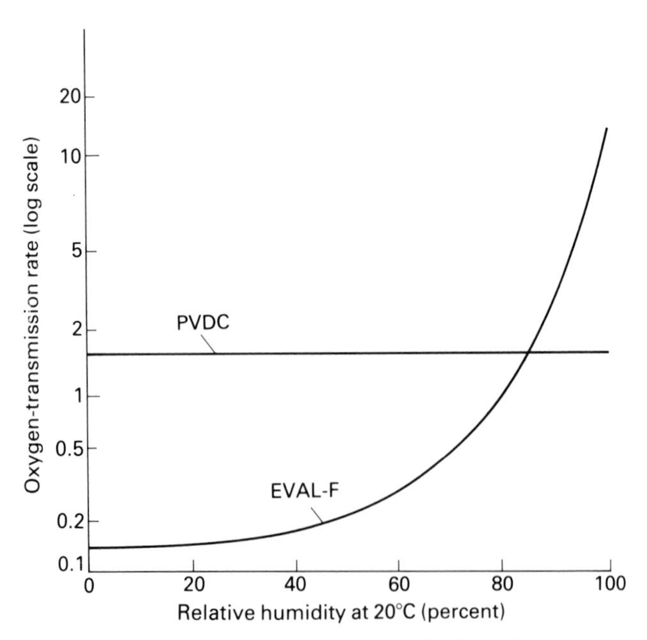

Fig. 10.5 Variation of the oxygen transmission rate (cm^3 m^2 day^{-1} bar^{-1} for 20 mm film) with relative humidity for the high barrier PVDC and EVAL films (from *Plastics Engineering*, 1984, May, 43, Soc. Plastics Eng. Inc.)

need to be sterilised in the container when it is filled, presents a difficult problem. Biaxially stretched PET is insufficiently form stable at 100 °C because the T_g of the amorphous phase is only 80 °C. Polypropylene has an adequate form stability but far too high an oxygen permeability. If a layer of EVAL is sandwiched between inner and outer layers of polypropylene then the oxygen resistance can be achieved so long as the EVAL remains below a 75% R.H. level. The problem is the sterilisation step, because the permeability of all polymers rises rapidly as the temperature increases (Fig. 10.6). Consequently the EVAL layer is at nearly 100% R.H. after sterilisation. If the EVAL layer is placed near the outside of the polypropylene sandwich then the water resistance of the outer polypropylene layer is lowered and the EVAL humidity can drop to an acceptable level within a week or so of filling the container. There are five layers in the container since there is a 0.6 μm thick layer of adhesive between the polypropylene and EVAL. The design of co-extrusion dies to produce a stable melt parison with five distinct layers is a formidable problem.

For a single material film, the rate at which water vapour permeates

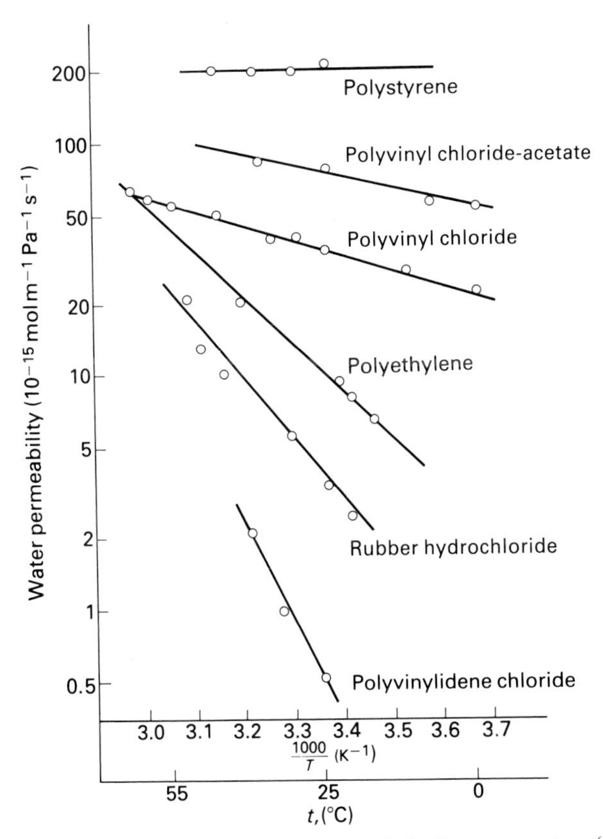

Fig. 10.6 Variation of water permeability with reciprocal absolute temperature (from Hennessy *et al.*, *The Permeability of Plastics Films*, Plastics and Rubber Inst., 1966)

through increases with temperature more rapidly than the permeability data in Fig. 10.6. This is because the vapour pressure of water in mmHg increases rapidly with temperature (17.5 at 20 °C, 55.3 at 40 °C, at 60 °C, 355.1 at 80 °C). From the definition

$$\text{relative humidity} \equiv \frac{\text{partial pressure of water vapour}}{\text{vapour pressure of water at that temperature}} \tag{10.16}$$

it follows that the water transport rate through the film is proportional to the permeability, to the water vapour pressure and to the relative humidity difference across the film.

10.1.5 Gas separation

With developments in the efficiency of gas separation membranes they have become competitive with other methods of gas separation. In Chapter 1 the separation of gaseous products of cracking naphtha was carried out by liquefaction and fractional distillation of the liquids at high pressure and low temperature. If gaseous separation can be carried out at ambient temperatures there are great savings in energy and the plant is more compact. The developments necessary have been in achieving high permeation rates while having mechanical durability. Equation (10.4) shows that to achieve high gas transport rates, the film thickness must be very small, the area of film must be large, and both the permeability and the applied pressure must be large. The contradictory requirements of having a thin film that can cope with a high pressure of up to 100 bar, have been met either by using porous fibres with a thin membrane skin, or by having porous layers that are cloth reinforced, with a thin skin of a polymer. Figure 10.7 shows the construction of a porous polysulphone fibre that has been spun from a water miscible solvent into water.

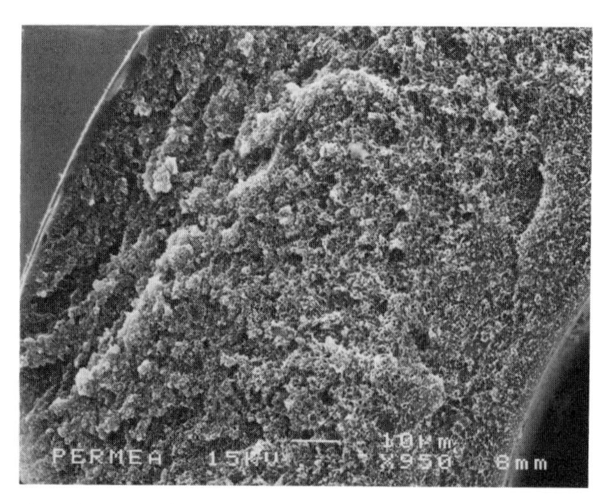

Fig. 10.7 Scanning electron micrograph showing close up of the wall of a prism-α hollow membrane used in gas separation applications. (Reproduced with kind permission of Permea Inc., St. Louis, USA.)

One area of application is in the provision of 99% pure nitrogen supplies from air, in competition with deliveries of liquid nitrogen or bottles of high pressure nitrogen. The selectivity α of a polymer membrane is defined as

$$\alpha \equiv \frac{P_A}{P_B} \tag{10.17}$$

where A and B are the two gases. When A is oxygen and B is nitrogen the values for most polymers lie in the range 3 to 5. Although this is inadequate to produce the required purity in a single separation stage, chemical engineers are used to multi-stage processes, and a second separation of the nitrogen rich output of the first stage will produce the required purity. Table 10.2 suggests that silicone rubber would be a good membrane material, but it is mechanically weak. Hence the need for a strong but porous support. The development of porous polysulphone layers or fibres has provided such a support.

One advantage of membrane separation systems is that they can be used for a variety of operations in the chemical industry. The permeability of gases increases as the size of the molecule decreases, so it is possible to separate small amounts of hydrogen from a mixture of gases. Applications are reviewed by Spillman (1989).

10.2 LIQUIDS

The diffusion of liquids into polymers is in general slower than the diffusion of gases. The diffusion coefficients are of the order of $10^{-13}\ m^2\ s^{-1}$. The major difference is that the equilibrium solubility can be much larger than that of gases, and the liquid content can change the diffusion constant, or even the physical state of the polymer. Semi-crystalline polymers are in general more resistant to organic liquids than are glassy polymers, so these will be dealt with separately.

Many hollow containers for liquids are made by the blow moulding of polyolefins, especially HDPE. They are used to contain a great variety of liquids because direct attack by environmental stress cracking or dissolution is relatively rare. High molecular weight grades with good ESC resistance can be blow moulded, and to improve the resistance still further MDPE copolymers can be used (see Chapter 8). There is, however, an affinity between polyethylene and aromatic or chlorinated hydrocarbons; they swell the amorphous phase of the polyethylene and have relatively high permeabilities. Table 10.4 lists the permeabilities of several common liquids. The corresponding figures for LDPE are a factor of ten larger, showing that diffusion through the amorphous rubbery phase dominates the permeability. Ethyl acetate is a representative constituent of foodstuffs. The high permeability of alkanes is not surprising considering the chemical similarity with polyethylene.

One application for blow moulded polyethylene containers is as petrol tanks for cars. There are considerable weight savings, corrosion is eliminated and the complex moulded shapes can fit into spaces above the rear axle.

Table 10.4 Permeability of liquids through HDPE of density 950 kg m^{-3} at 23 °C

Liquid	Permeability (g mm m^{-2} day^{-1} bar^{-1})
Toluene	37.5
n-Heptane	17.1
97 octane petrol	16
Ethyl acetate	1.6
Diesel oil	0.5 to 3
Methanol	0.15

Volkswagen used about 100 000 per year in the Passat model. Large containers have an advantage in terms of permeation because the surface area to volume ratio decreases as the volume increases (this also applies to carbonated drinks bottles, which is why the 2 litre PET bottles were introduced first). The container is designed to withstand a certain internal pressure; 3 bar for petrol tanks in spite of crash tests showing that internal pressures did not exceed 1 bar. Blow moulding does not produce a uniform wall thickness product; values ranging from 4 to 7 mm are found for a petrol tank. Consequently it might appear that the loss of petrol by permeation through HDPE tank walls would be insignificant. However the EEC regulations set a limit of 10 g emission over 24 hours at 23 °C, and US regulations the tighter limit of 2 g per 24 hour shed (sealed house emission determination) test, because hydrocarbon gas emissions can lead to photochemical smog. The constituents of petrol swell polyolefins, and the diffusion rate after 1 year is not acceptable, so some extra treatment is necessary. At present the polyethylene is sulphonated (treated with concentrated SO_3) to decrease the permeability of a surface layer by a factor of 10 (Fig. 10.8), or fluorinated to

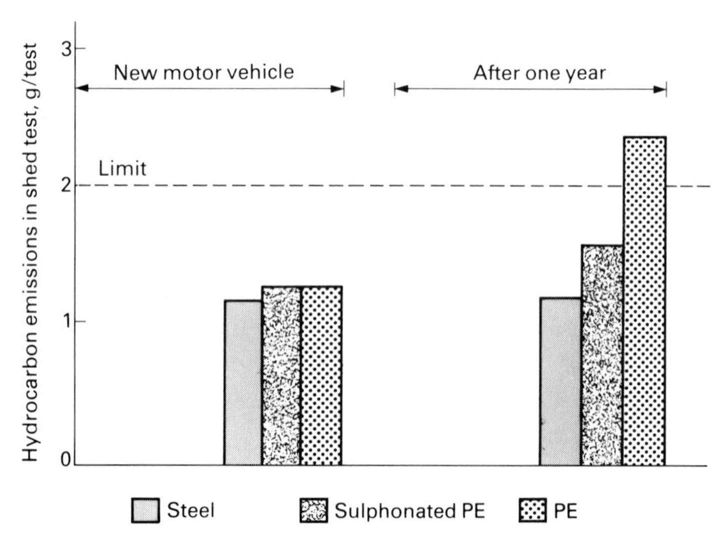

Fig. 10.8 Hydrocarbon emissions from steel, polyethylene, and sulphonated polyethylene fuel tanks (from *Automotive Engineering*, 1982, **90**, 68, Society of Automotive Engineers Inc.)

decrease the permeability by 97%. An alternative in future would be to use multilayer blow mouldings, or platelets of a less permeable polymer within the HDPE.

When polymers are used as food containers the additional problem of extraction of additives from the polymer arises. This is because certain food constituents diffuse into the polymer. For example fats or soils in many foods can diffuse readily into polyethylene. If the fat has a strong affinity for a polymer stabiliser or antioxidant the equilibrium concentration in the fat will be much higher than that in the polymer. There will be a two way diffusion process with the food component entering the polymer and the polymer additive entering the foodstuff. Fig. 10.9 shows some experimental results for a fat tricaprylin in contact with HDPE containing 0.25% of the hindered phenol antioxidant BHT. The results for different diffusion times at 40 °C are normalised by using $x/\sqrt{t}$ as the horizontal axis. The concentrations of both diffusants remain low and the results fit the theory for a constant diffusion coefficient of $5 \times 10^{-13}\ m^2\ s^{-1}$ for the fat and $1 \times 10^{-14}\ m^2\ s^{-1}$ for the antioxidant. There are two consequences of the extraction phenomenon. Firstly only certain non-toxic additives are permitted in polymers that are to be used as food containers. Secondly if a plastic container is re-used with another foodstuff then constituents of the first foodstuff may diffuse back out of the polymer into the second foodstuff. This is noticeable if polyethylene beakers or bottles are filled with orange squash, then re-used with water.

A third application area exists when the flow of a liquid through a polymer film is to be *maximised.* This arises in water treatment where seawater or brackish water can be purified by reverse osmosis, and in biomedical applications such as blood dialysis units for kidney patients. The design problem is to find a polymer of a high permeability for water, yet low

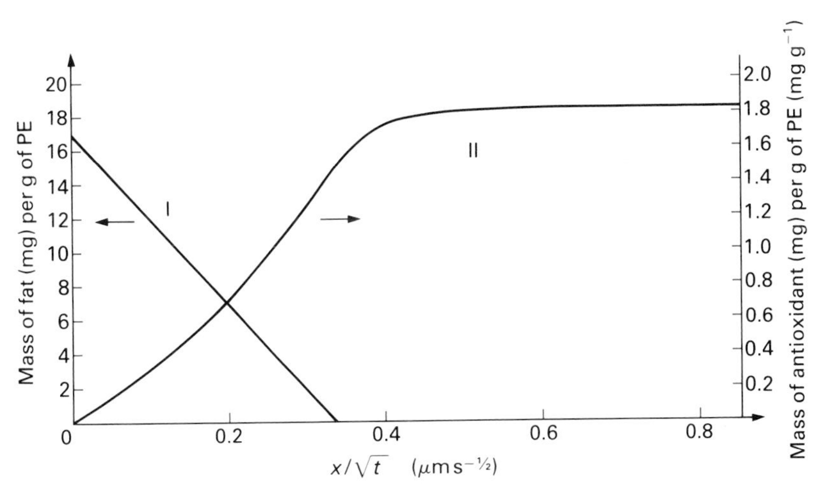

Fig. 10.9 Diffusion of a fat into HDPE and of an antioxidant additive from the HDPE, plotted against x/t where x is the distance from the surface and t the time (from *Angewa. Macromol. Chem.*, 1979, **78**, 179, Hüthig and Wepf Verlag)

permeability for the salt. To maximise efficiency the membrane area must be large and its thickness as small as possible consistent with a lack of pinholes. It is an advantage to apply a high pressure to one side of a reverse osmosis membrane. Because the membrane is thin, it is mechanically unable to resist the pressure unless it is supported on a backing structure. Fig. 10.10 shows the construction of a cellulose triacetate membrane on a porous cellulose nitrate–cellulose acetate support. To make the unit compact the composite membrane is spirally wound on to an inner cylinder, and the edges glued together. When a pressure of 70 bar is applied to the seawater side, NaCl rejection levels in excess of 99.7% can be achieved.

There are clothing applications where it is required to allow the passage of water vapour and to prevent the passage of water. The ingenious solution

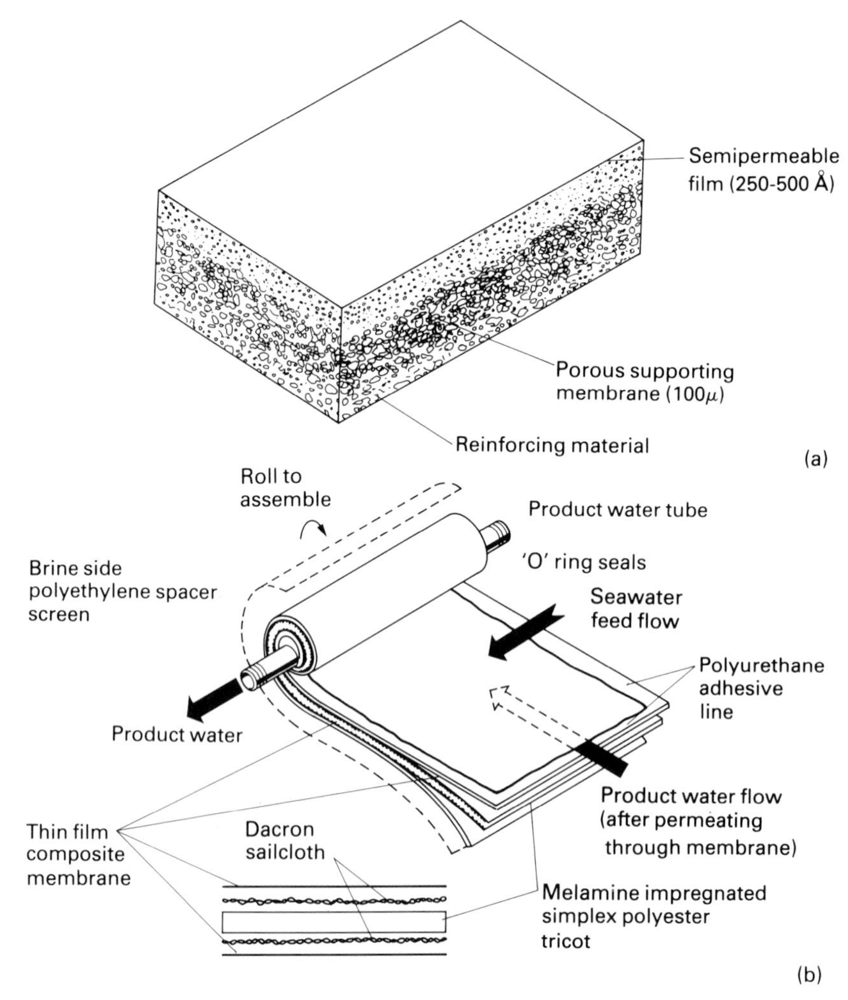

Fig. 10.10 (a) Cross section of a composite membrane for reverse osmosis of sea water. (b) Spiral winding of the membrane to make a compact unit (from Hopfenberg (ed.), *Permeability of Plastic Films and Coatings*, Plenum Press, 1974)

utilised in the Goretex fabrics is to stretch PTFE film in a manner that produces microscopic voids (Fig. 10.11). These allow the passage of water vapour from the skin but the very low surface energy of PTFE means that liquid water cannot wet the surface of the film and hence rain cannot be drawn through it by capillary action.

10.3 SOLIDS

In order to give some idea of how polymeric meshes or fabric can be used to prevent the passage of solids, some of the civil engineering applications of 'geotextiles' will be explored. The two main product types are both based on highly oriented polyolefins, the requirements being low cost and high strength. One type utilises polypropylene film that has been uniaxially drawn and fibrillated to produce a low cost substitute for fibres. This is woven into a coarse textile with a mass of between 100 and 300 g m^{-2} and a thickness of 0.3 to 0.7 mm (Fig. 10.12). The more recently developed Netlon products are based on the uniaxial or biaxial drawing of perforated HDPE sheets. These have a hole size that is an order of magnitude larger (up to 100 mm).

As well as having barrier properties, these geotextiles are very strong in tension in the plane of the product, so can be used for soil reinforcement. Soils have zero tensile strength, and steep sided soil embankments can fail by shear on surfaces at 45° to the vertical, especially if they contain clay and become waterlogged. Horizontal layers of geotextile, with a tensile strength of 50 to 100 kN per m width, can be incorporated into embankments at 1 m vertical separation, while the embankment is constructed. The plane of the geotextile is that in which the tensile principal stress of the soil acts, so it is efficient in its reinforcement action. Because the geotextile is buried it cannot be degraded by UV radiation.

Fig. 10.13 shows another application of the unidirectionally oriented

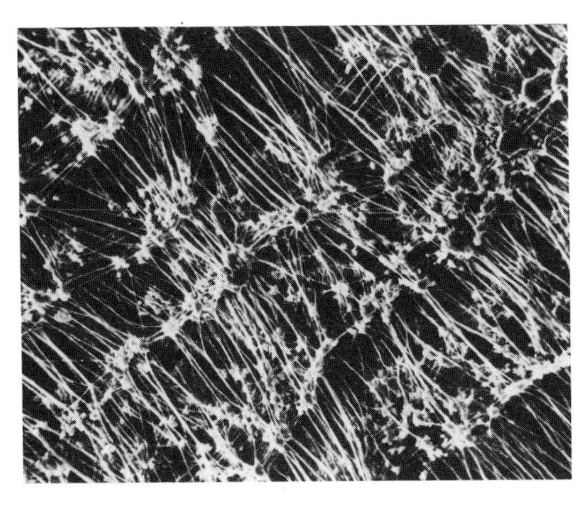

Fig. 10.11 SEM micrograph of Goretex fabric showing the pore structure that allows the passage of water vapour. (Courtesy of WL Gore & Associates (UK) Ltd.)

Fig. 10.12 Woven mesh of fibrillated polypropylene film used to allow water permeation but prevent the ingress of fine soil particles

Netlon product. Vertical rods couple the faces of the cellular structure, which can then be filled with granular material to form a 1 m thick stable 'mattress' at the foot of an embankment on top of soft soil.

Woven geotextiles prevent the loss of fine soil particles across the fabric plane. If the water flow is unidirectional across the textile a filter cake of fine particles builds up upstream of the textile and this aids the filtration process. By preventing road stone from being punched into soft underlying soil the total amount of road stone used can be reduced. The shear strength of the road stone layer is preserved without the ingress of fine soil particles. It also appears that geotextiles allow drainage in the plane of the fabric, so that in Fig. 10.13 water can drain to the sides of the embankment. This can aid the consolidation of a newly constructed embankment.

A third type of product that aids water drainage, but does not have the soil reinforcing element of geotextiles is the perforated corrugated pipe shown in Fig. 10.14. The hoop direction corrugations are produced by specially shaped cooling sections that move down the cooling section of an extrusion line on a caterpillar track. The corrugations provide a maximum resistance to diametral crushing by soil loads, and yet allow flexibility for coiling the pipe. Rectangular holes are punched in the small diameter regions and these allow the ingress of water into the drainage pipe. The elastic deflection δ that occurs when a line load of q N per unit length is applied to the pipe is given by

$$\delta = 0.0186 \frac{q D_m^3}{EI} \tag{10.18}$$

Since the mean diameter D_m of the pipe cannot be altered the deflection can only be limited by increasing the second moment of area I per unit length.

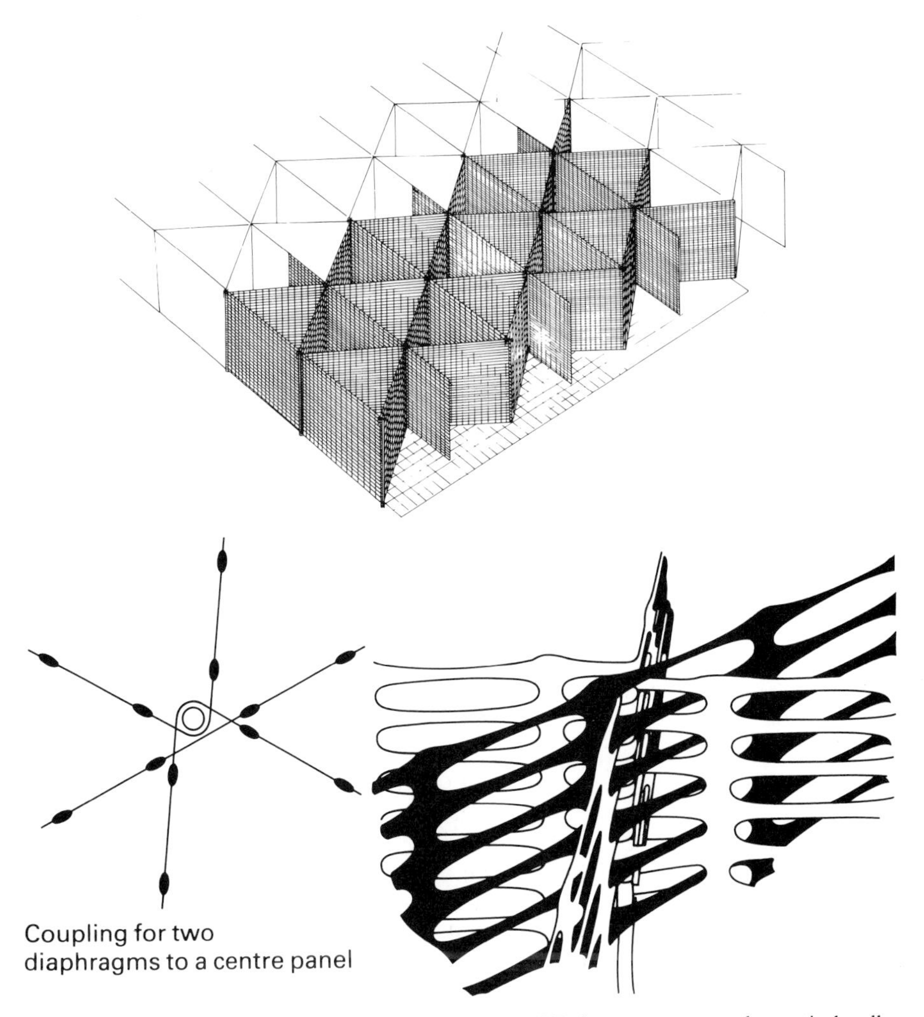

Fig. 10.13 Use of unidrectionally drawn perforated HDPE sheet to construct the vertical walls of triangular cells, that form a reinforcing mattress at the base of an embankment (from *Tensor Geocell Mattress* pamphlet, courtesy of Netlon)

This has a value

$$I = \frac{t^3}{12} \tag{10.19}$$

for a cylindrical pipe of thickness t. For the corrugated pipe the actual cross section in Fig. 10.14 can be approximated by a trapezoidal wave of wavelength L and crest length C. The I_c value of this shape is

$$\frac{I_c}{I} = 1 + \left(\frac{H}{t}\right)^2 \left(1 + \frac{4L}{C}\right) \tag{10.20}$$

if the vertical thickness remains equal to t. H is the height of the wave shape.

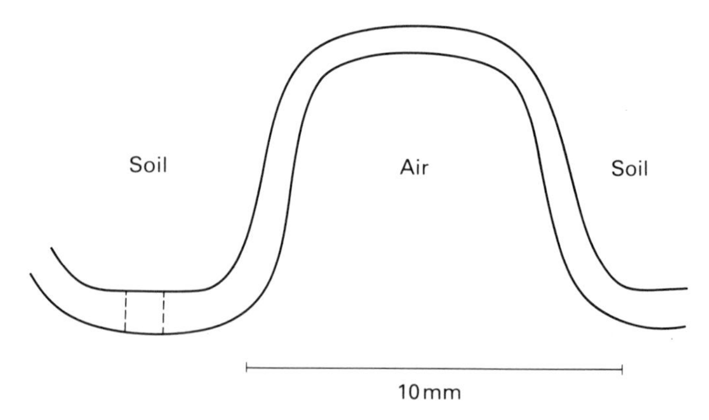

Fig. 10.14 Section through the wall of a corrugated PVC pipe, of diameter 150 mm, used for soil drainage. Slots of size 8 mm × 1 mm are punched at intervals in the base of the corrugation to allow water ingress

For the shape shown the stiffening factor calculated from equation (10.20) is approximately 450; this is equivalent to a materials savings of 87% compared with a cylindrical pipe of equal bending stiffness.

10.4 LIGHT

10.4.1 Refraction and reflection of light

The high optical clarity plastics are PMMA, PC and thermosetting diallyl glycol carbonate (tradename CR39). These glassy materials have a much lower Young's modulus ($3\ \mathrm{GN\ m^{-2}}$ as against $70\ \mathrm{GN\ m^{-2}}$) and tensile strength ($50–70\ \mathrm{MN\ m^{-2}}$ as against over $200\ \mathrm{MN\ m^{-2}}$) than conventional soda-lime glass. Usually only the infrared and ultraviolet components of the light are absorbed by the polymer, unless there are pigments present, or the polymer contains conjugated double bonds. The advantage of these polymers lies in their lower density and greater toughness, and the fact that they can be moulded to high precision, obviating the polishing stages needed with silicate glasses.

Table 10.5 Optical properties of glassy polymers

Material	Refractive index, n_Y	Dispersive power, D	Density, $\rho\ (\mathrm{kg\ m^{-3}})$	$\dfrac{\rho}{n_Y - 1}$	Stress optical coefficient $(10^{-12}\ \mathrm{m^2\ N^{-1}})$
PMMA	1.495	0.0189	1190	2400	4
CR-39	1.498	0.0172	1320	2650	34
PC	1.596	0.0333	1200	2010	78
PS	1.590	0.0323	1060	1800	9
Soda-lime silica glass	1.520		2530	4870	2.7

If the plastic is to be used in lens applications then the refractive index value and the dispersive power D are important. The refractive index is measured for particular yellow (587 nm), blue (486 nm) and red (656 nm) wavelengths and D calculated from

$$D = \frac{n_B - n_R}{n_Y - 1} \tag{10.21}$$

where n_B is the refractive index for blue light, n_R is the refractive index for red light and n_Y is the refractive index for yellow light. Table 10.5 shows that polycarbonate has a high refractive index, which allows lenses to be lighter, but the high dispersive power will increase chromatic aberrations.

The mass of a lens of a given diameter and focal length is proportional to the axial thickness of the lens and to the material density. If the radius of curvature of both surfaces of a biconvex lens is R then the focal length f is given by

$$f = \frac{R}{2(n-1)} \tag{10.22}$$

As the axial thickness of the lens is proportional to $1/R$, equation (10.22) shows that it will also be proportional to $1/(n-1)$. Hence the mass of the lens

$$m \propto \frac{\rho}{n-1}$$

Values of this quantity are given in Table 10.5, where it will be seen that plastic lenses allow considerable mass savings.

Fig. 10.15 shows the phenomena that occur when light meets a sheet of plastic at normal incidence. About 4% of the light intensity will be reflected back at the air/polymer and at the polymer/air interfaces; the reflected intensity R_0 for normal incidence is related to the incident intensity I by

$$R_0 = \left(\frac{n_1 - n_2}{n_1 + n_2}\right)^2 I \tag{10.23}$$

where the refractive index of the polymer $n_1 \cong 1.5$, and that of air $n_2 = 1$. If a high reflectivity is required (see the CD manufacture in Chapter 13) a conducting coating is required. Metals have a complex refractive index, with real part n_R and imaginary part n_I. This means that the light wave penetrates the metal with an exponentially decaying amplitude. The magnitude E of the electric vector varies with the distance y in the metal as

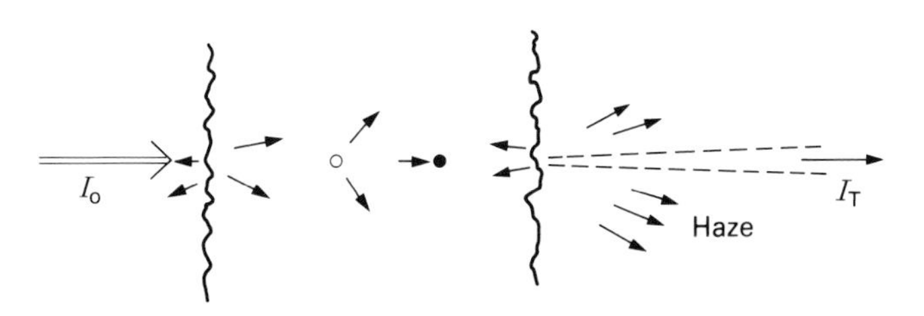

Fig. 10.15 Processes in a plastic film that reduce the transmitted light intensity

$$E = E_0 \exp(-\omega n_I y/c) \cos \omega(t - n_R y/c) \qquad (10.24)$$

where ω is the frequency of the wave and c the speed of light. For solid sodium at a wavelength of 589 nm the components of the refractive index are $n_R = 0.04$ and $n_I = 2.4$. This leads to a reflectivity for thick films of $R = 0.9$. For thin films the amplitude of the transmitted light decreases, according to equation (10.24), as $\exp(-2\pi n_I y/\lambda)$. This means that the metal only needs to be a few wavelengths thick for the reflectively to be high. Such thicknesses can easily be applied by vacuum evaporation, and the main concern is the protection of the layer from abrasion, with a transparent lacquer.

10.4.2 Light scattering

Light scattering can occur at the polymer/air interface, and internally in the polymer (Fig. 10.15). Different applications can tolerate different levels of light scattering. If the transmitted light just provides illumination, as in a roof light, or allows a liquid level to be inspected, as in a brake fluid reservoir, then a high level of light scattering can be tolerated. However if the light allows the performance of an eye–limb coordination task, like driving a vehicle, then a high level of optical clarity is required. Small angle deviations of the light path, caused by the lens effects of a non-planar plastic surface, will cause image distortion. High angle light scattering will cause glare from any bright light source in the field of view such as oncoming headlights at night.

Light scattering is marked in semi-crystalline polymers with a spherulitic microstructure, so unpigmented polyethylene appears milky and opaque. Light scattering is negligible when the inclusions in the matrix are of a diameter smaller than 10% of the wavelength of the light (Fig. 10.16). There

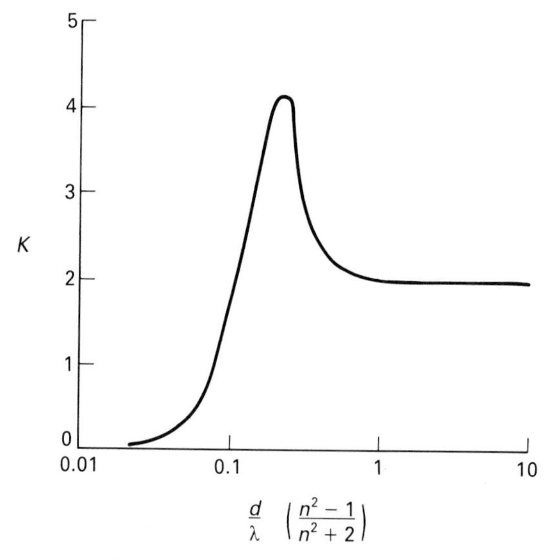

Fig. 10.16 Scattering coefficient for a single sphere of relative refractive index n versus the ratio of the sphere diameter d to the wavelength of the light λ

is a diameter at which the scattering coefficient peaks before it settles down to a constant value for large spheres. The scattering also depends on the difference $n_1^2 - n_2^2$ according to equation (10.17).

The mean refractive index n of a phase is related to the density ρ (and to the constant mean polarisability of the monomer unit α) by

$$\frac{\bar{n}^2 - 1}{\bar{n}^2 + 2} = C\rho\bar{\alpha} \qquad (10.25)$$

where C is a constant. For most semi-crystalline polymers the density of the crystalline phase is greater than that of the amorphous phase (Table 10.6) and the width of the crystal lamellae is of the same order as the wavelength of visible light. The exceptional case of polymethylpentene is one where the crystal has very open helical chain conformations, and the resulting semi-crystalline polymer is transparent. Nearly all rubber toughened polymers such as ABS are opaque because the phases differ both in density and polarisability.

Table 10.6 Densities of polymer phases at 20 °C

Polymer	Crystal density (kg m^{-3})	Amorphous density (kg m^{-3})
Polyethylene	1000	854
Polypropylene	940	850
Polymethylpentene	820	840

Even if a semi-crystalline polymer could be made 100% crystalline, there would be light scattering from neighbouring crystals that have different orientations. The anisotropy of bonding means that polymer crystals have a different refractive index n_c for light polarised along the covalently bonded **c** direction than it does for light polarised along the crystal **a** or **b** axes. Stretching the product aligns the **c** axes of crystals, and if crystallisation occurs on heating from the glassy state the high nucleation density results in a crystal size much smaller than λ. This is why the walls of PET carbonated drinks bottles are transparent.

The optical properties of polyolefin packaging films are important. If extrusion is carried out at too high a speed a surface roughness occurs on the molten extrudate and this will increase the light scattering. The average spherulite size in polyethylene film must be kept below the wavelength of light to minimise light scattering.

The scratching of glassy plastics (Section 7.2.5) will cause light scattering. The surface can be made more abrasion resistant by coating them with a hard layer of a highly crosslinked silicone thermoset. The layers are 5 to 10 μm thick and the tensile failure strain at 1.2% is smaller than that of the substrate. Consequently the presence of a brittle surface layer reduces the toughness of the product (see Section 8.3.3). This is less of a problem for spectacle lenses than it is for motorcycle visors, which are designed to cope with 145 km h^{-1} impacts of a 7 mm ball bearing.

10.4.3 Fibre optics

Fibre optics are the equivalent of pipes for light in that there is total internal reflection at the fibre/coating interface. Fig. 10.17 shows the propagation of a ray down the fibre. The refractive index of the fibre n_f must be less than that of the coat n_c so that the angle of incidence γ at the interface is greater than the critical angle θ_c, given by applying Snell's law

$$n_f \sin\theta_c = n_c \tag{10.26}$$

The amount of light that can enter the flat end of the fibre and be transmitted along it by total internal reflection is determined by the semi-acceptance angle α, given by applying Snell's law at the end face of the fibre

$$\sin\alpha = n_f \sin\beta \tag{10.27}$$

Assuming that there is a uniformly bright plane emitter at a fixed distance from the end of the fibre, the light gathering power of the fibre P is proportional to the square of $\sin\alpha_c$ and to the cross sectional area of the fibre. By equation (10.27) this gives

$$P \propto n_f^2 \sin^2\beta_c$$

and since $\beta + \gamma = 90°$ use of equation (10.26) gives

$$P \propto n_f^2 - n_c^2 \tag{10.28}$$

The right hand side of equation (10.28) is defined as the square of the numerical aperture of the fibre. The materials selection problem with fibre optics is to coat the fibre with a low refractive index that is durable and which protects the fibre. It might be thought that uncoated fibres would work well, but it would then be possible for light to pass between touching fibres, and the damaged caused by contact would cause losses.

There are several types of fibre optics and the requirements on losses determines the type of material used. Table 10.7 shows that the long distance telecommunication application has very stringent requirements on the transmission losses. These can only be met if the light travels axially down the fibre as a single mode wave (similar to a waveguide for cm wavelength radar waves). Fig. 10.18 compares the losses of the high purity silica fibres that are

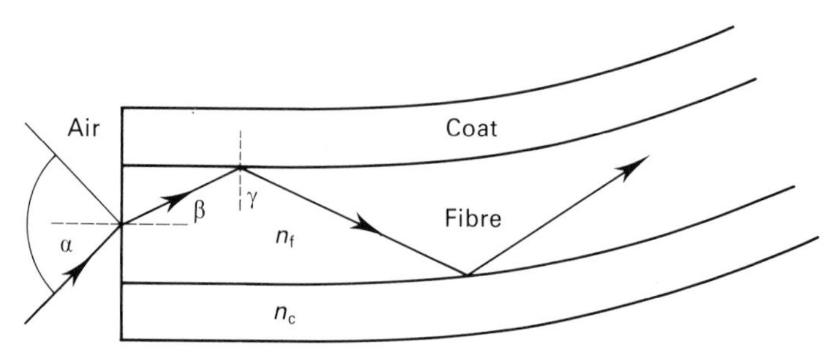

Fig. 10.17 Ray diagram for light passing down a fibre optic cable, showing the limiting ray that just undergoes total internal reflection. $n_f = 1.65$, $n_c = 1.50$, numerical aperture = 0.69, limiting $\alpha = 43°$

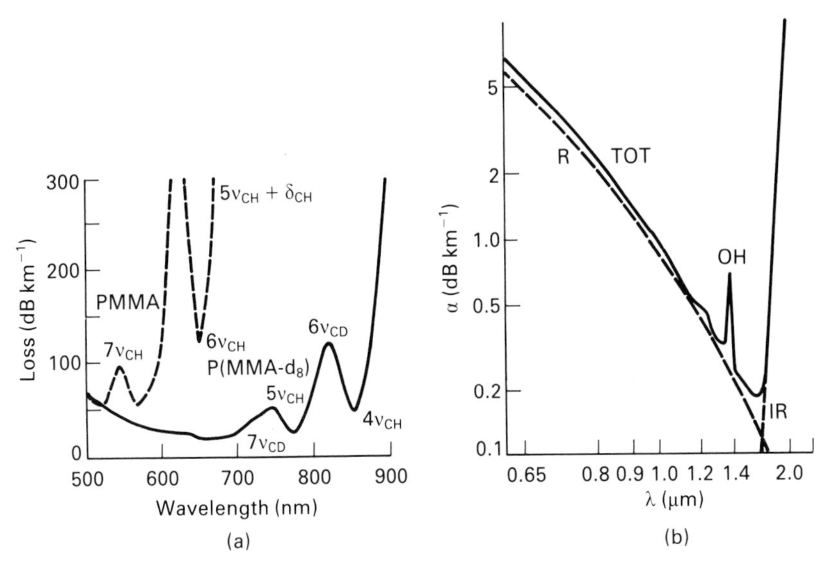

Fig. 10.18 Transmission losses versus wavelength for (a) PMMA and deuterated PMMA (from Kaino and Katayama, *Polym. Eng. Sci.*, 1989, **29**, 1209 and (b) pure silica single mode fibre. R is the Rayleigh scattering, IR the infrared absorption and OH the hydroxyl absorption (from *Philips Tech. J.*, 1989, **44**, 245)

used with that of ordinary and deuterated PMMA. The loss peaks are due to various molecular vibrations. In the PMMA these are harmonics of the C—H bond vibration, and the absorption becomes very strong in the infrared region. In the silica fibre there is a loss peak at 1.4 μm due to absorbed hydroxyl groups. There is a background effect of Rayleigh scattering, due to variations in the density of the material, which decreases as λ^{-4}. This explains why the telecommunications fibres operate with laser light at 1.30 μm or 0.85 μm, where the Rayleigh scattering is a minimum. Plastics are used as the outer protection of telecommunication fibres to prevent damage to the glass coating and to minimise bending of the fibres. A rubber interlayer is used between the outside of the 250 μm diameter glass and the 900 μm diameter UV-cured glassy polyacrylate secondary coating. The transmission losses of polymer cored fibres have been reduced by using fluorinated structures and avoiding the C—H bond. However the lower limit to the transmission loss is thought to be 5 dB km^{-1} at a wavelength of 0.65 μm.

The use of polymer fibres is preferred where flexibility is important. This could be, e.g., for supermarket bar-code readers, or endoscopes for observation in difficult-to-reach places. The medical use of endoscopes is growing since no tissue needs to be cut to reach the point of observation, and various cutting tools can be attached to the endoscope. Flexibility is achieved by having many fibres of small diameter—the theory of this is dealt with in Section 11.2.1. The PPMA fibres can be bent to much higher strains than could a silica fibre.

Table 10.7 Types of fibre optic system

Application	Type of transmission	Losses (dB km^{-1})	Core/coat diameter (μm)
Telecommunications	Single-mode wave	<0.4	8/125
Local area networks	Multi-mode waves	~0.6	50/125
Endoscopes	Ray	>500	250

10.5 THERMAL BARRIERS

The thermal conductivity of solid polymers is always within a factor of two of $0.3\ W\ m^{-1}\ K^{-1}$, with the higher values occurring in highly crystalline polymers. Thermal barrier materials have become increasingly important as the price of energy has risen, which has led to the development of low thermal conductivity building materials. If these materials can at the same time act as a barrier to water, and have some mechanical strength, so much the better. We shall consider the development of polyurethane foams and polystyrene foams for building applications.

In a closed cell foam there are a number of mechanisms that contribute to the overall thermal conductivity. The main ones are the thermal conductivity of the polymeric cell walls, the conductivity of the gas and convection and radiation in the cells. For foams of density $30\ kg\ m^{-3}$ the cell wall contribution is small. Convection inside the cells makes a negligible contribution for cell diameters smaller than 10 mm, and the radiation contribution is predicted to be linearly proportional to the cell size. Fig. 10.19 shows the effect of varying the cell size of polyurethane foams; there is clearly an advantage in minimising the cell size. With polystyrene foams the cells are rarely larger than 0.1 mm so radiation is expected to play a minimal part in the conductivity of the foam. The gas contribution can be estimated by following the changes in thermal conductivity with time. The fluorocarbon gas $CFCl_3$ used with polyurethane foam has a negligibly small diffusion coefficient through polyurethane. However, this is not true for the CO_2 gas initially present, or for oxygen and nitrogen which diffuse in from the air. Air has a higher thermal conductivity ($0.024\ W\ m^{-1}\ K^{-1}$) than the fluorocarbon gas ($0.009\ W\ m^{-1}\ K^{-1}$). Replacements of CFC gases with pentane or CO_2 to protect the ozone layer will have to take this into account.

Fig. 10.20 shows the change in the thermal conductivity value with time; the increase is due to the ingress of air and an increase in the total pressure of gas in the cells. The gas conductivity clearly dominates the total response. Heat loss calculations for buildings must be based on the long term value. Since polyurethane foam is invariably used with facings (paper, glass fibre, plasterboard, glass fibre reinforced concrete, etc.), the overall U value of the product must be calculated (in watts per m^2 area per degree temperature differential). This can be calculated from the thicknesses L_i and conductivities k_i of each layer using

$$\frac{1}{U} = \frac{A(T_i - T_0)}{q} = h_i + \frac{k_1}{L_1} + \frac{k_2}{L_2} + \ldots + h_0 \qquad (10.29)$$

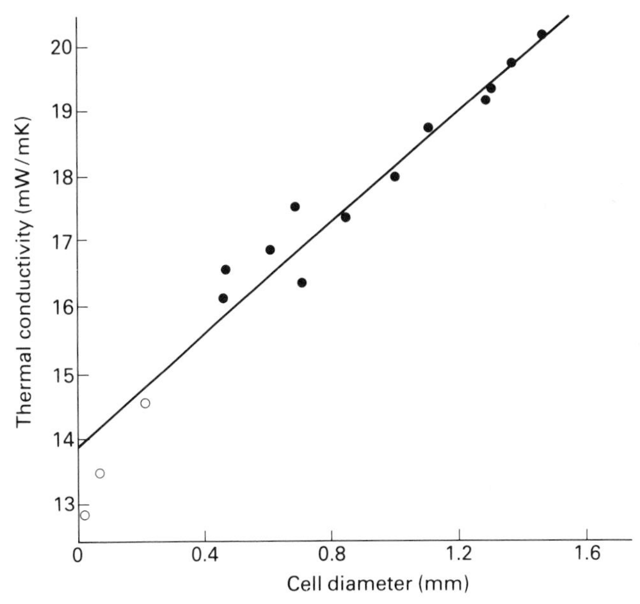

Fig. 10.19 Thermal conductivity of a polyurethane foam and of a fixed density versus the mean cell size (from J. M. Buist (ed.), *Developments in Polyurethanes*, Elsevier Applied Science, 1978)

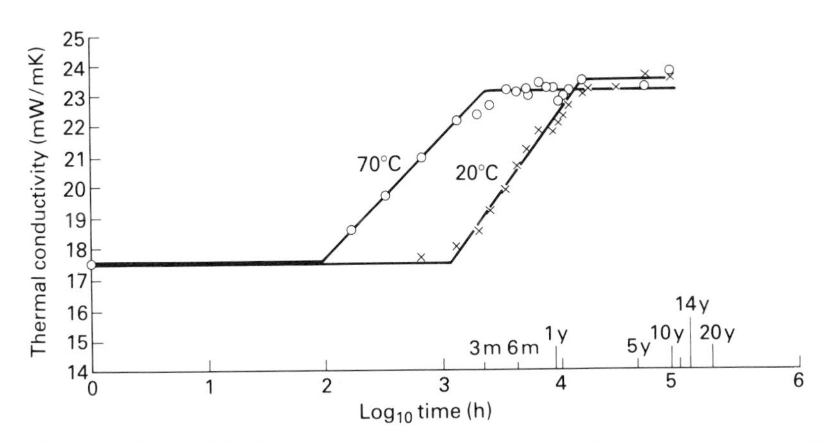

Fig. 10.20 Variation of the thermal conductivity of polyurethane foam with storage time (from J. M. Buist (ed.), *Developments in Polyurethanes*, Elsevier Applied Science, 1978)

The heat transfer coefficients h_0 at the outside is probably high enough when there is forced convection to contribute nothing to the overall U value. Current UK building regulations call for a maximum U value of $0.6\ \mathrm{W\,m^{-2}\,K^{-1}}$ for walls. This can be achieved using a 25 mm layer of polyurethane foam of density 30 to $35\ \mathrm{kg\,m^{-3}}$. At this level of insulation the losses from windows become significant—a single glazed wooden frame window has $U \cong 5\ \mathrm{W\,m^{-2}\,K^{-1}}$, whereas a double glazed PVC framed window has $U \cong 2.5\ \mathrm{W\,m^{-2}\,K^{-1}}$.

Polyurethane foam forms a good adhesive bond to most surface layers, and it has a relatively high Young's modulus. Consequently the sandwich structure performs efficiently in terms of bending stiffness per unit panel mass (Chapter 3). This has allowed the development of lightweight panels for use in roofs and walls. If thermal insulation is of paramount importance, as in cold stores, the foam thickness can be increased to 125 mm.

Polystyrene foam has somewhat different applications because of its physical form (expanded beads) and different properties (a much higher permeability to water and less effective adhesion to facing materials). The expansion gases pentane and steam escape fairly rapidly from the foam, so the thermal conductivity of the foam filled with air is about twice that of the best polyurethane foam—a 50 mm thick slab of foam has a U value of 0.5 to $0.6\ \mathrm{W\,m^{-2}\,K^{-1}}$. The bead-like form of the foam allows it to be mixed directly with concrete to form a low density ($850\ \mathrm{kg\,m^{-3}}$) and low thermal conductivity ($0.13\ \mathrm{W\,m^{-1}\,K^{-1}}$) concrete. The compressive design strength of $4\ \mathrm{MN\,m^{-2}}$ is still adequate because the mortar matrix takes the majority of the load so this expanded polystyrene concrete can be used to make thermally insulating blocks and walls. Loose polystyrene beads have been used to fill the cavities between the inner and outer leaves of house walls. If they are loose they are liable to settle or escape, so the beads are bonded together with adhesive. The water barrier properties of the bonded bead filling are inferior to the original air gap, with a 45% increase in water transfer.

11

Electrical properties

11.1 VOLUME AND SURFACE RESISTIVITY

The main reason for using plastics in electrical applications is that they are easily moulded insulators. With suitable additives they can be made to conduct electricity, but they cannot compete with metals as low cost conductors for long distances. The application determines whether resistivity is the key design property; in many cases the mechanical or thermal properties can be more important in determining the thickness of the insulating layer required.

The concept of a volume resistivity ρ implies that the resistance R of a bar of length L and constant cross section A is given by

$$R = \rho L/A \tag{11.1}$$

and also that when a constant voltage is applied there will be a constant current. However, for polymers, that have resistivities in the range 10^7 to $10^{16}\ \Omega$ m, the currents are extremely small, and they decay with time after a voltage step is applied. Consequently a time must also be specified at which the current is measured. Often it is 1 min for experimental convenience. The composition of the surface layers of the polymer may differ from that of the bulk, owing to the migration of organic anti-static additives. The surface conductivity of a connector can then have a dominant effect on the insulation resistance between two metal conductors. The concept of surface resistivity implies that there is a surface layer of thickness t and volume resistivity ρ, on top of an insulating substrate. The surface resistivity ρ_s is given by

$$\rho_s = \rho/t \tag{11.2}$$

It is also equal to the resistance in ohms between the opposite sides of a square of any size on the surface of the product.

The volume resistivity of polymers decreases with increasing temperature in a way typical of semiconductors. If the logarithm of the resistivity is plotted against the reciprocal of the absolute temperature, a straight line results (Fig. 11.1) except in cases where the polymer undergoes a phase change. Since the form of the graph is the same as that used by Arrhenius for thermally activated chemical reactions, it is tempting to inerpret the slope of

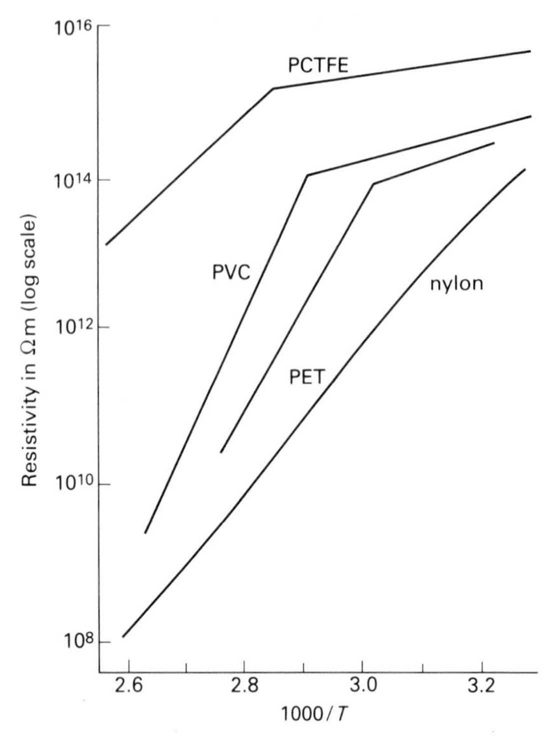

Fig. 11.1 Log of direct current resistivity versus reciprocal absolute temperature for polyvinyl chloride, polyethylene terephthalate, nylon 6,6 and polychlorotrifluoroethylene

the graph as the activation energy of the conduction process. If this is done the 'activation energy' for amorphous polymers above T_g is found to be between 0.2 and 0.5 eV. It is rarely known whether the charge carrier is an electron, or one of many possible ions, so the activation energy cannot easily be interpreted. The relationship between the resistivity and the concentration n of carriers of mobility μ is

$$1/\rho = qn\mu \tag{11.3}$$

The mobility is the ratio of the carrier velocity to the electric field, and is of the order of $10^{-9}\ \mathrm{m^2\ s^{-1}\ V^{-1}}$ in polymers. Consequently there only needs to be one carrier with an electronic charge q of 1.60×10^{-19} C for every 10^9 monomer units for the polymer to have a resistivity of $10^9\ \Omega$ m. Ionic impurity levels of this magnitude would have no effect on other physical properties.

Against the general background of high resistivity values a few polymers emerge with lower than usual resistivities. Polyamides have hydrogen bonds that lie in parallel planes in the crystal structure, so the resistivity is a factor of 100 smaller than that of non H-bonded polymers. At temperatures above 120 °C at least half of the conduction in nylon 6,6 has been shown to be due to protonic carriers, since hydrogen is liberated at one electrode. Nevertheless the resistivity is still high.

11.2 THE SPECTRUM FROM INSULATOR TO CONDUCTOR

The order of treatment of the insulation or conduction properties is that of increasing conductivity. This means that the insulation applications will be dealt with first, and the research on conducting polymers last. The divisions between the areas are rather diffuse. Fig. 11.2 shows the areas occupied by various materials on a conductivity–temperature map, and the main classification of polymers as low temperature insulators. It does not show the form of the materials, so it should be emphasised that the semiconducting polymers are thin films.

11.2.1 Low voltage insulation

It is convenient to divide applications into low voltage and high voltage areas. For domestic and electronic applications the voltages do not exceed 500 V, and the electric strength of the insulation is not critical. The function of the insulation is to separate two more conductors, and to provide mechanical support. The thickness of the exterior insulation (Fig. 11.3) is determined more by the safety aspects of abrasion and wear-and-tear than by the insulation values. The cross sectional area of the copper conductors determines the current carrying capacity of the cable, because the ohmic heating of the wire must be limited. The temperature rating of the insulation, 60 °C for plasticised PVC and up to 120 °C for crosslinked polyethylene, must not be exceeded (a heat flow analysis is given at the end of the next section).

The bending stiffness of a single conducting wire of diameter D can be reduced by using a cable which has n unbonded conductors of diameter d. Since the same cross sectional area of conductor is used in both cases $n = (D/d)^2$, and the total second moment of area I of the multicore cable is

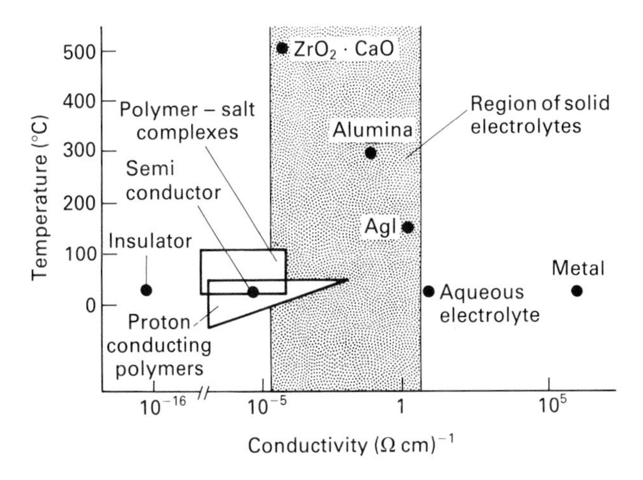

Fig. 11.2 The conductivity of solids and the temperature range of use (from J. M. Margolis, *Conducting Polymers and Plastics*, Chapman and Hall, 1989)

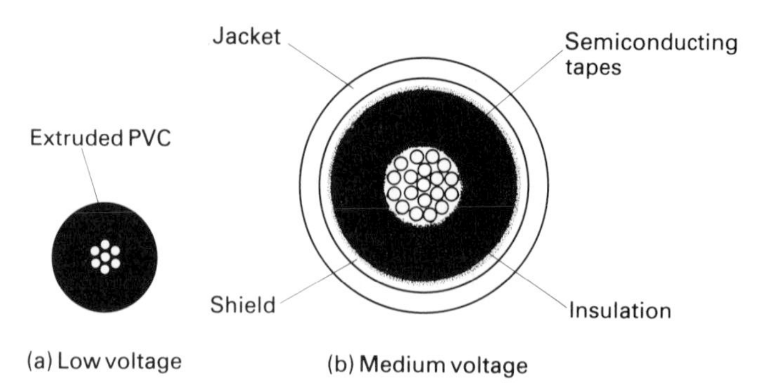

Fig. 11.3 Construction of a low voltage and a medium voltage cable. In the latter the semiconducting layers equalise the electric field and prevent corona discharges

given by

$$I = n\frac{\pi d^4}{64} = \frac{1}{n}\frac{\pi D^4}{64} \tag{11.4}$$

Therefore the bending stiffness of the cable is a factor of n smaller than the equivalent single wire. The insulating polymer sheath often has a diameter that is 3 times that of the cable, so although the polymer modulus may only be 0.1% of that of copper ($E = 100$ GN m^{-2}), the polymer can add significantly to the overall bending stiffness.

The temperature rating of the insulation may be limited by thermal degradation or oxidation (Chapter 9), which will cause the insulation to crack when the cable is bent. Alternatively the insulation may soften so much that it is penetrated by sharp objects. There are empirical tests of insulation failure, such as a chisel with a 0.125 mm tip radius pressed across the cable with a 3.5 N force. If electrical contact between the chisel and the conductor occurs within 10 min, the polymer fails the test. The test is repeated at higher temperatures until the insulation fails. The temperature rating of the insulation can be increased by crosslinking the polymer, so that it no longer melts on heating above T_m or T_g. Polyethylene is usually crosslinked by incorporating 1 to 2% of a peroxide which reacts, not when the polymer is extrusion coated on to the cable, but when the coated cable is treated with superheated steam in a long tube. Crosslinking increases the temperature rating from 75 °C to about 120 °C, and makes it possible for the insulation to survive momentary contact with a soldering iron.

Multiple pin edge connectors are used extensively in electronic circuits. The thermoplastic must position the many pins accurately for correct mating of the connector, and they must stand up to the temperature peaks when a soldered connection (at 300 °C) is made to one end of the pin. Glass fibre filled polycarbonate or polybutylene terephthalate is used, because it can be moulded accurately, and it survives ageing tests at 125 °C. When mains or higher voltages are involved, tracking can occur. Surface moisture absorption or ionic contamination causes the surface resistivity to be relatively low. The leakage current heats and dries the surface. If narrow dry bands form on the surface, due to the surface layer contracting, they have a higher resistance and sparks can occur across them. The sparks heat the polymer surface above 500 °C and carbonaceous degradation products form, eventually causing flash-over. Laboratory tracking tests involve dropping ammonium chloride solution on to the surface between electrodes, or exposing a sodium chloride contaminated surface to an artifical fog. PVC, which forms conducting conjugated structures on heating, performs badly in this test, polypropylene performs fairly well, whereaw PTFE with its excellent thermal stability, is outstanding.

11.2.2 High voltage insulation

The resistance of polymers to high voltages has been tested using many different test geometries (Fig. 11.4). There is no single electric strength value, independent of the thickness of the sample or the test geometry. This is because the interface between the conducting contact and the polymer plays an important part, as does the arcing in any air gaps. The electric strength is dependent on the size of foreign particles or voids in the polymer that happen to be in the highly stressed region. Consequently the results of many repeated determinations follow Weibull statistics (Fig. 11.4). This skewed distribution of strengths is found to apply to the tensile strengths of brittle ceramic materials, and to other cases where the weakest part of the product causes failure. The cumulative probability of failure $F(x)$ when an electric field strength x is applied is given by

$$F(x) = 1 - \exp[-(x/x_0)^b] \tag{11.5}$$

where x_0 is the value at which 63.2% of the failures have occurred, and the constant b describes the breadth of the distribution. Fig. 11.4 shows that for d.c. fields and low density polyethylene the cylindrical electrode geometry gives the highest x_0 values and the lowest distribution width b. The absolute electric strengths are high at 20 °C but they fall off rapidly at higher temperatures. These short term strengths are of no use for designing high voltage cables that are to be buried in the ground for many years. There is an exact parallel with the creep rupture phenomenon (Chapter 7) in that the time to failure increases as the electric field is reduced.

The construction of high voltage cables is shown in Fig. 11.3. The conductors are first extrusion coated with a layer of carbon filled polymer. This semiconducting layer equalises the electric field across the insulator and prevents corona damage. The main insulating layer consists of radiation or peroxide crosslinked polyethylene, or of ethylene–propylene rubber, and it is coated with a further semiconducting layer. The electric strength of such cables falls with time (Fig. 11.5) and the overall level is affected by defects in the polymer. Apart from metallic wear particles from the extruder screw or fibrous contamination that passes the filter screen in the extruder, there are the voids that can form in crosslinking. In particular the steam curing process can introduce dissolved water, that becomes supersaturated on cooling and nucleates as water filled voids. The breakdown process is associated with the growth of *trees*, named after the structures grown in laboratory tests from a

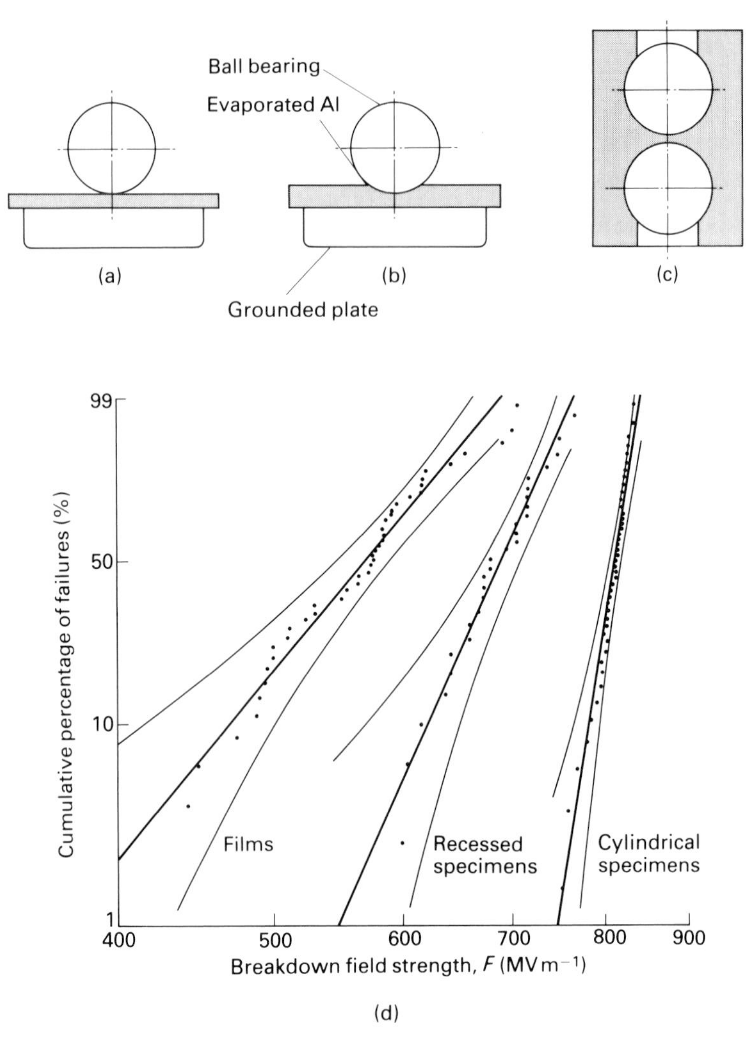

Fig. 11.4 Geometries for the determination of electric strength: (a) sphere on film; (b) sphere recessed into sheet; (c) cylinders embedded in plastic; and (d) the corresponding Weibull plot of the results of many tests on LDPE at 20 °C (from Seanor (ed.), *Electrical Properties of Polymers*, Academic Press, 1982)

charged metal needle tip (Fig. 11.6). Bow-tie shaped trees grow from spherical voids in cables. The void or needle tip acts as an electrical stress concentration which initiates the breakdown process. The breakdown processes are of two main types, electrical and electrochemical. In the former a corona discharge occurs in voids in the polymer, and as the hollow channels grow they become lined with decomposed polymer. The theory of such discharges predicts that as the spherical void diameter decreases the field strength required increases. Consequently if the polymer contains no voids larger than 25 μm, it should be possible to avoid such a breakdown. Electrochemical tree growth is the electrical equivalent of environmental

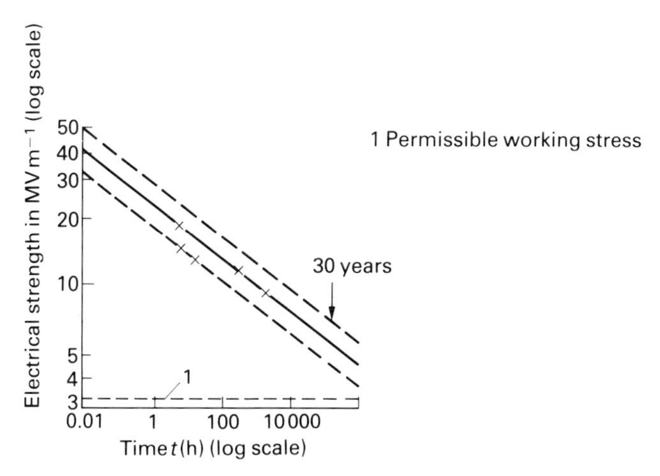

Fig. 11.5 Variation of the electric strength of polyethylene with the duration of the voltage application (from S. Y. King and N. A. Halfter, *Underground Power Cables*, Longman, 1982)

Fig. 11.6 Electrical tree grown from a needle tip at 10 kV in polyethylene (from Seanor (ed.), *Electrical Properties of Polymers*, Academic Press, 1982)

stress cracking (Chapter 9), happening at lower electric stresses than electrical tree growth. The mechanism of the slow growth of the water or fluid filled tree is not clear. One method of avoiding such tree growth is to use an impervious outer layer, such as an aluminium sheath, that prevents the slow diffusion of ground water into the polyethylene.

The design stresses for high voltage cable are only ~8 kV mm^{-1}. This is partly to avoid excessive temperature rises in the dielectric, which would promote the growth of trees. The conductor is surrounded by a series of materials with thermal conductivities k_i. If the materials are in the form of concentric layers, of inner and outer radii r_i and R_i, the temperature difference from the conductor to the outermost layer is

$$T_c - T_0 = \frac{W}{2\pi} \sum_i \frac{1}{k_i} \ln\left(\frac{R_i}{r_i}\right) \tag{11.6}$$

where W is the power dissipated in the conductor in watts per m length. The thermal conductivity of polyethylene at 0.2 W m^{-1} K^{-1} is low compared with sand bedding for the cable (0.5 W m^{-1} K^{-1} when dry), and the polyethylene layer has $R_i \cong 2r_i$. Consequently the dielectric is responsible for a significant fraction of the overall thermal resistance, and it has to operate at temperatures well above those of the soil.

11.2.3 Static electrification

Static charging is a particular problem with polymeric products, particularly if there is rubbing between dissimilar polymers. Polymer surfaces will charge just by contact with metals or other polymers, and it is possible to derive a work function, such that the polymer of a pair with the lower work function will acquire a positive charge after contact. There is a reasonable correlation between the order established from work functions, and those in the *triboelectric series*. The latter is found by various experiments involving the rubbing of, for example, textile fibres across each other. Table 11.1 compares the two lists, with the polymers that acquire positive charges at the top. Fairly large charge densities can build up by repeated contacts. For example the driving of conveyor belt rollers, that have nylon end mouldings, by a polyurethane rubber belt caused a voltage build-up to a level that operators received spark discharges. Similarly in a dry environment of less than 40% relative humidity, merely walking on a polypropylene carpet, or removing a sweater can cause the body to charge up to a voltage of 4 kV. As the capacitance of the body is around 200 pF the energy available for a spark discharge is 0.8 mJ. This is more than four times that necessary to ignite a petrol–air mixture. The problems are made worse by the widespread use of moulded polyurethane soles on shoes as this material has a high resistivity. If an LDPE bag is picked up from a surface under 15% humidity conditions the voltage generated (20 kV) could easily destroy a microchip, so additives are essential for packaging microelectronics components. Electric charges attract fine dust particles from the air which detracts from the appearance of plastics;

an examination of the underside of a polypropylene stacking chair will illustrate this.

Table 11.1 The triboelectric series and work functions

Polymer	Triboelectric series	Work function
Wool	Positive	4.08 ± 0.06
Nylon 6,6		
Cotton		
PMMA		
PET		4.25 ± 0.10
PAN		
PVC		4.85 ± 0.2
PE		
PTFE	Negative	4.26 ± 0.05

Most antistatic additives are polar waxes, which migrate to the surface of the polymer, and adsorb water if the humidity is not too low. A charge decay half time of 0.1 s or less is adequate protection against static electrification, and to achieve this the surface resistivity must be less than 3×10^{11} Ω/square. Such surface films are prone to be worn away, and they will not function adequately when the relative humidity is less than 15%. They can only be used in polymers such as polyethylene and polystyrene where the melt process temperatures are not excessive. The use of conducting fillers (Section 11.2.4) in the polymer is a more permanent solution to static electrification.

11.2.4 Electromagnetic screening of plastic mouldings

Plastics mouldings used as the housings for electronic equipment or micro-computers have the disadvantage that they do not prevent electromagnetic interference (EMI), because they are non-conductors. Radiation, such as from the thyristor switching of industrial heaters, can interfere with the operation of microprocessors, since the induced voltages can be large enough (>5 V) to be treated as signals by the computer. Conversely radiation escapes from the housing and may interfere with other equipment. These problems did not arise when sheet metal housings were used. Standards are being introduced for screening, for instance in Japan since 1988 a voluntary code calls for a 40 dB reduction of radiation in the 30 to 230 MHz frequency band, and a 47 dB reduction for 230–1000 MHz. The dB is a logarithmic measure of attenuation, and a 10 dB reduction represents a tenfold reduction of field strength. US regulations for emissions from commercial computers calls for electric field strength to be less than 30 $\mu V\ m^{-1}$ at a distance of 30 m from the computer.

There are two main mechanisms of attenuation of the signal, reflection from the surfaces of the material, and absorption inside the material. Unlike the scattering of light considered in Section 10.4.2, scattering of the longer wave length radiation is insignificant. The absorption A depends on the conductor thickness t and the frequency f in Hz according to

$$A\ (\mathrm{dB}) = 131.4t\sqrt{f\sigma_r\mu_r} \tag{11.7}$$

where σ_r is the conductivity relative to that of copper and μ_r is the permeability relative to copper. Table 11.2 shows that there is not much to choose between the non-magnetic metals. Magnetic metals have higher losses but corrosion problems prevent their use. The reflection loss R is proportional to

$$R\ (\mathrm{dB}) \propto 20 \log \sqrt{\frac{\sigma_r}{\mu_r}} \tag{11.8}$$

and Table 11.2 shows that it is much greater for the non-magnetic metals. There are various strategies of achieving the requisite attenuation, shown in Table 11.3. The most common method is to spray the inside of the moulding with a nickel containing paint. The disadvantage of this solution is the extra process step.

Table 11.2 EMI shielding properties of metals relative to copper

Metal	Specific conductivity, σ_r	Specific permeability, μ_r	Absorption loss, $\sqrt{\sigma_r\mu_r}$	Reflection loss (equation 11.8)
Copper	1	1	1	0
Aluminium foil	0.53	1	0.73	−2.8
Zinc	0.31	1	0.57	−4.9
Pure iron	0.17	5000	29.2	−44

Table 11.3 EMI shielding methods

Method	Thickness (μm)	Shield efficiency (dB)
Zn spraying	~70	60–90
Ni containing paint	~50	40–60
Al vacuum plating	2–5	40–70
Conducting filler	Dispersed in polymer	40–60

An alternative is to increase the conductivity of the polymer by incorporating conducting fillers, the most common being carbon black, and conducting fibres. The resistivity changes with the filler content in a non-linear manner, and when a conducting pathway is established across the polymer the resistivity drops dramatically (Fig. 11.7). It is essential to minimise the filler content, because most fillers bond weakly to polymeric matrices, and the toughness and tensile strength fall with increasing filler content. Fibres of a high aspect ratio are the most effective at creating conducting networks. Both brass and stainless-steel fibres have been used, but they are expensive, and cause wear in the moulds.

Fig. 11.8 shows the contributions to the EMI shielding as a function of the resistivity of the filled polymer. To reach a shielding of 40 dB with a 3 mm thick moulding requires the resistivity to be less than $10^{-2}\ \Omega$ m, and most of the screening is by reflection.

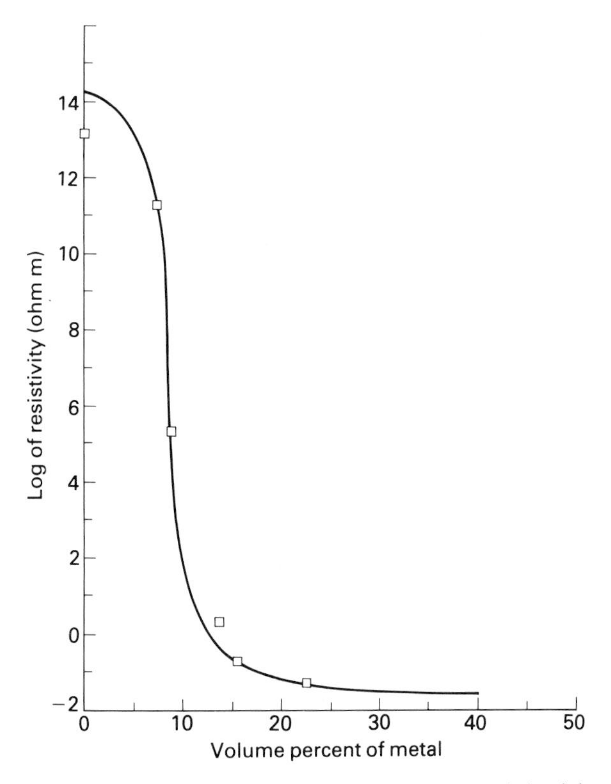

Fig. 11.7 Resistivity of polyethylene versus the volume percentage of aluminium fibres of 12:1 aspect ratio (from Seymour (ed.), *Conductive Polymers*, Plenum Press, 1981)

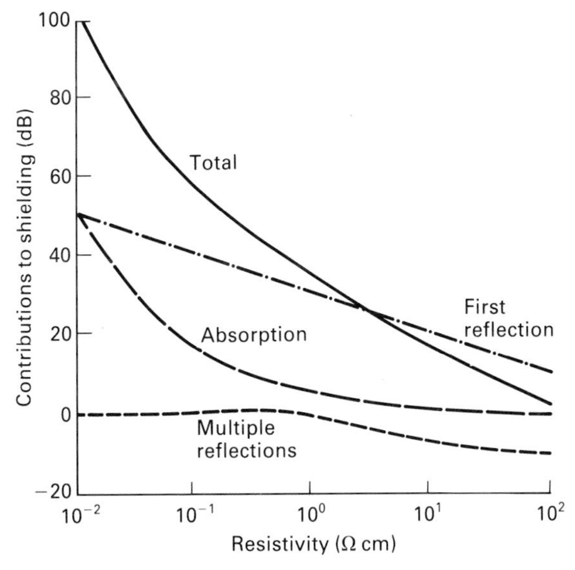

Fig. 11.8 Contributions of reflection and absorption to the EM shielding efficiency of a 3 mm thick filled moulding, at a frequency of 100 MHz (from Möbius, *Kunststoffe*, 1978, **78**, 31)

11.2.5 Conducting polymers

It seems illogical to go to great lengths to develop polymers with medium conductivities when there is a ready supply of metals of high conductivity. There is an academic interest in the mechanism of conductivity in one-dimensional conductors, which is different to the mechanism in metals. Few applications have been developed by 1992, but there is potential in these materials.

One type of conducting polymer is based on the oxidation of conjugated backbone structures. The simplest structure is that of polyacetylene (Fig. 11.9). The material is highly crystalline and the chain backbone is all-trans. In the neutral state it is not a conductor, so it must be oxidised to remove electrons from some of the π bonds. The radical ion that is formed on the chain is delocalised so can move along the chain, whereas the balancing anion from the oxidation process is unable to move. There are defects in the polyacetylene structure (missing double bonds) which also give rise to radical ions when they are oxidised. The charges must hop from chain to chain to produce bulk conductivity. Polyacetylene cannot be melted or dissolved, so it is impossible to process into useful forms. It is unstable on exposure to air, the conjugated structure being destroyed. Hence the search has been for more tractible and stable polymers.

Polypyrrole and polythiophene are ring containing conjugated structures, that can be polymerised as thin films by an electrochemical method (Fig. 11.10). The highest quality films are produced when there is an inert

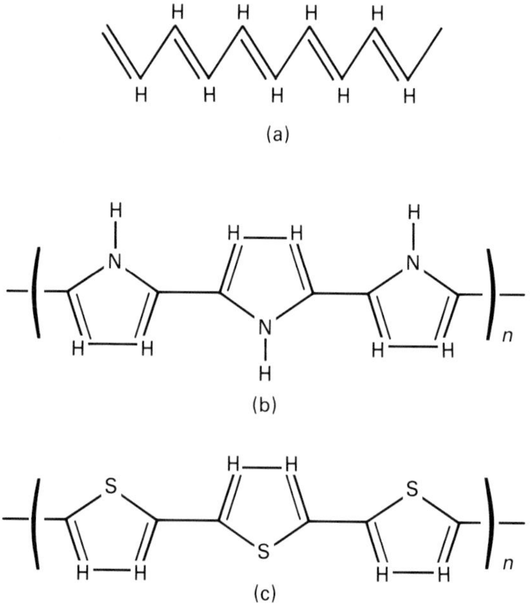

Fig. 11.9 Structures of (a) polyacetylene, (b) polypyrone and (c) polythiophene. The conjugated bonds between the C atoms are shown but not the C atoms themselves

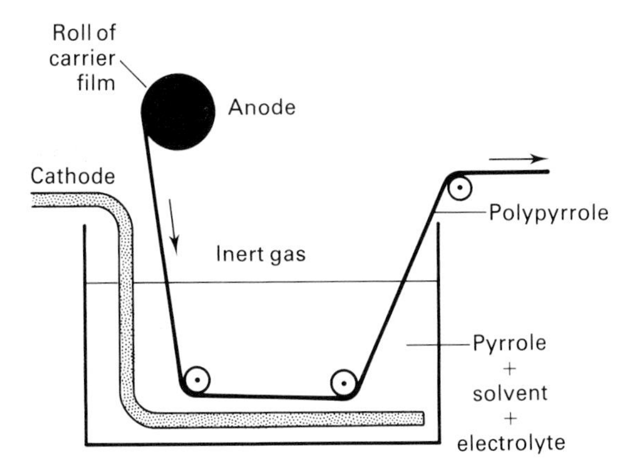

Fig. 11.10 Continuous belt process for the production of polypyrone films in an electrochemical cell

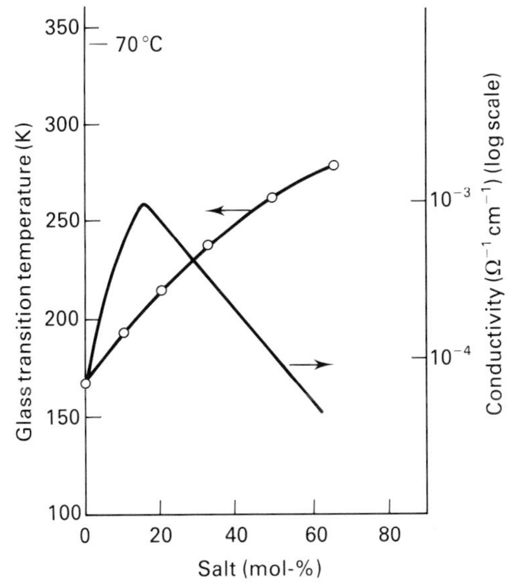

Fig. 11.11 Variation of conductivity and T_g with salt content for silver triflate in MEEP (from Margolis, *Conductive Polymers and Plastics*, Chapman and Hall, 1989)

atmosphere, the pyrrole concentrations is 0.01 to 0.2 molar, and there are an unreactive solvent, an electrolyte and a small amount (2%) of water present. Films of thickness 30 to 150 μm or coatings can be made and their resistivity ranges from 10^2 to 10^{-4} Ω m. This is still a long way from the resistivity of pure copper 1.7×10^{-8} Ω m, but it is adequate for many electrochemical applications.

The second main type of semiconducting polymer is solid electrolytes. These depend on the presence of mobile ionic salts to produce the conductivity. A typical system is polyether plus an alkali metal salt. Polyethylene oxide plus $LiClO_4$ has a resistivity of 10^2 Ω m at 110 °C. Although the polymer has a T_g of −60 °C it is still necessary to heat the solid well above ambient temperatures to achieve adequate conductivity. This is partly because the polymer–salt complex has a higher as the salt concentration increases. The temperature dependence of the conductivity can be described by

$$\sigma = \frac{A}{\sqrt{T}} \exp\left(\frac{B}{T - T_0}\right) \tag{11.9}$$

where A and B are constants and T_0 is close to the glass transition of the polymer–salt complex. The variation of conductivity and T_g with the salt content is shown in Fig. 11.11. The initial increase in conductivity is due to the increased numbers of mobile ions, but eventually the T_g of the mixture increases too close to the measurement temperature, and the conductivity falls according to equation (11.9). The use of these systems for batteries will be considered in Section 11.5.

11.3 DIELECTRIC BEHAVIOUR

11.3.1 Dielectric constants

When an alternating voltage is applied across a polymer its dielectric properties determine the current. The dielectric constant (or relative permittivity) ε^* is defined in terms of the electric field **E** vector and the electric displacement **D** vector by

$$\mathbf{D} = \varepsilon_0 \varepsilon^* \mathbf{E} \tag{11.10}$$

where the permittivity of free space $\varepsilon_0 = 8.85 \times 10^{-12}$ F m^{-1}. A more practical definition is in terms of a parallel plate capacitor with a polymeric layer between the plates (Fig. 11.12a). The capacitance C is given by

$$C = 4\pi A \varepsilon_0 \varepsilon / d \tag{11.11}$$

where A is the plate area and d the dielectric thickness. The applied voltage V varies sinusoidally with time

$$V = V_0 \exp(i\omega t) \tag{11.12}$$

V can be represented in an Argand diagram (Fig. 11.12b) by a vector of length V_0 that is rotating at an angular frequency ω. For an ideal capacitor of capacitance C the current I flowing is

$$I = C\frac{\mathrm{d}V}{\mathrm{d}t} = i\omega CV \tag{11.13}$$

so the current is represented in the Argand diagram as a vector rotating 90° ahead of the voltage vector. For a real dielectric the current, leading the voltage by an angle 90° − δ, is given by

$$I = i\omega\varepsilon^* C_0 V \tag{11.14}$$

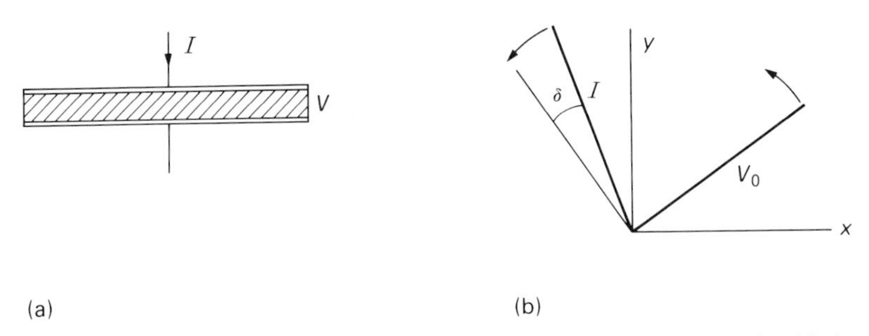

Fig. 11.12 (a) A capacitor with a polymeric layer between the plates. (b) Relationship between an alternating current i and the voltage V across a capacitor with a polymeric dielectric, represented on an Argand diagram

where C_0 is the capacitance with a vacuum between the capacitor plates. Consequently the dielectric constant ε^* is a complex number. It is an analogue of the dynamic compliance J^* in Section 6.4.1. We can write

$$\varepsilon^* = \varepsilon' - i\varepsilon'' \tag{11.15}$$

Data are usually presented in terms of the real part of the dielectric constant ε' as a function of frequency, and the ratio

$$\tan \delta = \varepsilon''/\varepsilon' \tag{11.16}$$

This is proportional to the ratio of the energy dissipated per cycle to the maximum energy stored (equation 6.21).

11.3.2 Polarisation loss processes

If there are chemical groups, free to move in the polymer, with permanent dipoles these will align with the electric field every half cycle, so long as the frequency is not too high. This *orientation polarisation* is expected in polar PVC where the C—Cl bond has a permanent dipole moment of 1.0 Debye (1 Debye $= 3.33 \times 10^{-30}$ C m). *Space charge polarisation* occurs in polymeric composites; charges accumulate at the interface of the two phases if the product of the dielectric constant and the resistivity $\varepsilon'\rho$ is not the same for both phases. This occurs both in rubber toughened polymers, and in polymers containing voids or inclusions.

In Debye's model of dielectric relaxation the polarisation process has a single relaxation time. The model has both an electrical circuit analogue and a viscoelastic model analogue (Fig. 11.13). In network theory terms the electrical circuit is the 'dual' of the mechanical model, because the voltages across the capacitor and resistor in series are added, whereas the forces on the spring and dashpot in parallel are added. We shall solve the mechanical analogue to illustrate the dielectric behaviour. The differential equation corresponding to the model in Fig. 11.13 is

$$e + \tau \frac{\mathrm{d}e}{\mathrm{d}t} = \left(\frac{1}{E_1} + \frac{1}{E_2}\right)\sigma + \frac{\tau}{E_2}\frac{\mathrm{d}\sigma}{\mathrm{d}t} \tag{11.17}$$

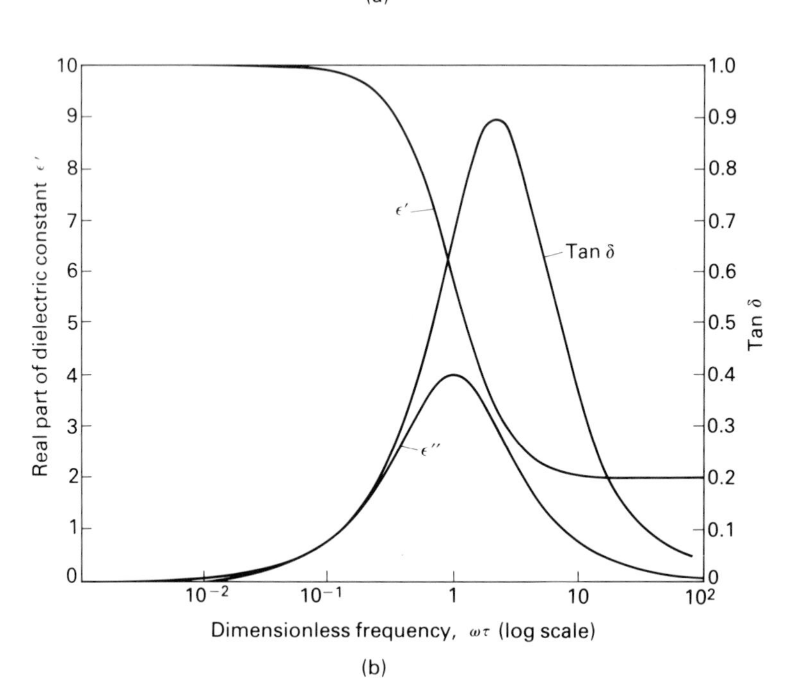

Fig. 11.13 (a) Spring and dashpot mechanical model and the equivalent electrical circuit that models a dielectric with a single relaxation time. (b) Predicted variation of the components of the dielectric constant with dimensionless frequency $\omega\tau$ for $\varepsilon_U = 10$ and $\varepsilon_R = 2$

where e is the strain, σ the stress and the retardation time $\tau = \eta/E_1$. Because the Voigt element is in series with a spring E_2 the compliances of the two can be added to give a complex compliance

$$J^* = \frac{1}{E_2} + \frac{1}{E_1(1+i\omega\tau)} \tag{11.18}$$

The dielectric equivalent of equation (11.18) can then be given in terms of the relaxed low frequency dielectric constant ε_R and the unrelaxed high frequency value ε_U as

$$\varepsilon^* = \varepsilon_U + \frac{\varepsilon_R - \varepsilon_U}{1+i\omega\tau} \tag{11.19}$$

Fig. 11.13b shows the variation of ε' and tan δ for such a model. Real polymers have a broader tan δ peak with a lower maximum value, indicating that there is a range of retardation times. Space charge relaxations occur at lower frequencies. If the damping peak of either of these processes coincides with the frequency of the electrical signal, then there will be a strong attenuation of the signal, or even significant heating effects. The electrical circuit in Fig. 11.13a can be used to model the signal attenuation and to explain why the d.c. resistivity of a polymer appears to change with time; when a constant voltage is applied there will be a polarisation current which decays as the retardation time is exceeded. There must, however, be a high resistance in parallel with the other elements to model the true d.c. resistivity.

11.3.3 High frequency insulation and capacitors

The dielectric for telecommunication cables needs to be non-polar to avoid orientation polarisation losses. Semi-crystalline polymers are used in preference to glassy polymers so that the cable can be bent without undue forces and without damaging the dielectric. Polyethylene is the most commonly used material for cost reasons. Fig. 11.14 shows the variation of tan δ with temperature at a 10 kHz frequency. The loss peaks are labelled as the α, β, γ, δ . . . peaks, starting at the highest temperature peak. The α loss peak at about 90 °C in HDPE is due to polar carbonyl groups (—C=O with a dipole moment of 2.3 Debye). These are introduced by degradation in melt processing (Chapter 9) and they can re-orient in the crystalline phase, some 40 °C below the melting point of the crystals. LDPE has a β peak at about 0 °C, due to the reorientation of carbonyl groups in the amorphous phase. Non-polar antioxidants can be used to minimise the oxidative degradation of polyethylene. This, and a reduction in the level of catalyst residues, has reduced the tan δ values below those in Fig. 11.14. As the electrical frequency is increased the loss peaks move to higher temperatures, so at the 30 MHz used in submarine cables the γ loss peak has moved to −28 °C, dominating the dielectric loss.

For mains frequency applications the use of a polar polymer for insulation is acceptable. Fig. 11.15 shows how the dielectric properties of PVC change with temperature for different amounts of diphenyl plasticiser. The imaginary

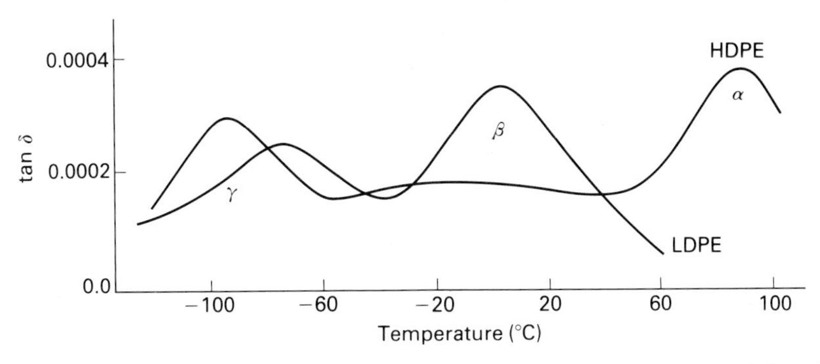

Fig. 11.14 Variation with temperature of the tan δ at 10 kHz of oxidised polyethylene (from M. E. Baird, *Electrical Properties of Polymeric Materials*, Plastics Institute, 1973)

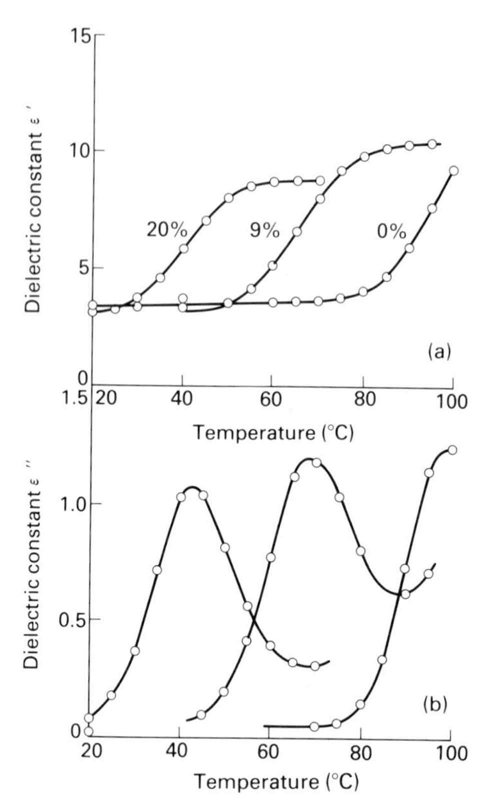

Fig. 11.15 Components of the dielectric constant of PVC at 60 Hz versus temperature, for (a) rigid PVC and (b) PVC plasticised with 9% and 20% diphenyl

part of the dielectric constant has a maximum value at the glass transition of the PVC. The amount of plasticiser used must not be too high otherwise there will be unacceptable losses at room temperature. On the other hand unplasticised PVC is too rigid to allow continual flexing of the cable.

Polymeric films have replaced paper as the dielectric medium in capacitors for high voltage or high frequency use. Equation (11.11) shows that the plate area A must be large and the dielectric thickness d must be small for the capacitance to be large. The former is achieved by rolling the plates up so the capacitor is cylindrical. The polymer must be drawn into a thin film; in this form it is inherently flexible so glassy polymers can be used. Polystyrene is non-polar, having a dipole moment of 0.3 Debye, and it has very low dielectric losses. Polyethylene terephthalate contains polar carbonyl groups, but can be drawn into film as thin as 1.5 μm, that can be handled without damage as tensile strength is 150 MN m^{-2}. The film is coated with a 15 nm thick vacuum-deposited metal layer, then slit into narrow tapes, and wound into capacitors. Neither type of capacitor can be used at temperatures above 75 °C. With polystyrene the oriented film begins to recover as the T_g of 100 °C is approached, and PET begins to crystallise above 100 °C (Fig. 11.16).

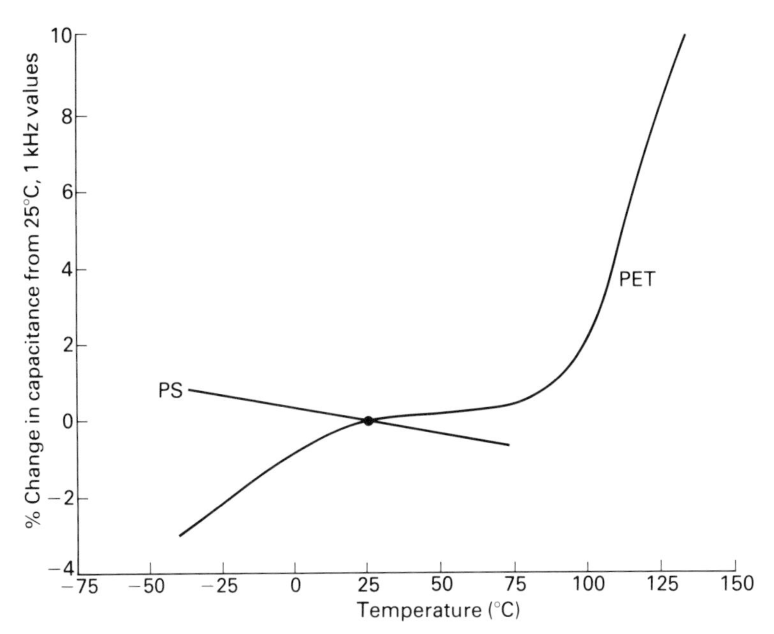

Fig. 11.16 Percentage change in capacitance with temperature for capacitors with polystyrene and polyethylene terephthalate dielectric (from P. Bruins (ed.), *Plastics for Electrical Insulation*, Wiley Interscience, 1968)

11.4 DEVICES

Currently there is a high level of research into the electrical uses of polymers, as this is a way of finding high value markets for polymers. Four product areas are described here to give an indication of the current applications. Most of the applications use the polymer in thin film form. It is the ability to process polymer into large areas of thin film, and the flexibility of these films, that allow the radical redesign of many electrical devices.

11.4.1 Film switches

The flexibility of polymer film allows it to replace mechanical switches in low voltage electronic equipment. Fig. 11.17 shows one design of a touch-sensitive switching panel. Films of polycarbonate of PET that are 125 to 250 μm thick are used both for the silk-screen printed outer layer, the two layers that carry the silver connections, and the spacer layer that has holes in it. The four layers are adhesively bonded together and positioned on a rigid base. Finger pressure causes the switch surface to bend elasticially through the cut out hole to close a circuit. Although these switches will operate for 10^7 or more operations, the only feedback to the operator is if there is a visual or aural signal elsewhere. The switches can be wiped clean and since the film is continuous it prevents the ingress of fluids or dirt into the electronics beneath.

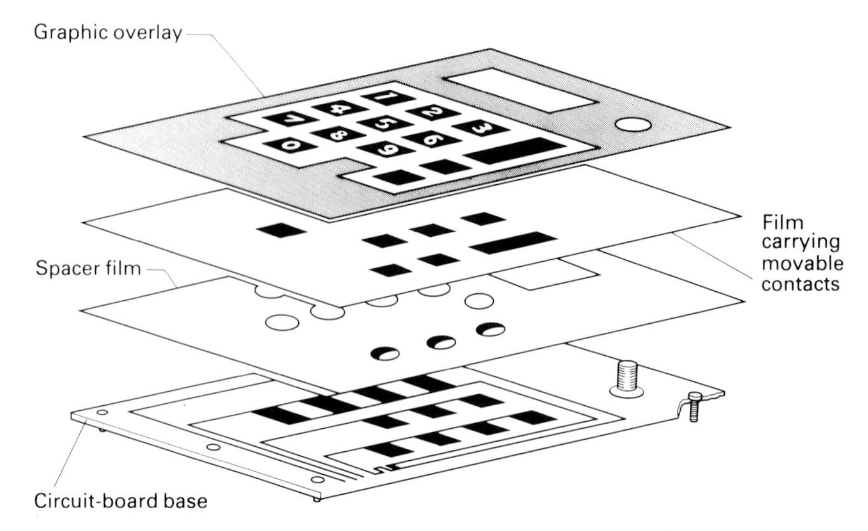

Fig. 11.17 Four layer polycarbonate film switch for a keyboard (from *Modern Plastics International*, McGraw-Hill, 1983)

11.4.2 Electrets

Several electrical devices are based on the permanent storage of charges in polymers. *Electrets* are the charge equivalents of magnets. The charges can be separated by distances of the order of 10 μm, or there can be permanently oriented dipoles in polar polymers. The polymer must be chosen to minimise the decay of charges by volume conductivity, or dipole relaxation. It must also be capable of fabrication as a thin film, so that flexible large area devices can be constructed. Useful devices can be made if the half-life for charge decay is measured in years. Fluorinated ethylene–propylene copolymer (FEP) is a semicrystalline polymer with many of the properties of PTFE, which can be melt processed. FEP film 25 μm thick is coated with 100 nm of aluminium on one surface, then a charge is injected into the other surface using a 20 kV electron beam. When the electrons strike the FEP surface about 20% cause secondary electron emission leaving the surface positively charged. The primary electrons penetrate about 5 μm before they are slowed to a point where they can be trapped (in conventional scanning electron microscopy, secondary electrons are used to form an image of the surface, and surface charging is avoided by coating the specimen surface with a thin layer of gold). The stability of charges can be studied by measuring the current as the temperature of the electret is gradually raised. In FEP measurable currents are observed above 125 °C; at 20 °C the charges have a half-life of 20 years. Microphones can be made by allowing sound vibrations to move an FEP diaphragm relative to another electrode. Fig. 11.18 shows that in a telephone the metallised surface of the electret is exposed to a sound source. The other electrode is a metallised ABS moulding with holes in it. Such microphones are insensitive to mechanical shocks and electromagnetic radiation, and are cheaper than the equivalent condenser microphones.

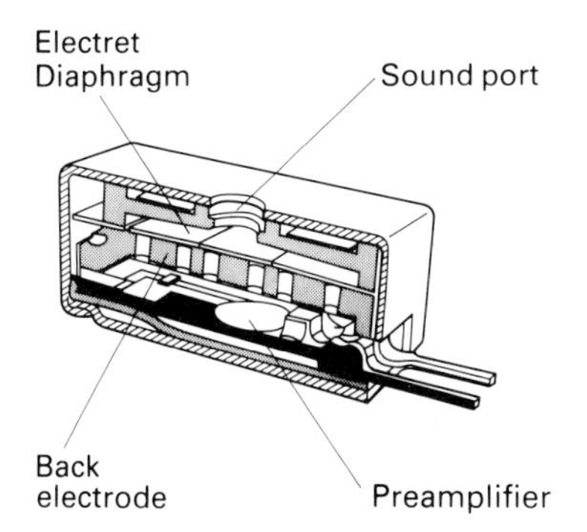

Fig. 11.18 Cross section of an FEP electret microphone for telephone applications (from *Bell System Technical Journal*, Copyright 1979, American Telephone and Telegraph Company)

11.4.3 Piezoelectric film

Polyvinylidene fluoride (PVDF) is a polar polymer; the CF_2 group has a dipole moment of 2.1 Debye. Both its microstructure and method of forming electrets are more complex than for FEP. It is a semi-crystalline polymer of 50% crystallinity and it has at least two crystalline forms. The type II crystals that form on the spherulitic crystallisation of unoriented PVDF have no net dipole moment because neighbouring polymer chains have opposite orientations of the polar CF_2 groups. If this material is drawn at 120 °C to a draw ratio of 4 or 5, then type I crystals are formed in which the polymer chains have an all-trans conformation, and all the dipoles are oriented parallel to the **b** axis (Fig. 11.19). The orientation of the crystals in the film is imperfect, there being almost complete alignment of the crystal **c** axes along the 1 axis (the draw direction), but as many **b** axes are in the positive 2 direction (the transverse direction) as are in the negative 2 direction. An electret is produced by heating the film to 100 °C, applying an electric field of 60 MV m^{-1} in the 3 direction (film normal) for 30 min, then cooling the film in the field. This preferentially aligns the **b** axes along the 3 axis, the average value of the cosine of the angle between **b** and the 3 axis being 0.84. The resulting film is piezoelectric, meaning that a charge density Q_3/A C m^{-2} appears on the upper and lower surfaces in the 3 direction as a result of stresses applied to the film. The largest piezoelectric stress coefficients are

$$d_{31} = \frac{Q_3}{A\sigma_{11}} \quad \text{and} \quad d_{33} = \frac{Q_3}{A\sigma_{33}} \tag{11.21}$$

for tensile stresses σ_{11} in the orientation direction, and for compressive stresses σ_{33} normal to the film. The magnitude of these coefficients depends on the degree of orientation of the crystal **b** axes and, as Fig. 11.20 indicates, on the measurement temperature. The major contribution to d_{31} is from the high Poisson's ratio ν_{31}, the ratio of the contraction in the 3 direction to the tensile strain in the 1 direction. The crystals act as rigid dipoles embedded in a deformable matrix, and the greater the contraction in film thickness the greater is the piezoelectric effect.

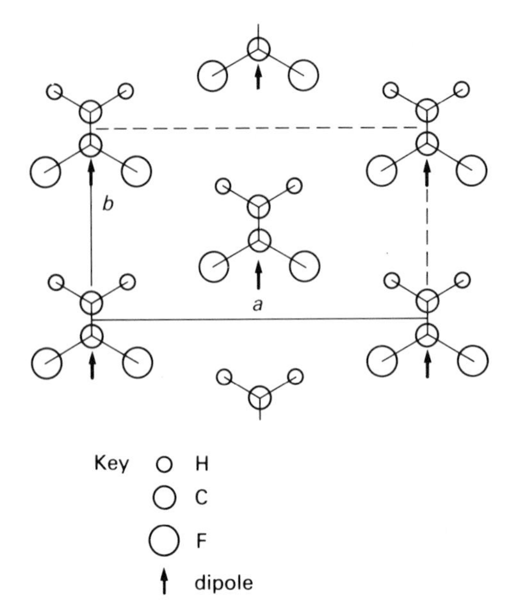

Fig. 11.19 Type I crystal unit cell of polyvinylidene fluoride, seen in the **a b** projection. The arrows represent the dipoles of the CF_2 groups. Large circles represent F atoms and small circles C atoms

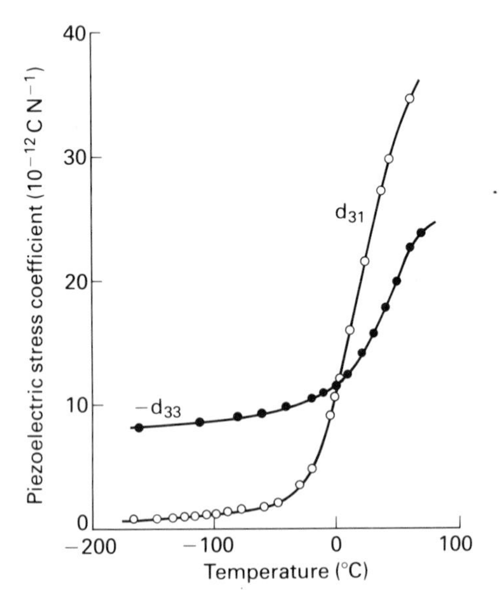

Fig. 11.20 Typical piezoelectric constants of a PVDF electret versus temperature (from J. Mort (ed.), *Electronic Properties of Polymers*, Wiley, 1982)

Quartz and ceramic crystals are alternative piezoelectric transducers. PVDF films do not have the thermal stability of quartz but they can be used for large area flexible transducers. The low mechanical impedance makes them ideal for coupling to liquids, hence their use in hydrophones and medical heart-rate monitors. The sensitivity to stress or strain makes it possible to construct pressure sensors (using the d_{33} coefficient), accelerometers by mounting a seismic mass on the film, etc.

11.4.4 Solid state batteries

Since 1972 polymer electrolyte batteries have been used in implanted cardiac pacemakers. The system used is lithium iodide with polyvinylpyridine. Although the conductivity of the Li ions in the LiI is poor, the current requirements are very small, and the major consideration is the high energy density that can be stored.

The research interest is in replacing NiCd rechargeable batteries using solid polymer electrolytes. Fig. 11.21 shows the cross section through a multilayer film battery. The electrolyte is the salt $Li^+[CF_3SO_3]^-$ dissolved in polyethylene oxide and the cathode is a composite of the electrolyte and V_6O_{13} and acetylene black. Since the multilayer films are flexible they can be rolled up into a confined space. The problems with such batteries are with the maintainance of the capacity after a large number of cycles. There is a delay while the battery is heated to its operating temperature, so it would be preferable to have a room temperature conducting polymer salt.

11.4.5 Gas sensors

The use of polymer films for the detection of chemicals by electrochemical

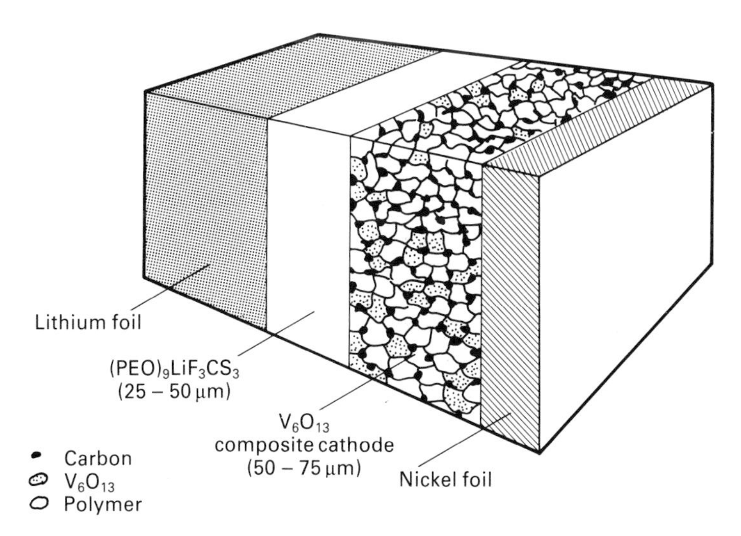

Fig. 11.21 Cross section of a polymer electrolyte battery (from Margolis, *Conductive Polymers and Plastics*, Chapman and Hall, 1989)

means will develop as the range of detectable gases increases. Biological sensors, made by incorporating biological molecules such as glucose into polypyrrole during electro-polymerisation, should be feasible. Currently only hydrogen sensors are marketed.

Blends of polyvinyl alcohol and phosphoric acid (H_3PO_4) can be used as the basis of hydrogen gas sensors. Fig. 11.22 shows the cell with platinum electrodes. The EMF of the cell depends on the difference of free energy between the electrodes. At the anode hydrogen at a partial pressure P_A is dissociated and dissolved into the polymer

$$H_2 \rightarrow 2e^- + 2H^+$$

and at the cathode where the partial pressure is P_C the reverse reaction occurs. Thermodynamic analysis shows (Margolis, 1989) that the potential E developed by the cell is

$$E \propto RT \ln\left(\frac{P_A}{P_C}\right)$$

so if a reference pressure of 1 atmosphere of H_2 is kept at one electrode the output of the cell is linearly proportional to the logarithm of the hydrogen concentration. The polymer is selective in only conducting protons—larger charged species are not able to diffuse through the film. The cell responds to a change in the hydrogen pressure on a time scale of 10 s. The only operational difficulty is the possibility of the temporary absorption of CO on the platinum surface. Once the CO is removed from the gas stream the CO desorbs and the sensor regains its sensitivity to H_2.

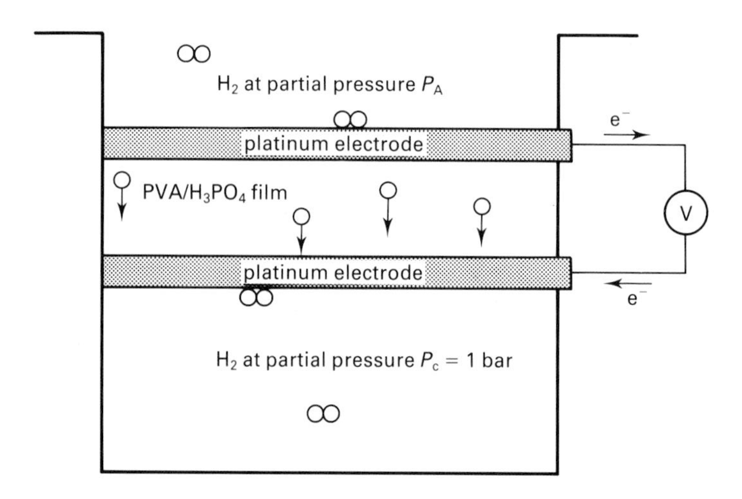

Fig. 11.22 A proton conducting polymer film as the basis of a hydrogen sensor

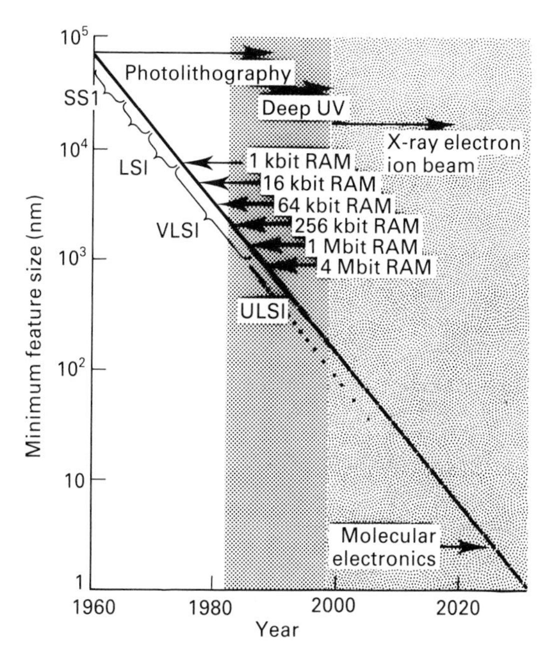

Fig. 11.23 Reduction in the size of the minimum width of silicon device features, as a result of the development of lithography. The current maximum size of random access memory devices is given alongside the trend line (from *Electronic and Photonic Applications of Polymers*, American Chemical Society, 1988)

11.4.6 Molecular electronics

If the progress in the miniaturisation of computer circuit features is maintained the predicted scale will reach that of molecules by the year 2030 (Fig. 11.23). The legitimacy of extrapolation on logarithmic scales for long periods of time is questionable. If the population of cars is extrapolated for 40 years there will be more cars than there is land to park them! Currently the main use of polymers in computer manufacture is in the photoresist processes in which UV light causes photopolymerisation. Once these masks for the etching and doping of the silicon wafers have performed their task they are stripped off chemically.

The potential for molecular scale electronic devices is intriguing, and one possible avenue is through the use of Langmuir–Blodgett (LB) films. These have been known since the 1930s. The principle, of the spreading of an organic film on a water surface until it is one molecule thick, is simple. The molecules need to have both hydrophilic and hydrophobic ends, and to be polymerisable once they form a monomolecular film. The current difficulties are that the LB films are not single crystals in contrast with silicon. Silicon also has the advantage of having a stable oxide film on its surface that is an insulator. Hence further development is necessary.

12

Material and shape selection

12.1 INTRODUCTION

There are several commercial materials selection packages for plastics, each of which covers slightly different data (Table 12.1). The investment necessary to develop one of these packages is considerable, and they are an asset in sales competition. They have partly replaced tabulations of the properties of grades, but the constant introduction of development grades by manufacturers for specific application areas means that they will not be entirely up to date. Nor is their purpose to provide all the technical service information that is available from the manufacturer. The advantages and disadvantages of using these selection packages will be explored in this chapter.

Table 12.1 Plastic selection packages

Package	Supplier	Coverage of materials
Campus	Consortium of manufacturers	German and US plastics
EPOS 90	ICI	ICI plastics
Peritus	Matsel	Metals, plastics, ceramics
Plascams	RAPRA	Range of UK suppliers

Another side of enegineering design is the use of certain features to improve the stiffness of products. Although there is not room here for a comprehensive treatment, the basics of bending and torsion stiffening can be explained. This will alert the reader to the possibilities and show why certain solutions are capable of efficient manufacture.

12.2 SELECTION USING A SINGLE PARAMETER

The first thing to appreciate is that the data used for selection are those which are readily available from manufacturers. This has two consequences: the information is incomplete and the quality of information for design purposes is limited. In the EPOS (Electronic Plastics On Screen) system the data are organised into a set of pull-down menus, when run on a PC. These are:

mechanical properties, thermal properties, electrical properties, tribology, desirability factors, miscellaneous, processing and additives. Inside each of these menus there will be up to 20 items; for instance in the first there is Tensile modulus, Notched Izod impact strength, . . . There is usually a single value for each property, although the creep data are more extensive. This aids the selection process but can lead the unwary user into traps. In earlier chapters the limitation of single values for mechanical data has been emphasised. A single value modulus refers to a specific time scale, to a specific strain (or a strain limit), to a test temperature and to a method of processing the test specimen. Usually the specimen is injection moulded as a tensile bar with end gating, so that the orientation will give high values of the strength and modulus. The discussion of impact tests in Section 8.5 will have alerted the reader to the need for tests on thicknesses of polymers identical with those in the product. Sadly these are not available, as the data are for standard 3.2 mm ($\frac{1}{8}$ inch) thick bars. There is no mention of fracture mechanics in the databases. This may reflect the difficulty of understanding the subject, or a lack of a generally agreed test method, but it is a serious drawback for design purposes.

The selection methods work by *ranking* the properties. Fig. 12.1 shows the range of tensile strength values for each family of thermoplastics in the EPOS database; the wide range reflects the effects of fibrous reinforcements on the strength. Seeing this or equivalent plots makes it easier to answer the questions 'minimum required value?' or 'maximum required value?' that the program poses. One method of selection is to enter the minimum (or maximum) values for the three or four key properties and see how many grades will meet these requirements. If there are initially 800 grades on the database it is unhelpful to be told that there are 344 grades that meet your requirements, as there is not room on the screen to list and examine them. The package can sort the grades under one requirement and list them in increasing order of that parameter. A useful sorting parameter is the price (the information is approximate), so the cheapest plastic that meet the requirements will head the list. If there are no grades that satisfy all your

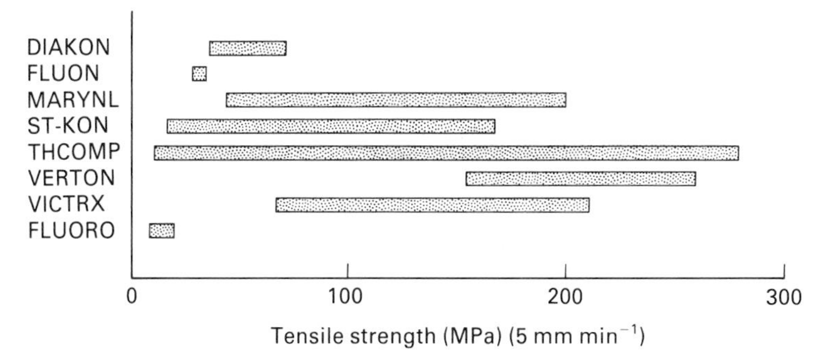

Fig. 12.1 The range of tensile strengths contained in the EPOS database. (ICI tradenames: Diakon = PMMA, Fluon = PTFE, Maranyl = PA, ST-KON = statically conducting compounds, THComp = specialist compounds, Verton = long fibre reinforced compounds, Victrex = PES and PEEK, Fluoro = PTFE)

requirements, you must relax one or more of them. The database may have information as to whether the grade is suitable for a particular process, but it will not tell you the cheapest process route or the cost of the manufactured product.

There are ways of showing visually how the properties of the plastic fit the selection criteria. In the PERITUS system this is by a plot that looks like a spider's web (Fig. 12.2). The properties are plotted as rays from a central point and the scales are normalised so that unit radius represents the minimum value that you specified. The plotted points are then joined to form a closed polygon. If you selected on 5 parameters the target figure would be a regular pentagon, and the data pentagon for an ideal material would just surround the target pentagon. A polymer with an excessively high strength would have a lop-sided data pentagon. This presentation alerts the user to the fact that there will never be a perfect match of the limited range of polymers to target specification. There is economic pressure to rationalise the number of grades manufactured, so that the remaining grades are sold in greater tonnages, and this will reduce the chance of a perfect match to requirements.

The needs to identify plastics for recycling and to design products that can easily be dismantled and separated into their components at the end of their lives will tend to reduce the number of plastics used. It will act against composite solutions; a PBTP bumper skin filled with polyurethane foam and attached to a steel subframe is far more difficult to recycle than a bumper with a PP skin, PP foam core and a glass-reinforced PP mounting beam at the back. This extra set of constraints could be contained in an 'expert system' but at present it is not part of the selection packages.

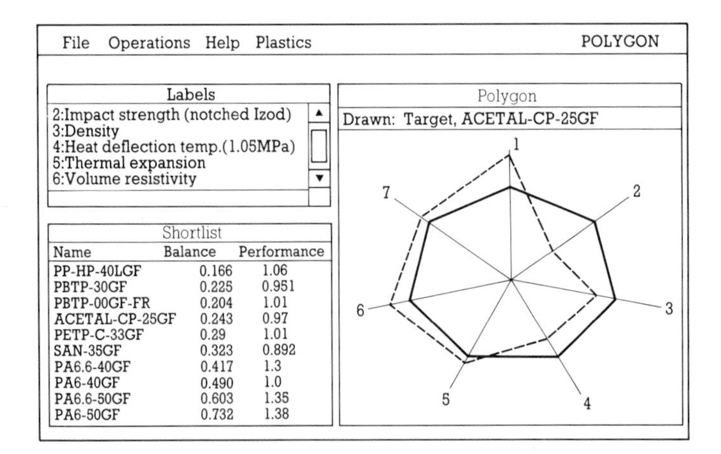

Name	Balance	Performance
PP-HP-40LGF	0.166	1.06
PBTP-30GF	0.225	0.951
PBTP-00GF-FR	0.204	1.01
ACETAL-CP-25GF	0.243	0.97
PETP-C-33GF	0.29	1.01
SAN-35GF	0.323	0.892
PA6.6-40GF	0.417	1.3
PA6-40GF	0.490	1.0
PA6.6-50GF	0.603	1.35
PA6-50GF	0.732	1.38

Fig. 12.2 Comparison of the properties of an Acetal grade (irregular polygon) with seven specified properties (regular heptagon) in the PERITUS database. The labels of five of the properties are shown in the inset. The shortlist shows the 'balance' (how close the polygon is to being regular) and the 'performance' (the average radius of the polygon) for this and rival polymers (courtesy of Matsel).

12.3 MERIT INDICES

When considering the selection of a panel material it is not appropriate to select on the basis of Young's modulus alone. The reasons are connected with the mechanics of bending of panels. In the bending of a flat panel the strain varies linearly through the thickness (Chapter 6). If the material is treated as elastic or linearly viscoelastic the stress variation is also linear, and the concept of the second moment of area I is valid. If however there is non-linear viscoelastic behaviour there is a more complex stress variation (Fig. 6.9) and the bending stiffness no longer depends on I.

For a panel of width w (this will be treated as a constant) and thickness t the value of I is

$$I = \frac{wt^3}{12} \tag{12.1}$$

Use of the beam bending formula, connecting the bending moment M the radius of curvature R and the Young's modulus E, gives

$$MR = EI = \frac{Ewt^3}{12} \tag{12.2}$$

The product is designed with a specified value of MR (the external loads fix M, and the limitations on the deflection give a maximum value of R), so if E is varied the required thickness depends on $E^{1/3}$. As the mass m of the constant width panel is directly proportional to t, then

$$m \propto \rho t \propto \frac{\rho}{E^{1/3}} \tag{12.3}$$

where ρ is the polymer density. To minimise the mass of the panel the figure of merit (which must be maximised) is $E^{1/3}/\rho$. This, or the value of $E^{1/3}$ divided by the volume cost of the polymer, is available on some of the packages as a selection criterion. To illustrate the results of using such a criterion, Table 12.2 shows a comparison between RIM materials, other thermoplastics and metals. This indicates that sheet moulding compound (SMC) or aluminium are the best materials for car door or wing panels, in terms of minimising the mass of the panel. If the raw material cost is to be minimised steel is preferred; however this does not include the processing cost of curved panels, which is highest for steel and lowest for RRIM.

Table 12.2 Panel materials with equal bending stiffness

Material	Thickness (mm)	Mass (kg m^{-2})	Material energy (MJ m^{-2})	Process energy (MJ m^{-2})	Total energy (MJ m^{-2})	Cost (£ m^{-2})
Steel	0.8	5.4	340	140	480	1.6
Aluminium	1.1	2.6	1060	150	1210	3.0
SMC	2.0	2.8	110	280	390	3.5
PC	3.5	3.5	820	140	960	9.2
RRIM	4.6	4.7	460	40	500	6.6

There are reasons why SMC is *not* used for the main structures of a car body. One is that the merit index is only appropriate for limited parts of a car structure. The main strength of the car body is in the hollow pillars that surround the passenger compartment. The bending stiffness of a square section thin-walled tube of width w and wall thickness t is given by the product of the Young's modulus with

$$I = \frac{1}{12}[(w+2t)^4 - w^4] \cong \frac{2}{3}w^3t \tag{12.4}$$

As the value of w is fixed by the design requirements that the door pillar, sills, etc., do not obstruct entry or vision, the figure of merit for such a thin walled tube is E/ρ. On this basis steel has a far higher figure of merit than SMC or thermoplastics. The production engineering problems of constructing a car with one technology for the panels and another for the tubular steel frame have been overcome for the Pontiac Fiero sports car and its successor, but the majority of motor manufacturers prefer to continue to use steel throughout. The total production costs for sheet metal panels (tooling and materials) are smaller than for plastics when the car is produced in high volumes, but they are lower for RRIM and SMC plastics when the production rate is lower.

12.4 PROPERTY MAPS FOR TWO PARAMETERS

With the EPOS package it is possible to produce graphs of one property against another. Fig. 12.3 shows for a number of polypropylenes the heat distortion temperature (HDT) plotted against the Izod impact strength. In the HDT test a bar is heated with a constant small bending load until a certain deflection is reached. It represents the point where the polymer is so floppy

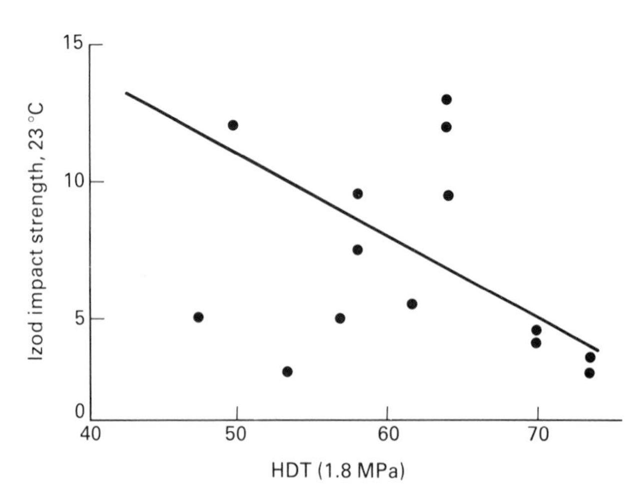

Fig. 12.3 A two parameter graph from EPOS: heat distortion temperature of polypropylenes versus the Izod impact strength

that it can hardly support its own weight. Although it indicates the limiting temperature for a short-time exposure (in a paint oven) it is of no relevance for long-term service where the maximum use temperature is the key parameter. There is a trade-off between the HDT and the impact strength, as the method used to improve the toughness of the polypropylene, using a greater rubbery copolymer content, leads to a lower crystalline melting point and lower stiffness. Hence there is no ideal polymer, and a compromise must be made in the selection process.

12.5 SHAPE SELECTION FOR BENDING STIFFNESS

When the creep modulus of plastics (the reciprocal of the creep compliance) is compared with the Young's moduli of metals it is found that there is a 1 : 100 ratio. For polypropylene E (1 h, 2 MN m^{-2}) ≅ 1.0 GN m^{-2}, whereas for aluminium $E = 70$ GN m^{-2} and for steel $E = 210$ GN m^{-2}. The creep moduli of plastics can be increased by the glass fibre reinforcement described in Section 3.8, but the values rarely exceed 5 GN m^{-2}. A steel panel used as the casing of a domestic appliance, or the skin of a car door, can be 0.6 mm thick and yet have adequate bending stiffness. If the replacement is a sheet of plastic of the same bending stiffness, the product Et^3 must be the same for both materials. Thus 3.6 mm of plastic of creep modulus 1.0 GN m^{-2} has the same bending stiffness as 0.6 mm of steel. Given that polypropylene has a density of 910 kg m^{-3}, compared with 8700 kg m^{-3} for steel, the plastic panel has 62% of the mass of the steel panel.

This approach does not maximise the bending stiffness to mass ratio of the product. The most common ways of modifying the product geometry to increase the bending stiffness are the use of corrugations, and ribs (Fig. 12.4).

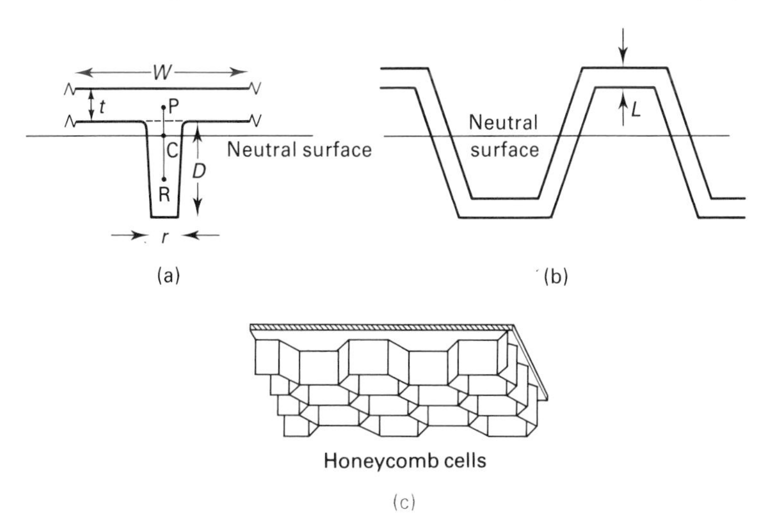

Fig. 12.4 Methods of increasing the bending stiffness of plastic panels. (a) Section of a ribbed plate showing the location C of the centroid; (b) section of a corrugated sheet; (c) honeycomb ribs that give isotropic stiffening

Ribs on the hidden surface of injection moulded products must have thickness r that is less than two thirds of the thickness t of the main part of the product, so that they solidify completely first and thereby avoid sink marks appearing on the surface of the product. As the section of the plate and rib is not symmetrical, the neutral surface, which passes through the centroid of the section, must be found first. In Fig. 12.5a there is a series of parallel ribs of spacing W. Each repeating unit of the ribbed plate, of area Wt, has a centroid at its midpoint P, and the rib of area Dr has a centroid at its midpoint R. The overall centroid is at a position C, where by the lever rule

$$Wt\,\mathrm{CP} = Dr\,\mathrm{CR} \tag{12.5}$$

This calculation neglects the taper angle of 1° or 2° that is needed on the rib to allow it to be easily ejected from the injection mould.

When a ribbed system is being designed there are several solutions that

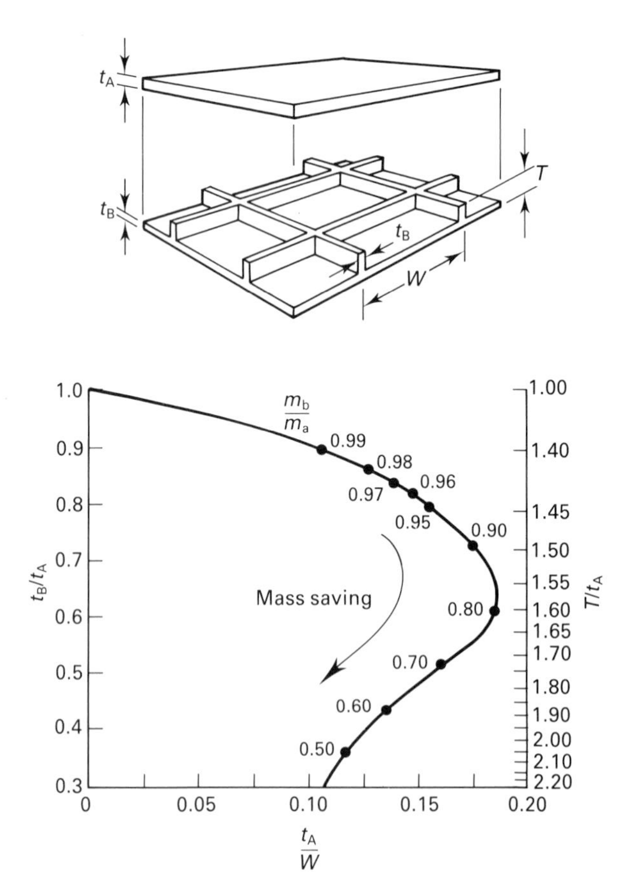

Fig. 12.5 (a) Replacement of a flat plate with a cross ribbed plate of the same bending stiffness; (b) the reduction in mass achieved by the change, and the rib dimensions required (from *Delrin Design Handbook*, Du Pont, 1980)

achieve the same required stiffness. Consider the replacement of a solid plate of thickness $t_A = 5$ mm by a cross ribbed plate of uniform thickness $t_B = 2$ mm. There will be several pairs of parameters (W, t_B) that meet the requirement of maintaining the bending stiffness. One of these pairs will have a minimum mass and this is shown in Fig. 12.5b. For example, the effect of a reduction in thickness of $t_B/t_A = 2/5$ can be found by reading across horizontally from the left-hand scale to the relative mass of 0.55 on the curve. The horizontal scale at the bottom of the chart can then be read to give $t_A/W = 0.125$. Substituting $t_A = 5$ mm means that the rib spacing $W = 40$ mm. Reading across to the right-hand scale shows that $T/t_A = 1.95$ so that the rib depth D, which is equal to $T - t_B$, will be equal to 8 mm. The stiffening in this cross ribbed design is only in the two rib directions; there is still a low bending stiffness along a direction at 45° to the ribs, and a low torsional stiffness. For an isotropically stiffened plate, which will need ribs in at least three directions, or the honeycomb structure of Fig. 12.4c, the mass saving will not be as high as shown by Fig. 12.5. On the other hand, the direction of the main bending moments may be known, so that the ribs can be placed in the appropriate directions. The edges of a product, such as stacking polypropylene chair, are natural places to incorporate ribs.

The maximum expected bending moment must not cause buckling of the rib in compression, if the bending moment is of a sign that puts the rib into compression (Section 7.2.3). The length L of the rib in equation (7.11) will be equal to the rib spacing W if there is cross ribbing as shown in Fig. 12.5. If this buckling calculation shows that the ribbed plate cannot meet the bending moment requirements the ribbing system must be redesigned with thicker or shorter ribs. The latter means having cross ribbing at closer intervals.

The addition of these stiffening features will add to mould construction costs, but very little to the materials costs or the cycle time of the process. The productivity of an injection mould for the ribbed plate will be controlled by the solidification time given by equation (4.6); it will be ~3 s for the ribbed plate and ~19 s for the 5 mm thick original design. This increase in the productivity of costly injection moulding tools is far more important than the materials saving of the reduction in mass of the moulding.

Fig. 12.6 shows the relative efficiency of different shapes of beam cross-sections. Each of these has the same cross sectional area and the neutral surface is taken to be horizontal. The I beam, of height equal to twice its width, is used for comparison and given a second moment of area 100%. The table shows that a ribbed plate with a rib depth equal to two thirds of the rib spacing has a relative efficiency of only 9%. I beams and hollow tubes are more efficient, but they are more difficult to mould in one piece; it would be impossible to reinforce a plate with a series of I beams behind it.

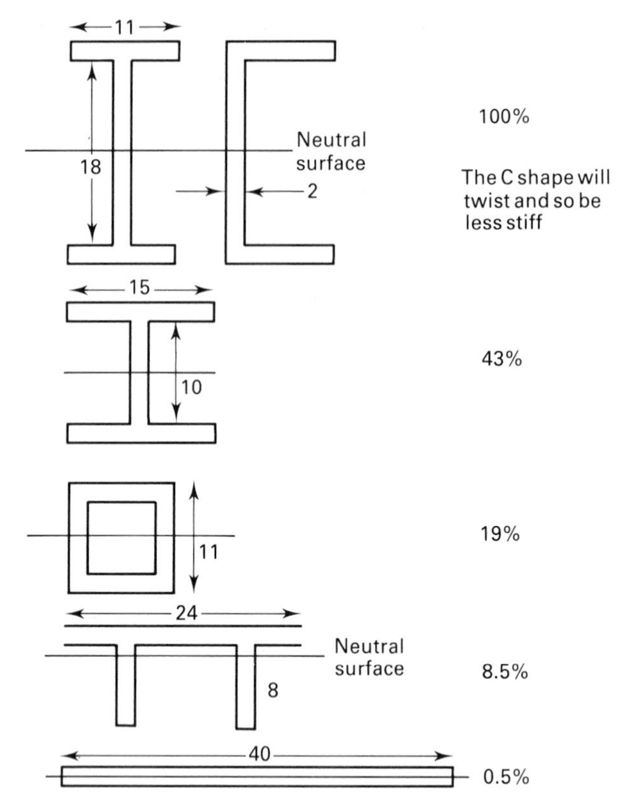

Fig. 12.6 Ranking of beams of the same cross sectional area and wall thickness for bending stiffness about the neutral axes shown

12.6 SHAPE SELECTION FOR TORSIONAL STIFFNESS

The theory of the distribution of the shear stresses in a twisted beam involves the solution of Poisson's equation

$$\frac{\partial^2 z}{\partial x^2} + \frac{\partial^2 z}{\partial y^2} = -2G\theta \tag{12.6}$$

where G is the shear modulus of the material, θ is the angle of twist per unit length and z is the potential function appropriate to the beam cross section. The solution of this equation is beyond the level of mathematics of this book. Luckily there is a soap bubble analogue that can be used to illustrate the stress distribution. The equation of equilibrium of the soap bubble, which is under a constant biaxial membrane stress T, is

$$\frac{\partial^2 z}{\partial x^2} + \frac{\partial^2 z}{\partial y^2} = -\frac{p}{T} \tag{12.7}$$

where z is the height of the bubble and p is the pressure differential across the bubble. The equations are identical in form, and the analogues are

shear stress τ ⇔ maximum slope of the bubble at the equivalent point

shear stress direction ⇔ contour line (z = constant) direction

torsional stiffness ⇔ volume contained inside the bubble

When the bubbles are examined for a ribbed plate (Fig. 12.7) the shear stress direction is seen to be nearly parallel to the surfaces, and the shear stresses are zero at the mid-section. The volume of the bubble is nearly the same as for a flat plate of total width $W+D$. The linear variation of τ across the section leads to the torsional stiffness being given by

$$\frac{T}{\theta} = G(W+D)\frac{t^3}{3} \tag{12.8}$$

To model the hollow walled tube the bubble is blown with a flat horizontal surface representing the inner shape of the tube. This horizontal surface is free to rise, so that the volume of the bubble is large. The shear stress τ is approximately constant in the tube wall, and it acts parallel to the wall. The volume of the bubble is proportional to the area A of the enclosed area of the tube and to the wall thickness t. The analytical solution for the torsional stiffness of a hollow tube of perimeter P, wall thickness t and enclosed area A

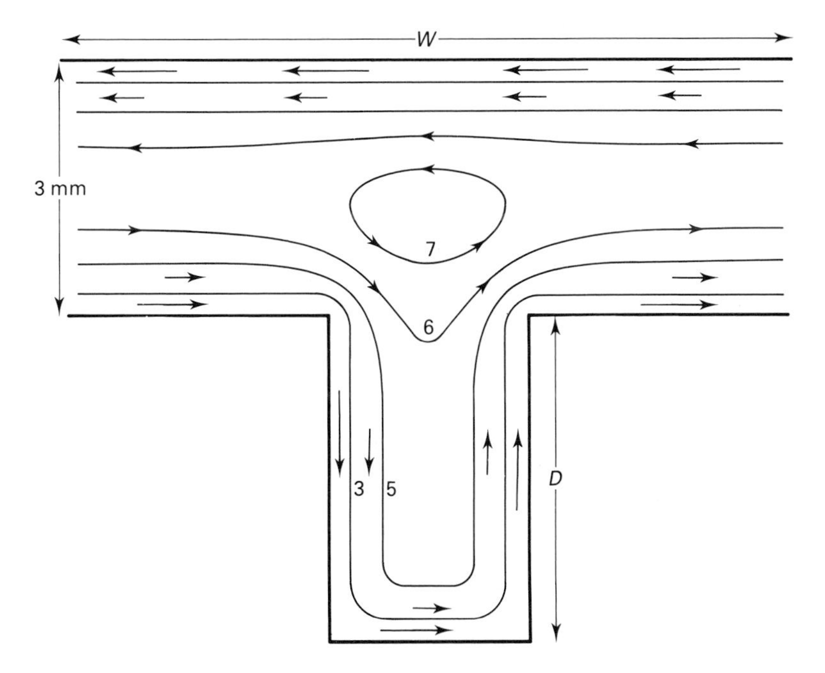

Fig. 12.7 Sketch of the contour patterns from a thermoformed bubble, which is an analogue of torsion of a ribbed plate. The numbers are the contour heights and the arrows show the direction and magnitude of the shear stresses

is

$$\frac{T}{\theta} \cong 4A^2G\frac{t}{P} \tag{12.9}$$

This reduces to $2\pi r^3 Gt$ for a circular section tube of radius r.

The bubble can be made by thermoforming, to produce a permanent display of the stress distribution. The shape of the beam cross section is cut from a sheet of metal or plywood and then a bubble of PVC, or other suitable thermoformable sheet, blown with a small air pressure on a thermoforming machine. The bubble is allowed to cool into the glassy state before being removed from the machine.

Fig. 12.8 shows the use of the above theory to rank the torsional stiffness of beams of constant cross sectional area. The optimum designs are hollow tubes of constant wall thickness, in which the shear stresses are uniform. In contrast, the torsional stiffness of all the open section shapes are uniformly low because the shear stresses must double back inside the cross section. One way of producing a stiff hollow section is to weld two mouldings together. If the joint is planar it may be possible to use hot plate welding (Section 13.2.4) but if it is non-planar there is the possibility of electrothermal welding. Fig. 12.9 shows the section through a car bumper where a woven copper braid has been placed between the two mouldings. When the mouldings are

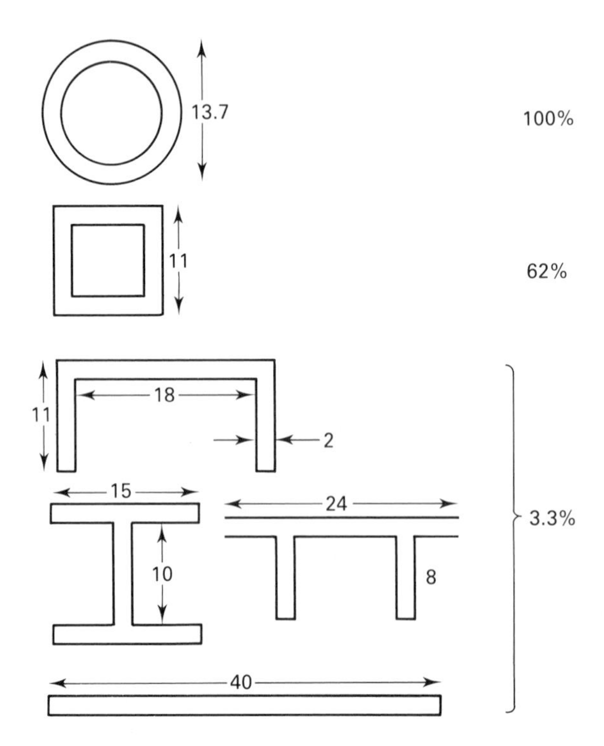

Fig. 12.8 Ranking of beams of the same cross sectional area and wall thickness for torsional stiffness

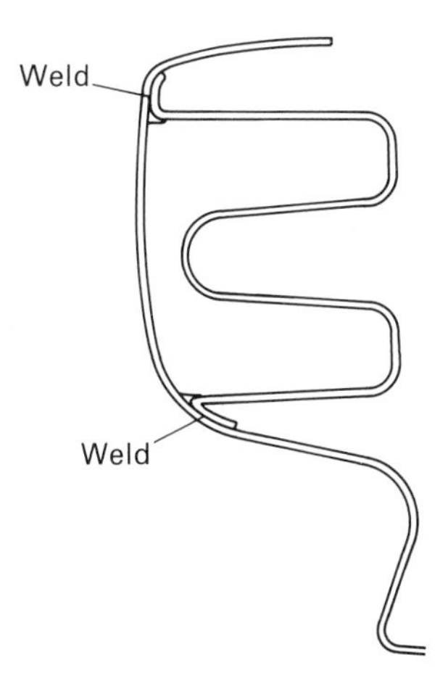

Fig. 12.9 Section through a car bumper. Two mouldings have been joined with an electrofusion weld, to form a hollow beam with high torsional and bending stiffness (from W. M. Risk, in *Advances in Exterior Body Panels*, Soc. Auto. Eng., 1987, p. 25)

clamped together and a current of hundreds of amperes is passed for a few seconds, a layer of 1 mm of plastic melts, flows through the spaces in the braid and forms a strong weld. The continuous weld allows the torsional stresses to flow uniformly around the section of the bumper. The alternative of mechanically fastening the two halves together is far less torsionally stiff because the shear stresses in the plane of the joint can only pass through the relatively small area of the fasteners.

12.7 STIFFNESS OF AN INJECTION MOULDED ACCELERATOR PEDAL

An accelerator pedal is subjected to bending loads but if it is not sufficiently torsionally stiff it can also deform by twisting, in a deformation mode that is reminiscent of buckling. The best shape for torsional stiffness is a hollow tube, but the pedal must have a bend in it which means that it is difficult to use movable cores, that must retract in a straight line before the injection moulding can be ejected. The I and sideways U beams have relatively high bending stiffnesses but their torsional stiffness needs to be improved. One solution is to mould a beam of which the cross section varies with position along its length (Fig. 12.10). The cross ribs on the I or U beam form a structure that takes tension and compression loads. They act like a pin-jointed framework (or a Meccano model) with some of the ribs being in tension and others in compression. In this way it is possible for the beam to have nearly the same torsional stiffness to weight ratio as a hollow box beam. Fig. 12.10 compares the bending and torsional stiffness of several versions of the design. The ribbed U section beam was preferred over the ribbed I beam because of its superior torsional stiffness, while its bending stiffness was only slightly less than the I beam. As the mass of the ribbed beam is close to that of the original

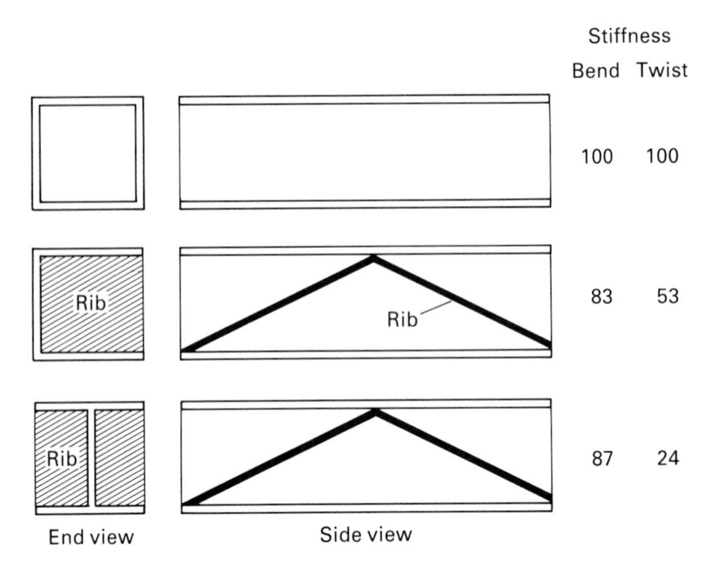

Fig. 12.10 The relative bending and torsional stiffnesses of ribbed beams where the cross sectional area changes along the length, compared with a hollow square section beam of constant cross section

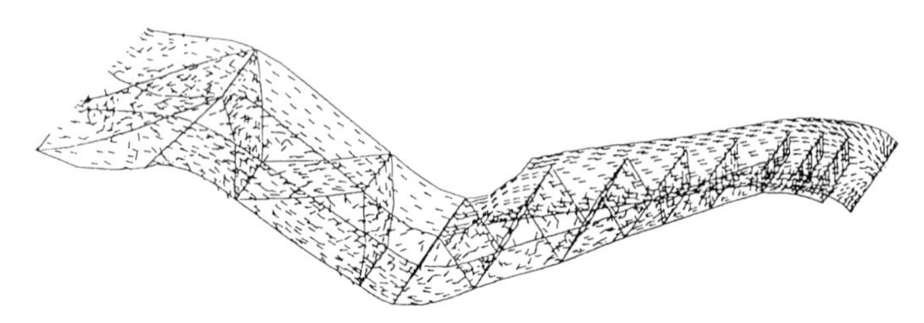

Fig. 12.11 Design of an injection moulded vehicle pedal. The arrows are the predicted flow directions when the mould is filled (from Brünings *et al.*, *Kunststoffe*, 1979, **79**, 254)

square cross sectioned beam, the stiffness to weight ratio of the designs is good. The major advantage is that these ribbed beams are relatively easy to injection mould, even if the beam is curved along its length.

In the final design the thickness of the top flange of the U beam was 1.5 mm, the vertical web of the U beam was 2 mm and the ribs were 1.5 mm thick. The U was on its side and half way along the pedal the direction of the U changed; this cancels out the slight twisting effect when the pedal is placed under a bending moment. The final version used V rather than X ribs, with the angle between the ribs and the flange being 15°. The pedal is moulded from glass-reinforced polyamide, the material being selected for its high strength and toughness (see Section 12.2). The flow orientation of the glass fibres along the flow directions as the mould fills adds to the modulus in this direction. Fig. 12.11 shows that the predicted flow directions are in the directions of the stresses in the ribs, so the microstructure is optimal.

13

Design case studies

13.1 INTRODUCTION

The three case studies are chosen to illustrate different areas of the rest of the book. The gas pipe case study involves the mechanical properties of creep, yield and fracture, and the process technology of welding. The mechanical design loads are well known, unlike in the second case of the bicycle helmet where there are unknown factors both in the type of impact experienced, and the tolerance of the brain and skull to impact. This second case emphasises the part played by the geometry of the loading, and the difficulties in meeting a set of conflicting requirements. The last case study is in the information storage area; it shows the interface between the electronics discipline, the optical properties of the glassy polymers and the injection moulding technology required to make the discs.

There are several problems in this chapter that should be tackled as the reader progresses. The answers to the problems are essential to the progress of the argument. The answers are provided at the end of the chapter, but it is strongly recommended that the reader attempts the problem first! This is a mean of revising concepts from earlier parts of the book.

13.2 PIPELINES FOR NATURAL GAS DISTRIBUTION

13.2.1 Introduction

The distribution of natural gas in the UK by the British Gas Corporation (BGC) is a major enterprise. There is a national grid which collects and distributes gas from the North Sea and the total network has a length exceeding 200 000 km, making connections to 14 million domestic and industrial consumers. A number of different materials have been or are being used in the distribution system (Table 13.1).

The last item in Table 13.1 may seem trivial, but it makes the point that the requirement of flexibility leads to a different choice of material. We shall consider in the case study the distribution of gas at the area and district level, where plastics have replaced a traditional material, cast iron. Gas at a

Table 13.1 Parts of the gas distribution system

Part of system	Gas pressure (bar)	Requirement	Material (old material)
National grid	7 to 70	Maximum flow High hoop strength	High strength steel
Local distribution	0.075 to 2	See later	Plastic (cast iron/steel)
Inside house	<0.075	Safe durable connections	Copper (lead)
On laboratory bench	<0.075	Flexible connection	Rubber, plasticized PVC

pressure of 70 bar from the national grid is received at local stations and reduced to a pressure of 7 to 16 bar in the local high pressure grid. This serves the local distribution system at 1 bar. Individual factories and housing estates take their gas supplies via the pressure governors which reduce the gas pressure to 40 mbar. Plastic gas pipe is being used for all the local distribution system.

13.2.2 Choosing a plastic

The various factors involved in choosing a plastic, specifying the particular grade that is best for the application and deciding which additives are necessary will be discussed. Three criteria—cost, strength and toughness—were used to make a decision about which plastic to investigate in more detail. The price of plastics fluctuates with that of oil, and it depends on the current balance of supply and demand. The figures in Table 1.1 can only be used for comparative purposes. The polymers that were considered were polyethylene, PVC and polypropylene, as engineering plastics would be too expensive for this application. ABS and polyamides are used for certain pipe or tubing applications where the quantity of material is less important than flexibility and durability.

The initial sorting by a materials selection package (Chapter 12) might be by the tensile yield strength in a room temperature test lasting the order of 1 min. Table 13.2 shows that such data would lead to PVC being preferred to polypropylene, with polyethylene the least favoured material. This selection could only be justified if the main design criterion was the prevention of the pipe wall yielding within a few minutes of installation! Certainly the amount of PVC needed would be significantly less than for polyethylene. The method of jointing the pipes differs between the polymers. It is impossible to make welds in PVC that are as strong as the original material, whereas the polyolefins can be welded easily. The rigidity of large diameter PVC pipe means that it must be installed as 12 m lengths with sealing gaskets between them. In contrast MDPE pipe can be welded above ground into a continuous pipe that is flexible enough to be pushed a section at a time into a trench, or be pulled into old cast iron mains in the relining process (Fig. 13.1).

The gas distribution application puts a high premium on toughness as a desirable material property. The explosive nature of gas/air mixtures will be

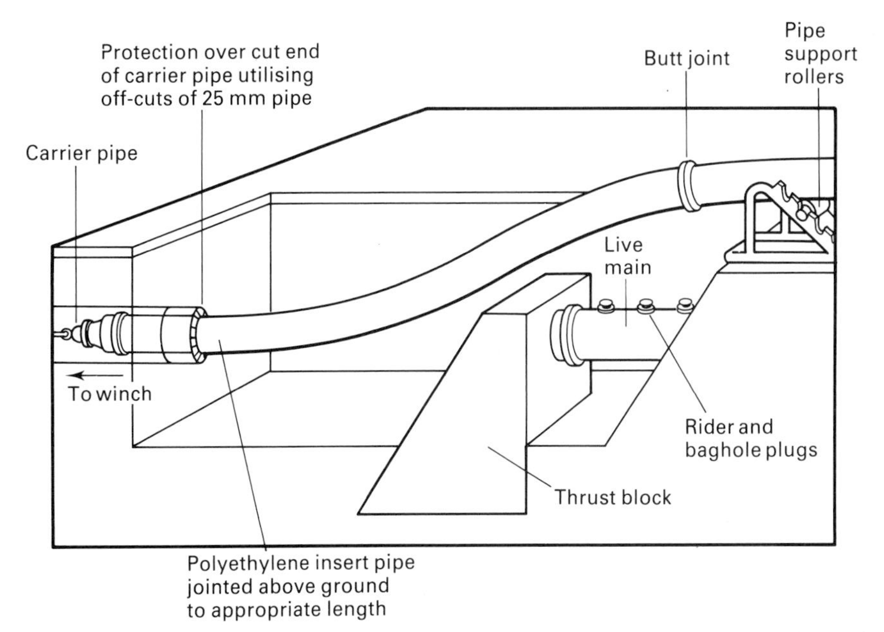

Fig. 13.1 Insertion of a welded length of MDPE gas pipe into an old cast-iron main without the need to excavate the old pipe (from *The Wavin Gas Handbook*, courtesy of Wavin Industrial Products Ltd)

familiar from newspaper accounts, whereas a leak in a water pipe, while inconvenient, rarely imperils life. It is not sensible to select a plastic on the basis of the plane strain fracture toughness K_{IC} alone (Table 8.1), because the thickness of the plastic should be below that necessary to cause plane strain fracture. Table 13.2 gives data for the fracture toughness and yield stress, measured in tests in which the loads were applied slowly. The two sets of data for the PVC reflect the difficulty of completing the fusion of the particle structure in thick walled pipe (Chapter 1).

Table 13.2 Toughness and yield stress of plastics

Plastic	Density ($kg\ m^{-3}$)	MFI (g per 10 min)	K_{IC} ($MN\ m^{-1.5}$)	At temperature (°C)	σ_y ($MN\ m^{-2}$)
Polyethylene	940	0.2	3.1	−35	22
Polyethylene	933	2	5.0	−60	15.8
Polyethylene	930	8	3.0	−60	16.4
Polyethylene	929	16	2.1	−60	14.5
Polyethylene	916	18	1.1	−60	9.4
PP copolymer		4	3.5	−70	26.4
PVC well processed		$K = 68$	4.0	−20	57.0
PVC poorly processed		$K = 68$	2.7	−20	57.0

Problem 1 Use the data in Table 13.2 to rank the plastics in terms of the transition thickness t_c for plane strain fracture (equation 8.16), and comment on why polyethylene is preferred to the other polymers.

The selection of the grade of PE is also determined by the performance in creep rupture strength tests on pipe (Section 7.3.1). These tests are carried out at 80 °C to accelerate both the ductile failure mechanism and the slow crack growth that leads to brittle failure (Fig. 7.14). Experimental data on HDPE shows that lowering the MFI (increasing the molecular weight) increases the creep rupture strength at long times. It delays the onset of the brittle fracture as it improves the resistance to ESC and slow crack growth.

The large market for pressure pipe has meant that grades of PE have been specially developed for this application. The first generation of HDPEs were not used in the UK. The second generation of MDPE copolymers have superior creep rupture times at the stress levels used for pipe design (Fig. 13.2). In the latest third generation materials the side chains produced by copolymerisation are no longer at random along the chain but grouped together. This means that a certain percentage of comonomer will have a greater effect in reducing the crystallinity. The consequent reinforcement of the amorphous phase in the polymer has a major effect on the brittle fracture mode. In Fig. 13.2 the data for the Solvay Eltex TUB 121 grade is shown; there is no sign of a steep brittle failure branch of the data.

Certain additions must be made to the polyethylene for the pipe to perform effectively. In the UK a yellow pigment is used so the pipe can be clearly differentiated from blue water pipe or red electric cable when the road is excavated at a later date. The pigment, usually cadmium yellow, must be efficiently mixed into the polyethylene, and dispersed so that there are no agglomerates exceeding 10 μm in size or streaks of pigment. Such agglomer-

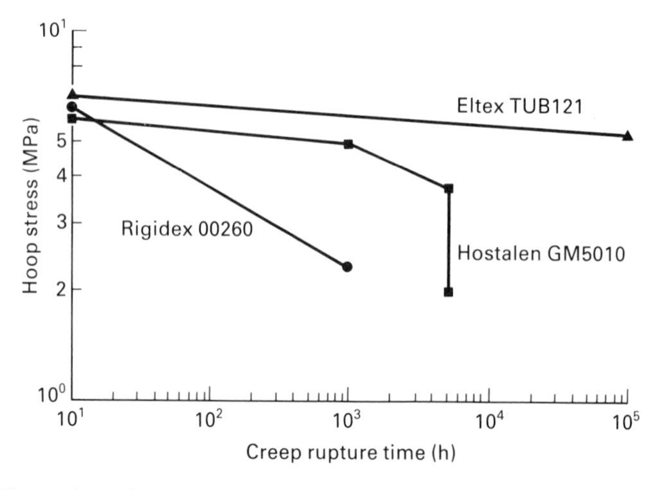

Fig. 13.2 Comparison of creep rupture data at 80 °C for the three generations of HDPE pipe material. (Rigidex, Hostalen and Eltex are the trade names of BP, Hoechst and Solvay, respectively)

ates can fracture internally and may act as crack initiation sites. The pigment, by absorbing light, partly protects the polyethylene from degradation during outdoor storage.

The gas pipe has to contain and transmit natural gas under pressure, so the gas should not be deleterious to the properties of the pipe, nor should it diffuse through the wall of the pipe at an excessive rate. Methane, and the other hydrocarbons in natural gas, diffuse through polyethylene at a rate that does not cause economic loss, or the build up of gas on the outside of the pipe. A rough calculation of the maximum daily loss from 1 km of pipe, of 90 mm outer diameter and wall thickness 8.5 mm, pressurised to 1 bar (the partial pressure of methane is 2 bar inside and 0 bar outside), is 4.4 litres. This is less than 1% of the losses that occur with jointed cast iron pipe, and can be compared with the daily throughput of gas, which is 2×10^5 litres for a pressure drop of 1 mbar km^{-1}.

13.2.3 Manufacturing the pipe and fittings

When the gas distribution network was planned the pipe sizes were standardized, since for each pipe diameter the necessary joints and couplings have to be made. An increased number of pipe sizes would make it easier to choose the diameter suited to the flow, but it would increase the costs of manufacture and stocking of the pipe. The pipe in the UK is made from either Rigidex 002–40 manufactured by BP, who are the sole UK manufacturers of polyethylene, or Aldyl A imported by Du Pont from the USA, or Solvay material. The range of pipe diameters is given in Table 13.3.

Table 13.3 PE pipe sizes

Outside diameter (mm)	SDR	Form	Length (m)	Mass per metre (kg)
20	9	Coil	50/100/150	0.12
25	11	Coil	50/100/150	0.17
32	11	Coil	50/100/150	0.27
63	11	Coil/straight		1.04
90	11	Coil/straight		2.10
125	11	Straight	6 or 12	4.04
180	11 and 17	Straight	6 or 12	5.43
250	17	Straight	6 or 12	10.44

The Standard Dimension Ratio (SDR) was defined in Chapter 7 by

$$\text{SDR} = \frac{\text{outside diameter}}{\text{minimum wall thickness}}$$

so the hoop stress in the pipe can be calculated directly from the SDR using

$$\sigma_H = \frac{p}{2}(\text{SDR} - 1) \qquad (13.1)$$

Problem 2 Use the data in Chapter 7 for the minimum creep rupture strength specified for PE gas pipe to calculate from equation (13.1) the SDR value needed if the gas pressure is 2 bar.

The pipe is clearly identified at 1 m intervals on both sides. Fig. 13.3 shows a typical legend, and the meaning of each of the symbols. If faults are found with a particular pipe, the labelling system makes it possible to identify all the rest of the pipe from that particular batch, and to find out the polymer used and the processing conditions.

The sockets are manufactured by injection moulding. It is usual in the injection moulding of thin walled products to use polyethylenes of high melt flow indices in the range 5 to 50, so that the mould can be filled easily. The sockets and gas fittings are exceptionally thick by injection moulding standards, up to 15 mm thick for the 125 mm o.d. pipe, so it is possible to fill the moulds using MFIs as low as 0.2. It is found to be important to match the MFI of the socket to that of the pipe to obtain fusion joints of optimum strength.

13.2.4 Installation and fusion jointing

The assembly of a gas distribution system either uses mechanical couplings, similar in principle to compression fittings on domestic water pipes, or uses fusion joints. Fusion joints cannot leak, unless a crack grows through the joint. They avoid the problems associated with the deterioration of seals in mechanical jointing systems. Socket fusion can be used to join pipes of sizes up to 125 mm, and involves the use of an injection moulded socket fitting that is a slight interference fit on the end of the pipe. The simple socket allows straight continuation of the pipe, whereas 45° and 90° elbows allow sharp corners, and equal tees allow branches to be added. Butt fusion can be used to make axial joints on pipes of diameter 63 mm and above, and is the only method used if the diameter is 180 mm or larger.

To make a fusion joint the polyethylene must be heated into the melt state.

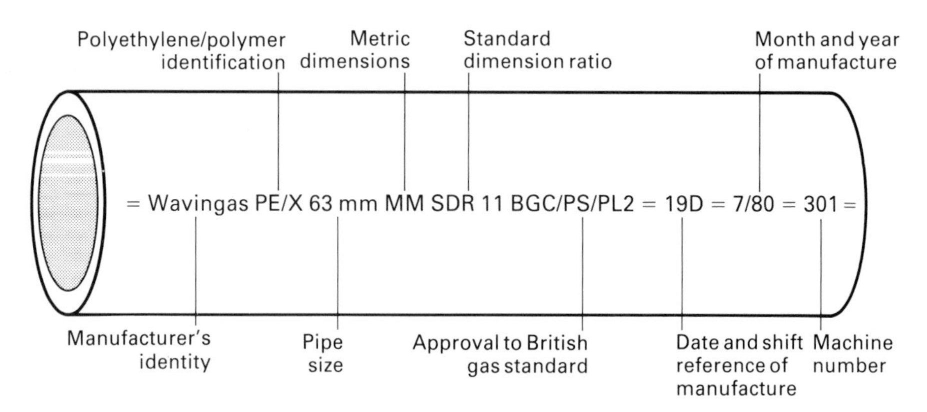

Fig. 13.3 Identification markings on gas pipe (from Wavin Gas)

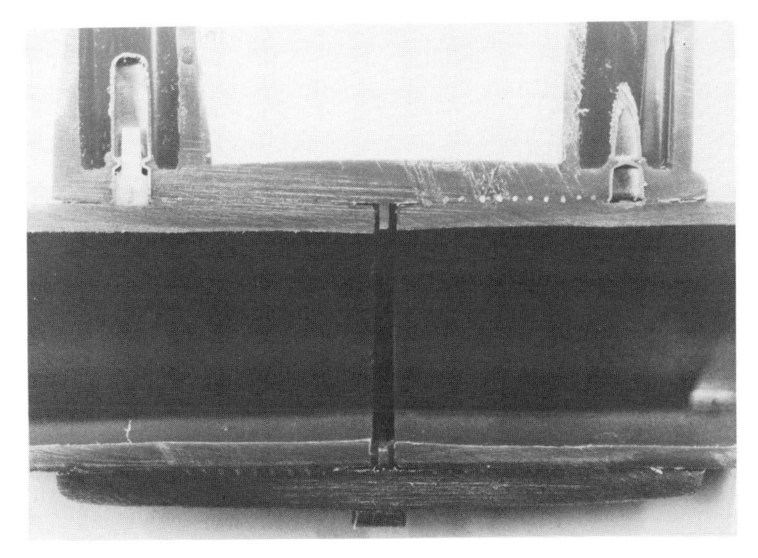

Fig. 13.4 Section of an electrofusion socket joint, in which the embedded copper wire is used to melt the surface of the joint prior to assembly

For socket joints it is convenient to have a copper heating coil insert in the injection moulding (Fig. 13.4), but for butt fusion an external heater is used. The polyethylene is pressed into contact with heated metal plates, coated with PTFE so that the melt releases from them when the pressure is removed. The plates must melt the polyethylene crystals; this is complete at 135 °C. The plate temperature should be hotter than 170 °C, to avoid the risk of the polymer surface crystallising before the joint is made. Cooling occurs by conduction to the underlying polymer, and by convection to the air. The plate must not be so hot that the polyethylene degrades rapidly, so it should not be above 270 °C. Experiments were needed to find the optimum temperature was 205 °C. The pressure is only applied for a short time to ensure the pipe is fully in contact with the heater plate. Fig. 13.5 shows how the temperature profile in the pipe changes with time, when the contact pressure is zero. It takes about 2 min to produce a 3 mm thick molten layer.

When a butt fusion joint is made the pipe ends are clamped together under an axial pressure. When this happens the molten layer flows radially acting as a high viscosity rubbery liquid. Even the partly molten layers behind the melt will deform fairly easily. With a very low welding pressure the melted layer is hardly distorted and it crystallizes in the spherulitic form. There is insufficient flow at the weld interface and a thin layer of crystals nucleate from the interface, producing a weak weld interface. The other extreme of a high pressure squeezes the melt until the two unmelted regions come into contact. The molecules in the melt are elongated in the direction of flow, and this causes excessive orientation of the **c** axes of the lamellar crystals in the direction of the flow. The crystal orientation makes the joint weak against the longitudinal stresses in the pressurised pipe. The strongest joints have microstructures between these two extremes (Fig. 13.6).

The procedure for the butt fusion of pipes has 5 stages:

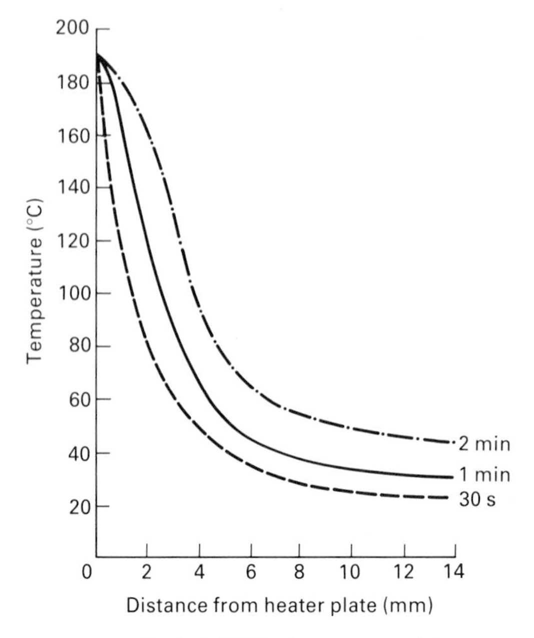

Fig. 13.5 The temperature profile in MDPE after contact with a heater plate at 210 °C for various times

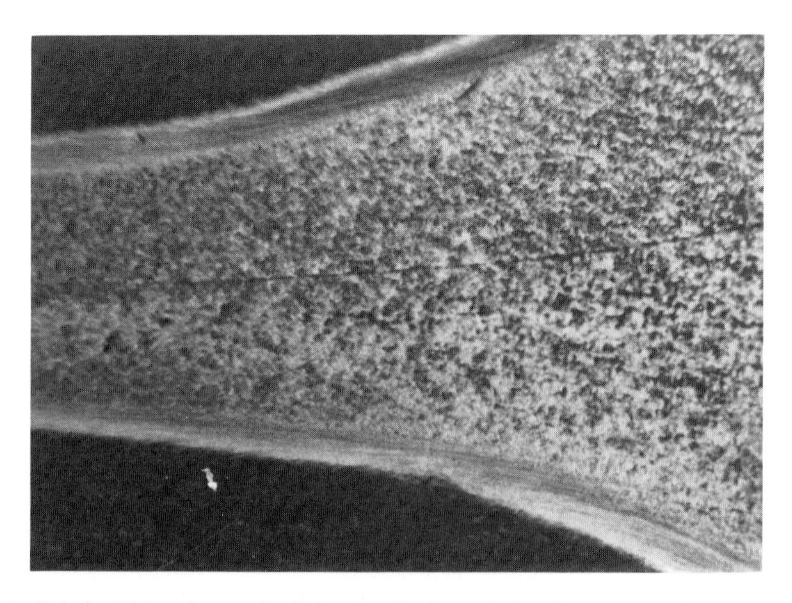

Fig. 13.6 Polarised light micrograph of a butt weld in 9 mm thick polypropylene sheet. Note the central weld line distinguishable by surface nucleated spherulites, the main weld pool on either side, and a layer of highly sheared material at the edges of the weld pool

(i) The two pipes are clamped into a machine which ensures that they are in axial alignment. A set of rotating blades machines the ends of the pipes so that they are clean and parallel.
(ii) A double-sided flat heating plate at a temperature of $205 \pm 8\,°C$ is inserted into the gap between the pipes. A pneumatic ram pushes the pipes into contact with the heater, with a compressive stress of $0.15\ MN\ m^{-2}$ over the end surfaces. This stress is maintained until a 2 mm wide bead of molten polymer forms in contact with the heated plates.
(iii) The pressure is reduced to zero, and heating is continued for the appropriate time (120 s for 125 mm diameter pipe).
(iv) The carriage is opened, the heating plate removed, then the carriage is rapidly closed to form the weld. The compressive stress of $0.15\ MN\ m^{-2}$ is maintained for a cooling time of 10 min (125–180 mm pipe).
(v) After removing the pipe from the machine the bead is checked for completeness, and that its width falls in the range 7 to 11 mm.

The butt joints have beads at the inner and outer pipe surfaces. The outer beam may be cut off, but the inner one is difficult to remove. The bead meets the pipe wall at a relatively acute angle, and the resulting stress concentration will be exacerbated if there are differences in the wall thicknesses of the two pipes to be joined. It was found by testing butt jointed pipes at 80 °C that any step of greater than 10% of the pipe wall thickness produced an unacceptable reduction in the creep rupture life. This explains the requirement for wall thickness limits.

13.2.5 Soil loads on the buried pipe

The creep strain due to the internal pressure in a pipe have been analysed in Section 6.3.1. There is a requirement that the hoop strain should not exceed 3% after 50 years in use. This relates to the need to butt weld new pipe to older material, without there being a mismatch in dimensions. It overlooks the fact that the polyethylene begins to recover its original dimensions once the gas pressure is removed, but of course this process would not be complete in a time less than the creep time.

Problem 3 An HDPE pipe, for which the tensile creep data is given in Fig. 6.6, must not have a hoop creep strain greater than 3% after 50 years. If the gas pressure is 2 bar, calculate the maximum SDR that can be used. Poisson's ratio can be assumed to be equal to 0.4 and constant.

We now look at the creep due to soil loads. When a gas pipe is installed in a trench, and the trench is back-filled and the road surface or paving stones replaced, the soil exerts forces on the pipe. It is difficult to calculate the exact magnitude of these forces, as they depend on the degree of support of the back fill by the walls of the trench, and on the relative stiffnesses of the pipe and the soil. We want to be certain that the pipe will not be significantly distorted in shape in the period before the gas pressure is applied. Fig. 13.7 shows three approximations to the loads experienced. In Fig. 13.7a the entire

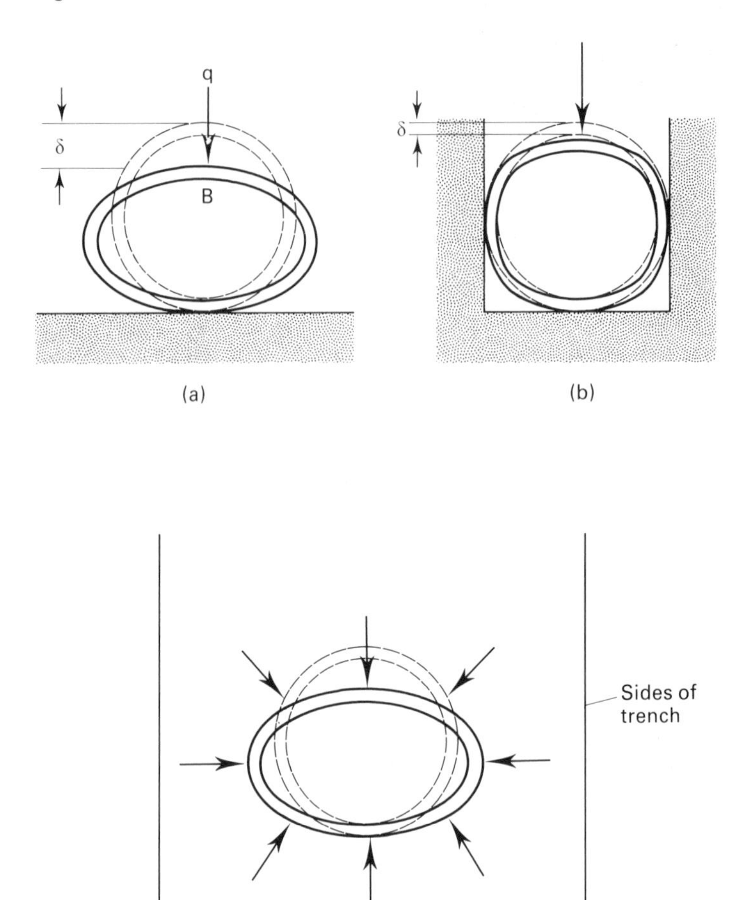

Fig. 13.7 Diametral compression test on pipe (a) without side support and (b) with smooth rigid walls for side support. (c) Loads due to backfill of a trench with soil

weight of the back fill directly above the pipe acts as a diametral load on a pipe that is free to expand sideways. Although this is easy to analyse it is more severe than the real situation. In Fig. 13.7b the sides of the pipe cannot expand sideways as they are constrained by the smooth walls of a rigid box. This is the test situation specified the BGC test for resistance to external loads. Finally Fig. 13.7c shows what happens with buried polyethylene sewer pipes. Initially the polyethylene pipe expands sideways and compacts the soil, but within about 1 year an equilibrium situation occurs in which the loads on the pipe surface become a uniformly distributed pressure, and creep ceases.

The pipe in Fig. 13.7a is a *statically indeterminate* structure, which means that the bending moments cannot be calculated from the external loads alone. Consider first the bending moment distribution in the pipe wall as a result of the applied load per unit length q. If the pipe is cut in half horizontally the bending moment is given as a function of the angular distance θ from the top of the pipe of diameter D by

$$M = \frac{qD}{4}\left(1 - \cos\theta\right) \tag{13.2}$$

In cutting the pipe wall we have released the bending moment m which keep the walls vertical at these points. The value of m is obtained by an analysis of the stored elastic energy in the pipe wall as $m = -\frac{1}{2}qD(\frac{1}{2} - 1/\pi)$, so the maximum bending moment (per unit length) has a value

$$M_{\text{max}} = 0.159qD \tag{13.3}$$

under the applied load at B. Since the elastic bending stresses reach a maximum value of

$$\sigma_{\text{max}} = \frac{6M}{t^2} = 0.95\frac{qD}{t^2} \tag{13.4}$$

on the inner and outer surfaces of a beam of thickness t, we are in a position to calculate the maximum stress given the applied load q. We can then determine the creep modulus that corresponds to the maximum stress, and predict the time dependent reduction in the vertical diameter of the pipe using

$$\delta(t) = 0.223q\left(\frac{D}{t}\right)^3 J(\sigma_{\text{m}}, t) \tag{13.5}$$

Problem 4 An empty pipe is placed at the bottom of a trench and a depth of 1 m of soil placed on top of it. If the soil density is 2000 kg m^{-3}, and the weight of the soil can be considered to act as a concentrated force as in Fig. 13.7a, what is the maximum allowable SDR for the pipe if the vertical deflection after 1 hour is not to exceed 10% of the pipe diameter? The polyethylene creep data can be taken from Fig. 6.6.
Hints: the force exerted by the soil per unit length of pipe is

$$q = Dh\rho g$$

where h is the depth of soil of density ρ and g is the acceleration of gravity = 9.8 m s^{-2}. Start with an assumed SDR of 20, calculate the maximum stress using equation (13.4), then find the creep compliance for this maximum stress. If the deflection found from equation (13.5) is too large, revise the value of the SDR and repeat the calculation.

The analysis of the deflections of a pipe surrounded by soil is based on a semi-empirical formula due to Spangler. The increase in the horizontal diameter Δx, which is approximately equal to the decrease in vertical diameter $-\Delta y$, is given by

$$\Delta x = \frac{0.15P_{\text{v}}D}{(t/D)^3E + 0.091E'} \tag{13.6}$$

where P_{v} is the vertical soil pressure on the pipe. E' is the *modulus of soil reaction*, a kind of material hardness, defined by

$$E' \equiv \frac{P_H D}{\Delta x} \tag{13.7}$$

where P_H is the horizontal soil pressure on the pipe. Typical values for the modulus of soil reaction are:

Gravel—uncompacted	20 MN m^{-2}
Sandy clay—uncompacted	2 MN m^{-2}
Sandy clay—compacted	<9 MN m^{-2}

Note the similarity between equations (13.5) and (13.6); the different numerical constant is because there is a uniform vertical pressure in one case and a concentrated load in the other. The Spangler analysis shows that for most soils there is no risk of long term soil loads causing excess deformation. We have also neglected the effect of the gas pressure, which will act against the deformation from the vertical soil loads.

13.2.6 Fracture mechanics of the pipeline

In Section 8.5.3 the likelihood of runaway crack growth in pressure pipe was analysed. This requires a value of K_{IC} measured under the same conditions of temperature and strain rate as those the pipe experiences. The K_{IC} data given in Table 13.2 were from slow loading in a tensile test machine, and the polyethylene was cooled to well below room temperature to ensure that the fracture was plane strain. More recent experiments, in which the load was applied for the order of 1 ms to the polyethylene at room temperature, show that the K_{IC} values appear not to change with temperature. However, this cannot be assumed to be universally true for polymer fracture.

When a pipe of a certain diameter and SDR is manufactured it is vital to know if plane strain fracture is possible. The rate of application of the load will affect the yield stress, and hence the thickness at which there will be a changeover from plane stress to plane strain fracture. Even if the thickness is such that a plane strain fracture can occur, this fracture cannot propagate for a long distance unless the K value supplied by the internal gas pressure is higher than K_{IC}. A crack could be initiated by a local impact by a mechanical excavator but this crack will arrest within a short distance if $K < K_{IC}$.

This problem has become less severe with the development of super tough grades of medium density polyethylene. The plasticity at the crack tip in these materials is so extensive that it spreads to the far boundary of the pipe wall before the crack can propagate. This means that the concept of K_{IC} as a fracture criterion no longer applies. In its place it is necessary to calculate the pressure necessary for through-section yielding. The consequence is that the presence of small cracks is no longer able to reduce the failure pressure from that for ductile creep rupture by more than a small percentage.

13.2.7 Summary of the design requirements

When the design calculations 2 to 4 are completed (see the answers at the end of the chapter), the critical (smallest) safety factor can be found, and hence the upper limit on the design SDR can be set. In these calculations it was assumed that the gas pressure was 2 bar.

Problem	Design criterion	Conclusions
2	Ductile creep rupture at 20 °C	$\sigma_H < 10$ MN m^{-2} at 50 years so SDR < 101
3	50 year hoop strain <3%	SDR < 41
4	Soil loading of 20*D* kN m^{-1}	SDR < 20 so diametral deflection <10% at 1 h. Long time: soil supports sides of pipe
	Creep rupture data at elevated temperatures, extrapolated to 20 °C	Select a suitable MDPE of low MFI
	Slow crack growth (Section 9.5)	Eliminate foreign particles >0.1 mm in the PE, monitor flaw sizes at welds
	Runaway crack growth (Section 8.5.3)	If $K = 2.1$ MN m$^{-1.5}$, and SDR = 21, $D < 0.70$ m

The conclusion is that the SDR should be less than 20 to ensure safety against all the failure modes listed, and that the polymer grade selected should meet certain fracture mechanics requirements. The melt flow requirements of manufacturing the pipe and sockets place a lower limit on the MFI that can be used. This rules out increasing the toughness of the polyethylene by using a very low MFI. The initial design by British Gas used high safety factors (see Table 13.3), because of a number of uncertainties, including the size of any flaws in fusion joints, the possibility of accidental damage to the pipe during installation and bending stresses in the pipe at the junction of service pipes to the main. Confidence has grown in the MDPE pipe system since there have been no failures (it was reported that in 1989 in Hungary a crack propagated 750 m while a newly constructed PE gas system was being proof tested at 50% above the working pressure). One way of cutting back on excessive safety factors is to increase the allowable pressure in existing gas pipes.

13.3 DESIGN OF RIGID FOAM PROTECTION IN CYCLE HELMETS

13.3.1 Biomechanics criteria for the protection of the head

In this case study we are faced with far more unknowns than in the first one. The reasons why plastics are used are the low density of the materials and the possibility of economic mass production. There are many applications of polymer foams for injury prevention, for instance the fascia padding in cars. The helmet area is chosen here because of the author's research experience. For bicycle helmets the outer shell plays a minor part in the energy dissipation, so the design process can be concentrated on the foam liner (Fig. 13.8). For motor cycle helmets, where the thermoplastic shells are 4 to 5 mm thick, the deformation of the shell can absorb the order of 40 J when the helmet hits a hemispherical anvil of radius 50 mm.

The purpose of bicycle helmets is to reduce head injuries and deaths in 'accidents'. The word 'accident' is a misnomer as there may be culpable parties, whose behaviour could have been modified by training or by the enforcement of the traffic laws. The psychological factor of perceived risk may cause the user of a safety product, be it a seat belt or a protective helmet, to behave in a more reckless way than if he did not wear the product. He may increase his speed until his perceived risk is the same as before, or he may drive as if he was invulnerable. The intended social benefit of the product, of reducing hospital costs and the costs of supporting disabled citizens, need to be evaluated by epidemiological studies. These have shown that the wearing of motorcycle helmets has decreased serious injuries and deaths by approximately 30% (comparing States in the USA with and without helmet-wearing laws). The compulsory wearing of bicycle helmets in Australia after July 1990 has caused a significant reduction in head injuries, but some of the reduction can be put down to other causes, such as a reduction in the number of cycles on the roads.

The injuries to the head can be classified into three types:

(a) Skull fractures. These may be caused by rigid objects penetrating the skull. The outer shell of a helmet may prevent such penetration, by spreading out the force applied to the head. Some minor skull fractures do not cause brain injuries, and it could be argued that this is one of nature's mechanisms for absorbing impact energy.

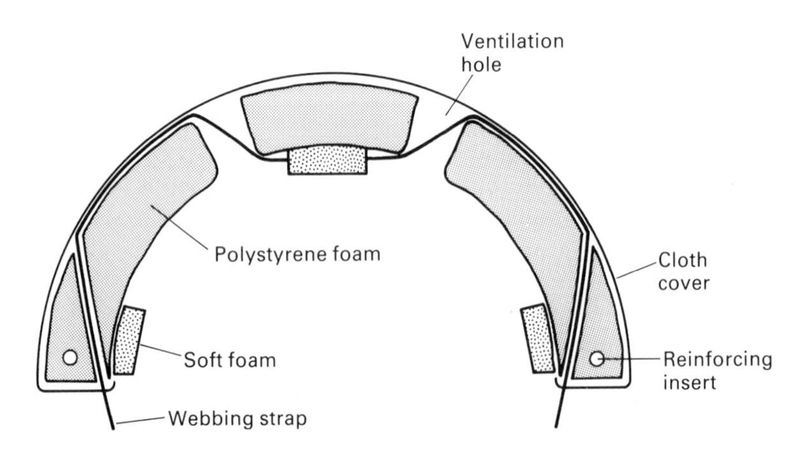

Fig. 13.8 The components of a bicycle helmet, seen in cross section. The shell may be omitted or replaced by a soft cloth cover. The comfort foam pads in the interior are used to fit the range of head shapes to a limited range of liner sizes

(b) Linear acceleration of the brain. In helmet or car crash standards it is the linear acceleration of the head that is used as the criterion of protection. The current UK bicycle and motorcycle standards set an upper limit of 300 g for the acceleration of the rigid headform used in testing. The foam liner of the helmet, by providing a stopping distance for the head, reduces the linear acceleration. A direct blow to the skull can cause brain swelling and bleeding (haematoma) below the impact point, on the opposite side of the skull (a contra-coup injury), or distributed in various parts of the brain. There is experimental evidence that blows to the sides of the head can cause more severe brain injuries than frontal blows with the same acceleration levels.
(c) Rotational acceleration of the brain. When the heads of animals were subjected to high levels of rotational acceleration, while the skull remained undeformed in shape, it was possible to produce concussion, or permanent brain damage of a diffuse nature. There are no tests for rotational acceleration in helmet standards, because it is not perceived that this is a major cause of injury. Different shell materials (ABS or fibreglass) produce approximately the same sliding force at a given sliding velocity for oblique impacts on a road surface, so it is not clear what a limit on the rotational force could achieve. A situation which does produce a high level of rotational acceleration is when a rider falls from a moving horse and his/her helmet indents a muddy field.

It is difficult to be sure that a specific closed head injury is solely due to linear acceleration, as linear and rotational acceleration of the brain usually occur at the same time. In this case study we examine the design methods and materials selection for reducing linear acceleration of the brain, under the constraints of mass, size and cost. It should be made clear at the start that helmets cannot prevent all head injuries, and hence the aim is to minimise the social costs of injuries to the population of road users.

13.3.2 Geometry of the helmet/impacted object interface

Although neither the human skull nor the outer surface of a helmet is exactly spherical, it will be assumed that the section where the impact forces act is spherical. This allows a simple geometrical analysis of impact geometry (Fig. 13.9). The most common object hit is the flat road surface, which can be treated as a rigid body. In the analysis the shell of the helmet is assumed to be of negligible bending stiffness. The error resulting from this assumption will emerge when experimental data are compared with the predictions of the theory.

The foam is assumed to crush at a contant yield stress σ_y when compressed. The contact area is a circle of radius a when the maximum deflection of the foam outer surface is x (Fig. 13.9). Applying Pythagoras's theorem to the triangle OAB gives

$$R^2 = (R - x)^2 + a^2$$

If the amount of linear crush x is much less than the radius of curvature R of the spherical outer surface, the x^2 term in the expansion of the equation can

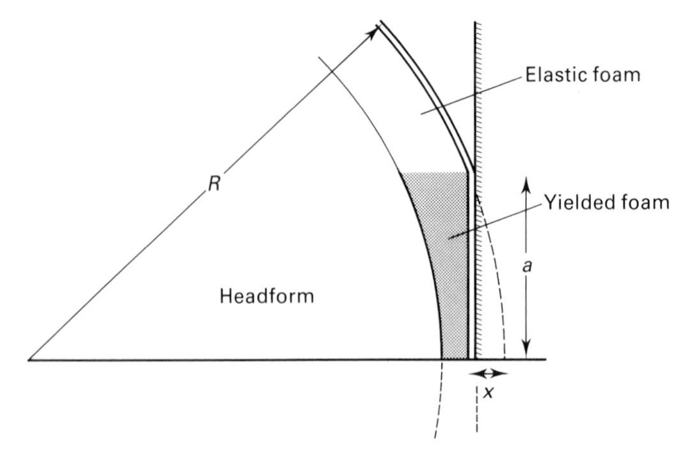

Fig. 13.9 The geometry of the contact between a spherical helmet shell and the rigid flat road surface (from Mills and Gilchrist, *Accident Anal. Prev.*, 1991, **23**, 153)

be ignored, and the contact area A is given by

$$A = 2\pi Rx \tag{13.8}$$

The force F transmitted by the foam is

$$F = 2\pi R\sigma_y x \tag{13.9}$$

so long as the strain is increasing. Once the foam begins to unload the force drops rapidly as the cell walls do not fully recover from their buckled state. Substituting typical values of $R = 160$ mm and $\sigma_y = 0.7$ MN m^{-2} for the side of a cycle helmet liner into equation (13.9) gives an effective foam loading constant of

$$k = F/x = 700 \text{ N mm}^{-1} \tag{13.10}$$

A more refined calculation, using the measured stress–strain properties of the foam and the strain distribution across the contact area, predicts a linear force–distance relationship until the deflection exceeds 80% of the foam thickness.

The radius R of helmets varies from about 100 mm at the front to about 170 mm at the sides, a consequence of the oval shape of human heads. The loading constant predicted by equation (13.10) will change according to the impact site. If the impact is on to a hemispherical anvil, then equation (13.8) for the contact area must be modified. Fig. 13.10 shows the experimental data for the impact striker force as a function of the central deflection of the helmet outer shell.

The loading part of the responses is linear, in agreement with the simple theory. The slopes of these graphs are compared with the theoretical predictions in Table 13.4. The measured yield stress of the 68 kg m^{-3} density foam in the Centurion helmet is 1.06 MN m^{-2} (Mills, 1991). The theoretical value is 59% higher than the experimental value for the impact on a flat surface, probably because the headform radius is slightly smaller than that of

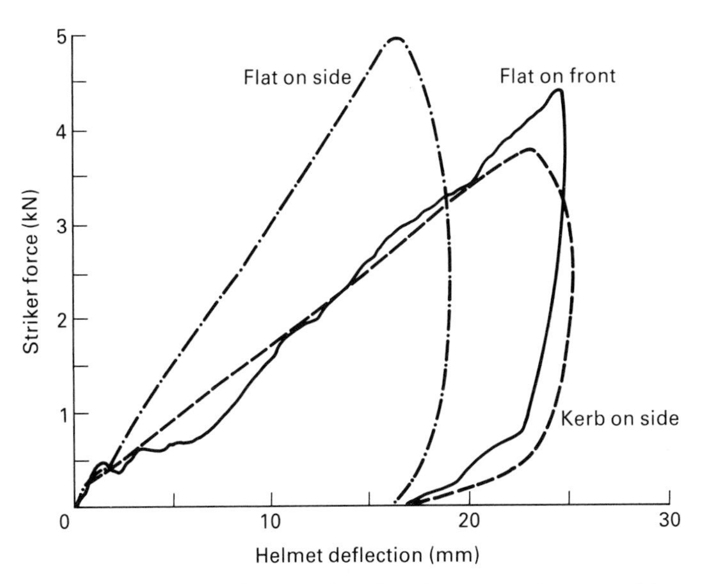

Fig. 13.10 Force deflection curves for a bicycle helmet impacted at the side and front on to flat and kerbstone anvils (from Mills, *Br. J. Sports Med.*, 1990, **24**, 55)

the interior surface of the foam liner. The theoretical slope for the impact on the hemispherical anvil is only 73% of the experimental value. This suggests that the shell, by refusing to conform completely to the shape of the hemispherical anvil, increases the area of foam crushed compared with a 'soft shell' helmet. The slope of the loading curve for the frontal impact on a flat surface is lower (309 N mm^{-1}) than for a side impact, because the radius of curvature of the helmet is lower at the front.

Table 13.4 Loading curve slopes for impacts on a Centurion bicycle helmet

Impact conditions	Experimental (kN mm^{-1})	Theoretical (kN mm^{-1})
Side on flat anvil	711	1130
Side on hemisphere	350	257

13.3.3 Design of a helmet liner for a particular impact velocity

There is a limit to the energy absorption of a foam in compression. Opposite cell faces begin to touch, and the compressive stress rises rapidly as the compressive strain approaches $1-R$, where R is the relative density of the foam. Section 7.4.2 described the gas-pressure hardening that occurs when these materials are compressed. The behaviour described by equation (13.8) will cease when the foam 'bottoms out'. The thickness of the foam must be chosen so that it does not bottom out during the impact. As the foams have relative densities close to 0.05 this means that the foam cannot be compressed by more than 80% of its thickness before marked non-linearities occur in the force deflection relation.

Most standards specify an impact velocity that the helmet and headform have before they strike a rigid fixed anvil. In BS 6863 this is $4.57\ \mathrm{m\,s^{-1}}$ corresponding to a free fall from 1 m. The kinetic energy of the helmet of mass 250 to 500 g is much less than the kinetic energy of the headform. For an average-size head, of mass approximately 5 kg, the kinetic energy is 50 J for a 1 m drop. Calculations will be made for a higher 100 J impact energy, to show what protection level is possible. The EC standard due to appear in 1993 may have a drop height of 1.5 m, and a pass/fail criterion of 250 g headform acceleration. Helmet designers allow a margin for material variability, and tests at high and low temperatures, so the target maximum acceleration is 200 *g* at 20 °C. This is equivalent to saying that the force on the 5 kg headform must not exceed 10 kN. In Fig. 13.11 the energy under the force–deflection curve is equal to the energy input. If the force just reaches 10 kN as the headform decelerates to a momentary halt, the deflection at this point x_{max} must be chosen so that

$$0.5 \times 10\ \mathrm{kN} \times x_{max}\ (\mathrm{mm}) = 100\ \mathrm{J} \tag{13.11}$$

Problem 5 Use the above analysis to determine the minimum thickness of foam in a bicycle helmet to keep the head acceleration below 200 *g* for an impact on a flat surface with 100 J kinetic energy. What then should the foam yield stress be, if the radius of the helmet exterior at the impact site is 100 mm?

There cannot be an optimum design as the shape of the surface struck

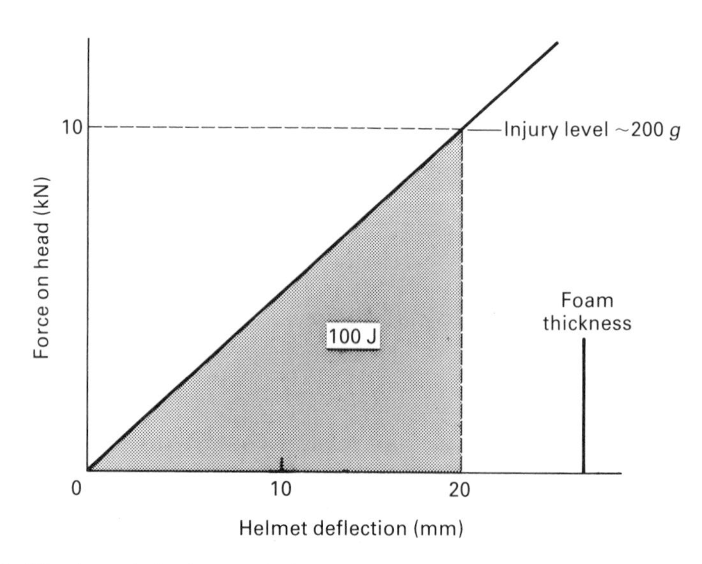

Fig. 13.11 The force deflection graph for a bicycle helmet impact showing that the impact energy (the area under the loading curve) must be absorbed without crossing the injury force limit, or the foam deflection limit when the foam bottoms out

varies. The foam that is ideal for an impact with a flat surface has a too low a yield stress to be ideal for an impact on a kerbstone. The geometry of the foam at the side of the helmet must be altered if the loading stiffness is not to be higher in this large R region; one way of decreasing the stiffness in this region is to incorporate ventilation holes and thereby reduce the contact area of the foam.

As the impact velocity increases the thickness of the foam must increase with the square of the velocity, and the yield stress of the foam must decrease in proportion to keep the head acceleration below 200 g. The thickness of the foam is limited by the mass of helmet that is comfortable to wear, the necessity not to restrict the field of vision of the wearer and increased aerodynamic drag with increasing helmet size.

13.3.4 Choice of foam

The requirements are for a helmet liner of minimum weight and low production cost. Equation 7.24 shows that the yield stress of closed cell foams varies with the 1.5th power of the relative density of the foam. In order to minimise the foam density the polymer chosen must have the maximum yield stress in the bulk state.

Problem 6 Polystyrene has a bulk yield stress at high strain rates of $\sim$120 MN m^{-2} whereas polypropylene has a yield stress of $\sim$60 MN m^{-2}. If a bicycle helmet liner is to be designed as in Problem 5, which foam should be used? Consider the density of the required liner.

There are several processes for moulding helmet liners of complex shapes. The cheapest process is to pour a self-foaming liquid such as a rigid polyurethane into a mould. However this is a slow process and the density of the liners is double that of polystyrene liners of the same yield stress, because of the low bulk yield stress of polyurethane. The main process used is the fusion of prefoamed beads of polystyrene or polypropylene, using pressurised steam in a mould similar to an injection mould. Although the polypropylene liners cost approximately twice as much to manufacture the foam is less brittle and recovers better after an impact. This may be a consideration for products such as skate board helmets which suffer a large number of minor impacts. Polystyrene foam only recovers by a small amount after an impact (Chapter 7) so the helmets must then be destroyed.

13.3.5 Summary

The above calculations show that bicycle helmets can be designed for impacts of up to 100 J kinetic energy. The mass of the helmets can be as low as 200 g and they are comfortable to wear. Design for higher impact energy levels would lead to helmets of an unacceptable size, so it is impossible to protect riders for the most extreme impacts.

Polystyrene foam is the best foam for a helmet that offers single-impact protection, as the mass of the liner is minimised. This material has the drawbacks of limited surface durability, but a thin shell of a thermoplastic or fibreglass will reduce abrasion of the exterior of the helmet. Polystyrene foam is capable of brittle fracture if bent too far. This means that the shape of ventilation holes needs careful design, as does the means of attaching the retention straps that pass under the rider's chin. If this is not done the forces of a crash could cause the helmet to fracture into several pieces.

13.4 DATA STORAGE ON POLYCARBONATE COMPACT DISCS

13.4.1 Information storage on plastic discs

The familiar compact disc for music reproduction is one of a family of information storage products, the others being the CD-ROM used to store text for retrieval by a computer and the larger laser-vision disc used to store video signals. In this case study we shall concentrate on the selection of the plastic, the features of the moulding process required to produce this high precision product and the optical properties of the disc. Fig. 13.12 shows the 'light pen' that reads the information stored in the pits in the CD surface.

The development of the CD and its introduction on to the market in 1982 required the technical advances of

(a) Solid state lasers of small size. They must be small enough and light enough to fit into the moving pick-up head of the CD player. A high light intensity is needed at the information tracks, and a focused laser was the only practical solution for this. The AlGaAs laser operates at a wavelength of 780 nm and is of 1 mW power. Layers of this material are grown epitaxially on GaAs crystal substrates, then doped to be n or p type. The result is that the laser light emerges from the thin layer in the centre of a GaAs sandwich in a plane polarised form. The laser is less than 1 mm long and it forms a negligible part of the total mass of the light pen. A solid state laser that operates at a shorter wavelength (closer to 400 nm, within the transparency window of plastics) would allow a higher density of information storage, but such lasers have not been developed yet.
(b) Servomechanisms that can keep the light pen focused on the information track, as it moves past at 1.25 m s^{-1}. The vertical position of the light must be kept within limits of 2 μm so that the track is in focus, and the radial position must be kept within 0.2 μm so that the pen is centred on the track. This must be done with a response time measured in ms. The resonant frequencies of the suspension systems are 45 Hz in the focusing direction and 900 Hz in the radial direction, and the systems can respond to frequencies just below these limits. The disc is not perfectly flat nor are the tracks perfectly circumferential—the radial position of the track can vary by 300 μm as the disc rotates, and focusing is required over a range of 1 mm.
(c) Digital signal processing techniques to reduce noise in the replayed music. These include error correcting codes so that any faults in the disc, or scratches of dirt on the surface of the disc, do not lead to the type of clicks that mar the

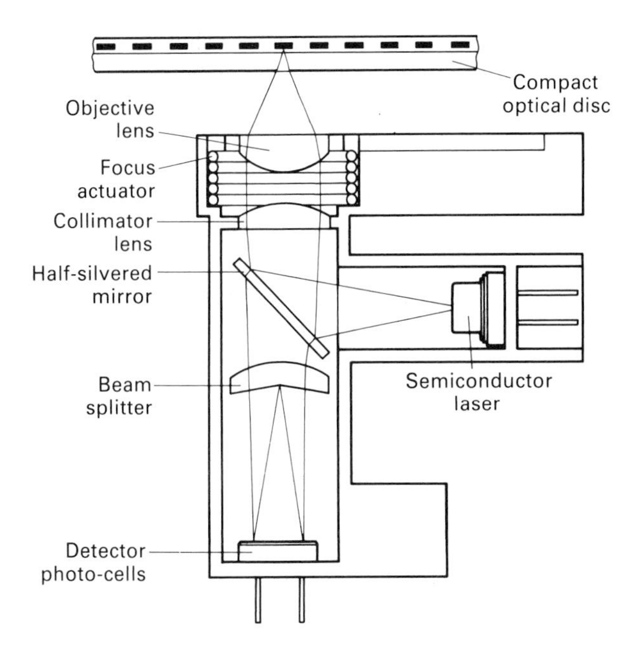

Fig. 13.12 Light from a solid state laser is focused on to the series of pits in the compact disc. The diffracted light is detected by a series of photo-cells. (From Thomas, *Philips Tech Rev* (1988), **44**, 51)

response of old vinyl LPs. Digital signal sampling as 16 bit binary numbers (1 part in 64 500) allows a much better signal to noise ratio than with LPs (90 dB compared with 60 dB for the LP signal, and only 30 dB for the channel separation with the LP). The use of a computer buffer to store a section of the signal before replay allows the time base of the signal to have the stability of the computer clock. This cures the problems of wow and flutter detectable when sustained notes are reproduced on a turnable when the rotation speed varies. In the CD player there is feedback control to the rotation motor from the detected signal.

13.4.2 The optical design for information storage

When music is recorded in stereo via microphones there are two channels of analogue signals. The human ear is not sensitive to frequencies above 20 kHz, so each channel is sampled at a frequency of 44.1 kHz. This allows some leeway above the 22 kHz at which aliasing of the signal would occur. Since each digital sample of the sound intensity is digitised to 16 bit accuracy there are 1.41 Mbit s^{-1} of information to be recorded. The process of coding for error correction increases the rate by a factor of 1/3 then finally each block of 8 data bits is modulated into 17 channel bits for optical recording. Part of this modulation ensures that the length of the pits is between certain limits. The end result is that 4.332 Mbit s^{-1} of information must be stored. Hence if the velocity of the disc past the light pen is 1.25 m s^{-1}, the bit length is 0.3 μm.

The SEM image of the surface of an uncoated CD (Fig. 13.13) shows that there are pits of depth approximately 0.12 μm, that the pits are 0.6 μm wide, spaced by 1.9 μm in the radial direction. The length of the pits varies between 3 and 11 channel bits (0.9 and 3.3 μm) and there is always a distance of at least 3 channel bits between transients. The ramp at the end of the pits must be shorter than 0.3 μm in length. The small size of the pits compared with the 0.78 μm wavelength of the light, and the 1 μm diameter of the focused spot used to read the disc, means that diffraction theory must be used to understand the 'reflection' of the light. Simple explanations of the light beam not being reflected when the light hits a pit are erroneous, especially when the beam is out of focus. The regular spacing of the pits in the radial direction means that the disc acts as a diffraction grating, as can be seen from the colours observed when white light is reflected from its surface. This diffraction grating property means that several pieces of information are contained in the diffraction pattern. By using four detectors to scan different parts of the pattern, these signals can be combined in different ways to achieve the following:

$1+2-3-4$ gives the radial tracking error signal—this is a left-right asymmetry signal;
$1-2-3+4$ gives the axial focusing error signal—this is a measure of the edge slope of the signal, and is a maximum when the pits are in focus;
$1+2+3+4$ gives the audio signal.

The position of the detectors (Fig. 13.14) is such as to measure the first lateral diffraction peak, which occurs at 18° on either side of the normal to the

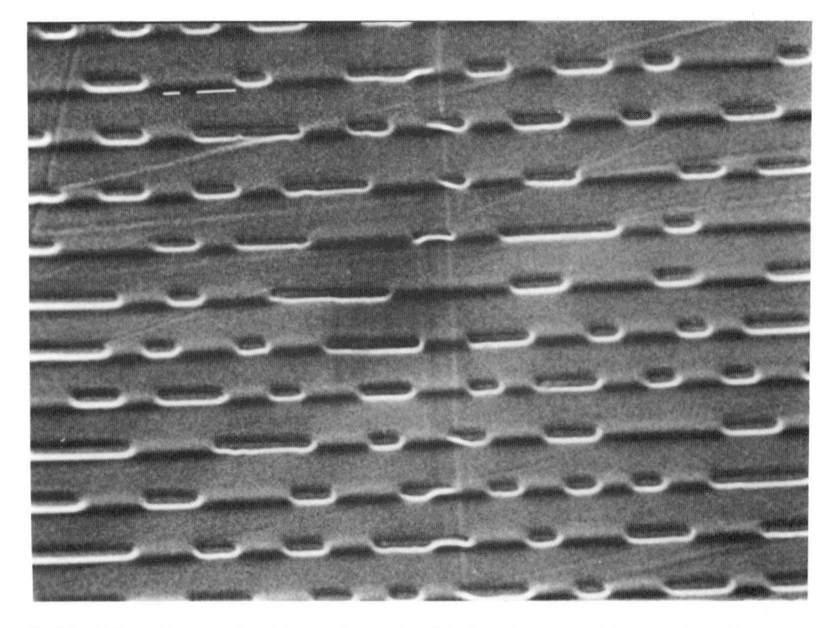

Fig. 13.13 SEM micrograph of the surface of a CD showing part of the spiral track

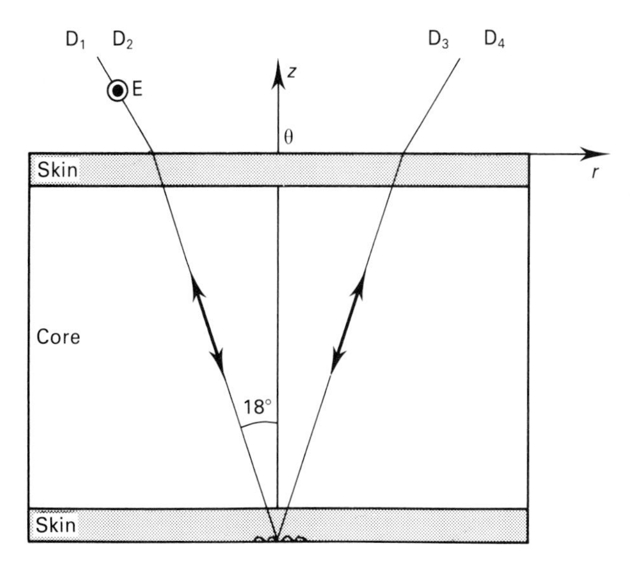

Fig. 13.14 A radial section through a CD showing the rows of pits as a diffraction grating, the positioning of the detectors at the first diffraction peak and the oblique path of the plane polarised light that causes the optical path to depend on n_r as well as n_θ

disc surface. The pits of the CD have an equal reflectivity to the rest of the disc, but they form a phase object. The light that is reflected from them is advanced in phase by

$$\phi = 2\frac{nh}{\lambda} \tag{13.12}$$

where h is the depth of the pit and n the refractive index of the plastic. The detection system can be compared with a phase contrast optical microscope, operating in reflection, with the light passing through the phase object twice. This means that a very shallow pit is possible of $h = 0.12$ μm, compared with a transmitted light disc which would need pits of twice the depth. The limitations of the injection moulding process mean that pits of depth to width ratios greater than 0.2 are very difficult to mould successfully, so the pit size would have to be increased for a transmitted light disc. The phase shift caused by the pit makes the diffraction pattern shift sideways and hence the detectors are no longer at the first diffraction peak. The detection technology with the light beam only being focused on the protected side of the disc means that the unprotected side, where the beam is 0.7 mm wide, can have minor scratches without there being significant deterioration of the signal.

The reflectivity of the surface of the disc is achieved by the vacuum evaporation of a thin layer of aluminium on to the disc under clean room conditions (not more than 100 particles of size greater than 0.5 μm in each cubic foot of air) followed by a coating of lacquer that is rapidly UV cured. This side is then overprinted with the label, further protecting the pits underneath.

13.4.3 The requirements on the plastic, and its selection

Table 13.5 shows some of the qualitative requirements on the plastic used. This allows the elimination of some of the contenders. Polystyrene, which is a low cost material, has too low a resistance to crazing and stress cracking, and the birefringence of a moulded disc would be high because the melt is relatively elastic and has a high stress optical coefficient (this is defined in equation (8.9) and a table of values given in Chapter 8). Extruded PMMA sheet is used for the protective layers of the laser vision discs, but it is not possible to injection mould this polymer with sufficient surface detail. PVC has too low a heat distortion temperature, and its lack of thermal stability makes the injection moulding of high definition surfaces rather difficult.

Table 13.5 Qualitative comparison of glassy materials for compact disc production

Property	Glass	PMMA	PC	PS	PVC
Heat distort temperature	5	3	4	3	2
Birefringence	5	4	2	1	3
Toughness	1	3	5	2	4
Solvent resistance	5	3	3	1	3
Processing	1	4	3	4	2

The scale is 5 excellent to 1 poor.

It is necessary to consider quantitative criteria to explain the selection of the best material. The disc design is asymmetric, with an impermeable aluminium coating on one side, so dimensional changes caused by diffusion of water into the polymer make the disc bow. The water absorption of PMMA at saturation relative humidity is 2.1 weight %, compared with 0.4% for PC, and the diffusion constants as 23 °C for water are 0.5×10^{-6} and $4.8 \times 10^{-6}\,mm^2\,s^{-1}$ respectively. PMMA will expand considerably over a period of tens of days as the water diffuses through the 1 mm thickness of polymer. This does not matter with the laser-vision disc which is of symmetrical construction, so it will not bend as it expands. The requirement for the CD is that the warping of the surface is less than 0.6°, otherwise the laser spot cannot be focused properly. This rules out a PMMA disc with one side sealed.

The transmission of the plastic at a wavelength of 780 nm must be >90%, which is not a critical criterion for polycarbonate. The optical path difference $\Delta p_{r\theta}$ between rays polarised in the radial (r) and circumferential (θ) directions, traversing the disc once, is specified to be less than 50 nm. As the disc is 1.2 mm thick this means that the birefringence $\Delta n_{r\theta} < 4.2 \times 10^{-5}$. As the laser light is polarised in the θ direction, it would appear that only variations in n_θ with the position r,θ matter, and it is not clear how the condition on $\Delta n_{r\theta}$ arises. However, Fig. 13.14 shows that the diffracted light returns through the disc obliquely to the z axis. This means that the path length of the light is affected not only by the refractive index in the direction of polarisation n_θ, but also by the value in the radial direction n_r. If the difference between these refractive indices is too large then the laser light cannot focus properly on the pits (it approaches as a cone of light) and the diffracted signal is affected, reducing the signal to noise ratio of the reproduced music. The birefringence

$\Delta n_{r\theta}$ is affected by the processing conditions and the molecular weight of the polymer, which will be discussed in the next section.

13.4.4 Optimising the processing of the polycarbonate

There are several requirements of the process. The first is to reproduce accurately the shape of the mould surface. This requires both a polymer melt of low viscosity and a high mould pressure for a period while the surface of the disc is solidifying. The viscosity requirement could be met by using a polymer of sufficiently low molecular weight, and processing the melt at as high a temperature as possible. For glassy polymers properties such as impact strength and stress cracking resistance depend on the M_N value exceeding twice the entanglement molecular weight. Hence a low molecular weight may mean that these mechanical properties fall below the minimum acceptable values. It was not possible to use a regular polycarbonate with acceptable mechanical properties for CDs as the viscosity is too high at the 340 °C processing temperature. Consequently a 'special polymerisation technique' (Siebourg, 1986) was used, and the melt flow index was increased to 60 g per 10 min at 300 °C using a 1.2 kg mass, while retaining 80% of the Izod impact strength of the standard polycarbonate. The mould must be filled rapidly or the surface roughness of the type shown in Fig. 5.14b could occur. The mean surface roughness value for injection mouldings is of the order of 1 μm, so the CD requirement that $R_a < 15$ nm is out of the ordinary. The surfaces of many non-CD moulds are etched to impart a pattern that disguises scratches on the plastic moulding. Here the requirement is for an extremely smooth mould, patterned with bumps of 120 nm height, that must be reproduced in detail. Dust must be kept away from the polycarbonate in all stages of transport and drying prior to moulding, and the plasticising stage of the injection moulding machine has a special homogenising screw with starve feed.

The birefringence criterion can only be met by minimising several effects that were discussed in Chapter 5. The first is that the flow into the mould cavity should have axial symmetry, so a film gate allowing flow in all radial directions is used in preference to the common 4 arm gate. The film gate is slightly more difficult to remove cleanly, but it does not start the flow along certain radii. The second condition is to reduce the birefringence of the moulding to a minimum. Earlier in the book it has been shown that birefringence can be caused by molecular orientation in the skin or in the core of the moulding, and by residual stresses in the moulding.

Problem 7 What strategies can be used to reduce the molecular orientation in the skin of injection mouldings, and to keep the skin as thin as possible?

The flow geometry affects the pattern of orientation. There is shear flow in the *rz* plane with the high shear strain rate region close to the disc surfaces. As the disc of melt expands in size elements closer to the core extend in the circumferential θ direction. This extensional flow is tensile in the θ direction and compressive in the *r* direction. The complexities of this radial spreading flow mean it is impossible to predict the orientation distribution from the viscous and elastic flow properties of the melt. The solution to Problem 7 provides a method of reducing the skin orientation, but does not address the orientation in the core of the disc. The core of the moulding does not solidify until after the flow ceases so there is some time for the core orientation to relax (see equation 5.2) before solidification takes place. Fortuitously the core orientation turns out to be equally biaxial in the r and θ directions, so it is less important than the other contributions. The methods of reducing the core orientation are to have a melt of low elasticity, to fill the mould relatively slowly, and to have slow solidification so that there is plenty of time for the core orientation to relax. It is possible that during the relaxation the form of the core orientation changes to being equally biaxial.

The residual stress contribution to the birefringence is related to the pressure history in the mould cavity. Fig. 13.15 shows the mould pressure history for a CD moulding. There is a step at about 80 bar which represents the maximum pressure at the end of the filling stage, then a steep rise to a packing pressure of 450 bar, which is necessary to obtain a good impression of the bumps on the mould surface. The packing pressure is reduced to zero after 1 s. The steep fall in the melt pressure at this stage shows that melt is flowing out of the cavity, and that the gate is not yet frozen. This reduction in mould pressure occurs while the solid layer at the surface of the moulding is only about 0.12 mm thick.

Proglem 8 Section 5.3.4 describes the residual stresses found in injection mouldings—what is the effect of having the majority of the moulding solidify while the melt pressure is zero?

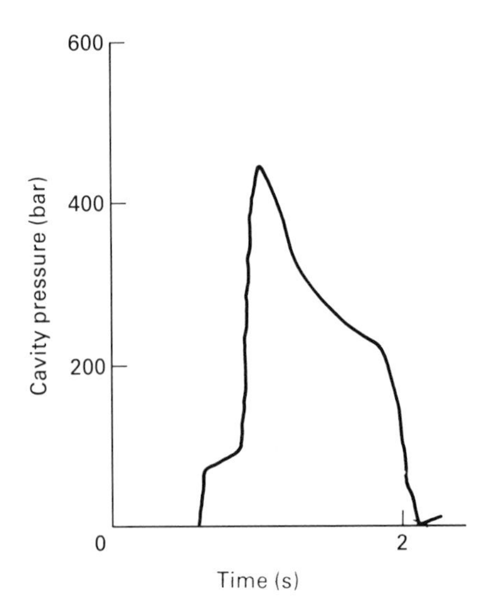

Fig. 13.15 The mould pressure history for the injection moulding of a CD (from Anders and Hardt, *Kunststoffe*, 1987, **77**, 25)

Residual stresses in a flat product should be biaxial, so that for the CD the residual stresses will be equally biaxial compression in the r and θ directions near the surface and equally biaxial tensions in the core. These residual stresses will give rise to a birefringence, according to the stress optic law (equation 8.9), that is proportional to the difference between the principal stresses. Therefore the birefringence $\Delta n_{r\theta}$ due to residual stresses should be zero both at the surface and in the core. The path difference for a ray travelling in the z direction will be the integral of the birefringence

$$\Delta p_{r\theta} = \int_0^t \Delta n_{r\theta}\, dz \tag{13.13}$$

Hence it is possible for compensation to occur between effects that occur in the skin and in interior layers. When we wish to study the magnitude of the effects it is best to measure the path difference Δp along an axis in which the value of Δn does not vary. This rules out the direct measurement of $\Delta p_{r\theta}$ because of the skin core effect along the light path. The radial symmetry of the flow means that Δn_{rz} does not vary along the θ axis and the fact that the flow velocity vectors are along the r axis means that $\Delta n_{z\theta}$ does not vary rapidly along the r axis.

The birefringence effects in CDs has been investigated by Wimberger-Friedl (1990), who cut thin sections of the disc and measured the birefringence for light travelling in both the θ and r directions. Fig. 13.16a shows the distributions of the birefringence Δn_{rz} as a function of the radial position and the layer. The values are much higher than in the important $r\theta$ plane, and they vary markedly with position. The peak A near the surface is due to the shear flow in the rz plane while the skin forms during mould filling; the peak could be removed by insulating the surface of the mould with Teflon. The level of $\Delta n_{rz} \approx 4.5 \times 10^{-4}$ at the midplane seems to be due to the extensional flow of the core. When a short shot moulding was examined (Fig. 13.16b), in which the mould was just filled but no feeding pressure was applied, there is seen to be a nearly uniform value of $\Delta n_{rz} \approx 4 \times 10^{-4}$ for all but the skin. Therefore the feature B in Fig. 13.16a is a negative peak, that can be attributed to the packing pressure applied for 0.5 s. From the theory above a layer that solidifies under a high packing pressure will have a compressive residual stress. When the birefringence distributions $\Delta n_{\theta z}$ were measured they were found to be almost identical with Fig. 13.16a. Using the relationship

$$\Delta n_{r\theta} = \Delta n_{rz} - \Delta n_{\theta z}$$

shows that most of the effects cancel out in the $r\theta$ plane. When the two birefringence distributions were subtracted, the variation of $\Delta n_{r\theta}$ through the thickness of the CD was found. Fig. 13.16c show that there is a residual shear flow orientation effect in the skin at A and a packing pressure effect at B. By adjustment of the process parameters these two effects were made to cancel. The result is that the path difference Δp calculated by equation (13.13) is less than 10 nm for radii between 25 and 55 mm on the disc. Beyond 55 mm the path difference begins to rise rapidly. This is due to the edge effect, as there is more effective cooling at the perimeter of the disc. The outer rim of the disc

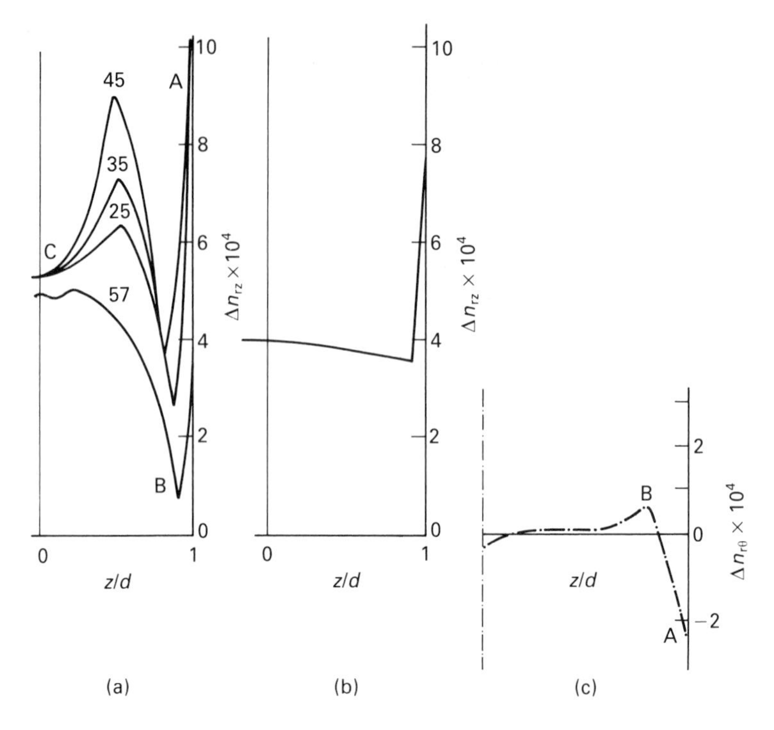

Fig. 13.16 Birefringence variation with distance Z from the midplane of a CD in the rz plane (d is the half thickness). (a) Δn_{rz} for a standard disc at various radial distances; (b) Δn_{rz} for a short shot when the mould cavity pressure <40 bar; (c) $\Delta n_{r\theta}$ for a standard disc. The integral of this distribution gives a path difference of ~10 nm (from Wimberger-Friedl, *Polym. Eng. Sci.*, 1990, **30**, 813)

will solidify before the rest of the disc and this will cause residual compressive stresses in the θ direction in the rim of the disc. The outermost 5 mm of the disc are not used for recording, one of the reasons being this deterioration in the optical quality.

13.4.5 Summary

The CD has very stringent moulding requirements together with optical birefringence limits that could only be satisfied with extensive development. This case study shows that apart from the development of the means of information storage at high density (2×10^9 pits on the disc) there was a plastics technology development. It was essential to mass produce the discs for a low cost using the established technology of injection moulding. However this required significant advances in the cleanliness of the process, and in keeping bubbles out of the moulding. We have seen that common effects of molecular orientation and residual stresses have had to be reduced in magnitude, then balanced against each other, in order to meet the <50 nm path difference requirement. Once the discs have been produced they are not subjected to wear, as occurs with the movement of a stylus in an LP track, or to the ferrite layer coating of a PET tape that passes through a tape recorder. Therefore CDs should be a long life storage medium for information as well as a means of improving the quality of sound reproduction.

ANSWERS TO PROBLEMS

1 When equation (8.16) is used to calculate the transition thickness the values obtained are 40/400/66/42/27 mm for the polyethylenes, 35 mm for the PP and 10/4 mm for the PVCs. The pipe wall thicknesses of gas pipe are in the range from 4 mm up to 40 mm depending on the diameter. There is a risk that if PVC is not processed in an optimal way that the wall thickness could be greater than the critical thickness, so fast brittle fracture could be possible. With PP the value of t_c is comparable to that for the MDPE (the first polymer in the table) so its use could be possible. The much lower melt stability of PP (see Chapter 9) means that the extrusion of thick walled pipe would be difficult without the inner surface suffering molecular weight loss by oxidation. The data for PE shows that for most pipe sizes there should be no possibility of plane strain fracture.

2 The British Gas specification if for a 50 year creep rupture stress greater than 10 MN m^{-2}, and it is suggested that there is a safety factor S in the design to allow for unexpected variations in the pipe properties, dimensions or defects caused by the installation. If a safety factor of $S = 2$ is used then the hoop stress in equation (13.1) must be less than 5 MN m^{-2}. With the substitution of the pressure we obtain $(\text{SDR} - 1) \times 0.1 < 5$, so $\text{SDR} < 51$.

3 When the pipe is placed under pressure there will be hoop and longitudinal stresses in the wall given by equation (7.16). This means that the longitudinal stress is half the hoop stress. The hoop strain for a viscoelastic material is given in equation (6.14).

With the substitution of $\nu = 0.4$ this is

$$e_{\text{H}}(50\,\text{yr}) = (1 - 0.4/2)\,\sigma_{\text{H}} J(50\,\text{yr})$$

As explained in Section 6.3.1, the creep hoop strain will be smaller than that in a tensile creep test (Fig. 6.6) by a factor $(1 - \nu/2)$ if the hoop stress is the same as the tensile stress. Therefore, we examine Fig. 6.6 to find the tensile stress that will cause a creep strain of $3/0.8 = 3.75\%$ after 50 years. The stress is approximately 4 MN m^{-2} using the extrapolation of data shown. For a hoop stress of 4 MN m^{-2} to be caused by a pressure of 2 bar the SDR of the pipe must be 41 by equation (13.1). Therefore, the maximum SDR that meets the 50 years creep strain requirement is 41.

4 The soil load per unit length of pipe is $q = Dgh\rho = 20\,000\,D$ N m^{-1}. This value of the loading q can be substituted in equation (13.4) to find the maximum stress in the pipe:

$$\sigma_{\max} = 0.95\frac{D}{t^2} \times 20\,000\,D = 19\,500\left(\frac{D}{t}\right)^2 = 19\,500\,(\text{SDR})^2$$

Starting with a value of SDR = 20 this gives a stress of 7.6 MN m^{-2}. The creep strain after 1 h is from Fig. 6.6 approximately 2.2%. Therefore, the condition that the diametral deflection δ is to be <0.1 D leads from equation (13.5) to

$$\frac{\delta(1\text{ h})}{D} = 0.223 \times 20\,000 \times 20^3 \times \frac{0.022}{7.6 \times 10^6} = 0.103$$

By chance this result is very close to the required value, so the maximum allowed SDR is 20. The conservative nature of the pseudo-elastic calculation means that there is an in-built safety factor in this result.

5 The condition of equation (13.11) can be met by a 25 mm thickness of foam, and a loading slope of 10 000/20 = 500 N mm^{-1}. Hence for a frontal impact on a helmet of radius R = 100 mm on to a flat surface the yield stress of the foam must be 0.8 MN m^{-2}, by equation (13.8).

6 The density of a polypropylene foam must be 59% higher than the density of polystyrene foam to achieve the same yield stress, according to equation (7.24). In reality the difference in density is less, reflecting the orientation of the crystalline phase in the cell walls of the polypropylene. Other factors such as the cost of the materials and the processing will be important. In Section 7.5.2 the plastic collapse of polystyrene foams was described. The reader may suspect that the response of semi-crystalline polypropylene foam is different to that of glassy polystyrene foam. Hence the foams may differ in properties other than the compressive yield stress.

7 The molecular orientation during the flow depends on the melt elasticity of the polymer, and increases with the flow rate. If the molecular weight of the polymer is kept to a minimum then the melt elasticity is also minimised. The polycarbonate has been chosen partially because it has a low melt elasticity, compared with polystyrene for example. The flow rate can be controlled on an injection moulding machine, and for CD manufacture the mould is filled in between 0.2 and 0.4 s, which is a low flow rate. The thickness of the skin depends on the rate of solidification in the mould, and it can be reduced both by having a very hot melt at 340 °C and by keeping the mould temperature above ambient. With a mould temperature of 95 °C, the calculated thickness of the skin is a maximum near the gate of 4% of the disc thickness. This calculation was carried using a finite difference method (Mills, 1982) for polycarbonate, which has a mean thermal diffusivity of 0.11 mm^2 s^{-1}.

8 Equation (5.7) involves the pressure p in the mould as the layer of melt solidifies. While the cavity pressure is zero all the layers that solidify have the same reference length, and hence the same value of residual stress. This means that only the 0.12 mm thick surface layer, that was solid before the cavity pressure fell to zero, will be under compression, and the rest of the moulding will be under a small residual tension. In this way the residual stress in the surface layer can be kept reasonably low.

Appendix A
Diffusion of heat or impurities

MOLECULAR MODELS FOR DIFFUSION

The starting point for a molecular theory of diffusion is the analysis of a random walk of an atom or molecule. In Chapter 2 we dealt with the possible shapes of a one-dimensional polymer chain, and showed that [equation (2.13)] the number of distinguishable chains of end to end length r is

$$W = A \exp\left(-\frac{r^2}{2nl^2}\right) \tag{A.1}$$

where A is a constant. The probability that a chain, chosen at random, has a length r, is proportional to W. The function must be normalised, so that the sum of the probabilities of all chain lengths is unity. Since the integral

$$\int_{-\infty}^{\infty} \exp(-Ax^2)\, \mathrm{d}x = \left(\frac{\pi}{A}\right)^{1/2} \tag{A.2}$$

the probability $P(r)$ that a chain has a length r is given by

$$P(r) = \frac{1}{(\pi 2nl^2)^{1/2}} \exp\left(-\frac{r^2}{2nl^2}\right) \tag{A.3}$$

This solution can be adapted for diffusion as follows: the diffusing species is assumed to take ν steps per second, so in a time t has traced out a walk of $n = \nu t$ steps. The step length l becomes the jump length in the microstructure. The one-dimensional polymer chain becomes a planar diffusion problem, in which the concentration C only varies in the x direction, and is constant in the yz plane. The equivalent of equation (A.3) then is

$$C(x) = \frac{M}{(\pi 2\nu t l^2)^{1/2}} \exp\left(-\frac{x^2}{2\nu t l^2}\right) \tag{A.4}$$

describing what happens to an initially planar source of impurity atoms (Fig. A1). M is the total amount of the impurity per unit cross sectional area,

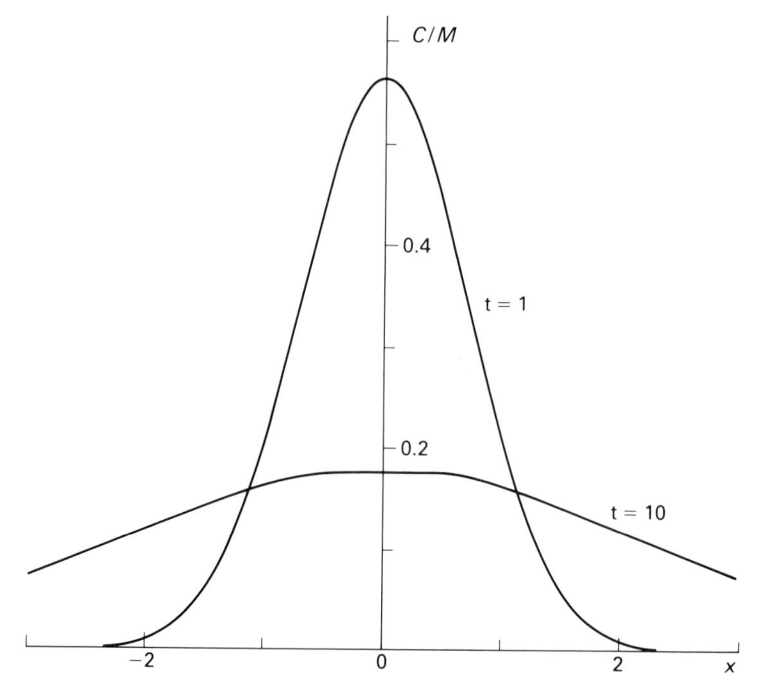

Fig. A1 Diffusion from a planar source in an infinite body at times 1 and 10 s. The distance x is in units of $2\sqrt{D}$ or $(\nu l^2)^{1/2}$

and the diffusion distance x replaces the random walk length r. With increasing time the gaussian impurity distribution spreads out and diminishes in intensity.

DIFFERENTIAL EQUATIONS FOR DIFFUSION

The diffusion of impurities or heat are governed by differential equations, that can be derived from the molecular models just described. We shall only consider a one-dimensional problem; two and three dimensional diffusion problems are analysed in texts as Crank*.

Consider two layers in a solid a distance Δx apart (Fig. A2). Let the concentration of impurity atoms be C_1 and C_2 in the two layers, so the concentration gradient is

$$\frac{\Delta C}{\Delta x} = \frac{C_2 - C_1}{\Delta x} \tag{A.5}$$

If Δx is chosen to be equal to the diffusion step length l then the numbers of impurity atoms per unit area in the layers are $C_1 l$ and $C_2 l$ respectively. In a time interval $\Delta t = 1/\nu$ half of these will jump to the left and half to the right, so the net flow of atoms from layer 1 to layer 2 is

* Crank, J., 'The Mathematics of Diffusion,' 2nd Ed., Oxford University Press, 1975.

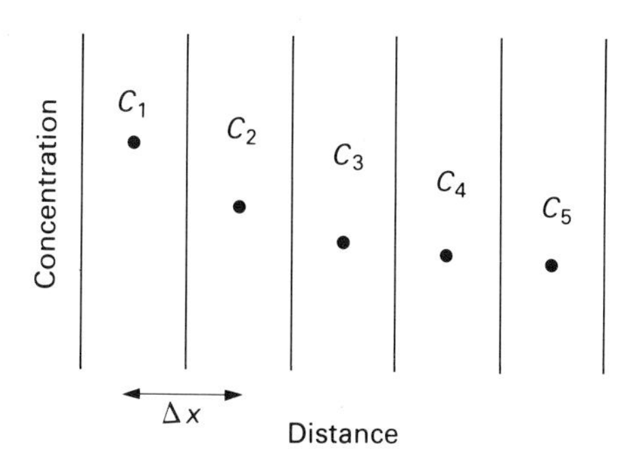

Fig. A2 Finite difference approximation to a concentration profile. The concentrations are C_1, C_2, C_3, . . . , in layers Δx thick

$$\Delta n = \tfrac{1}{2}C_1 l - \tfrac{1}{2}C_2 l$$

Hence the flow rate

$$\frac{\Delta n}{\Delta t} = \tfrac{1}{2}\nu l(C_1 - C_2) = -\tfrac{1}{2}\nu l^2 \frac{\Delta C}{\Delta x} \tag{A.6}$$

In the limit as Δt and Δx tend to zero, equation (A.6) can be written as

$$F = -D\frac{\mathrm{d}C}{\mathrm{d}x} \tag{A.7}$$

which is known as Fick's first law. F is the flow rate of atoms per unit cross sectional area, and the diffusion coefficient

$$D = \tfrac{1}{2}\nu l^2 \tag{A.8}$$

gathers together the three constants (number of step directions, frequency, step length).

The equivalent of Fick's first law for heat flow is used as the definition of thermal conductivity k. It is

$$Q/A = -k\frac{\mathrm{d}T}{\mathrm{d}x} \tag{A.9}$$

where Q is the heat flow in watts down a temperature gradient $\mathrm{d}T/\mathrm{d}x$, and A is the cross sectional area.

A second differential equation is needed for the analysis of non-steady impurity or temperature distributions. It can be derived from equation (A.7) or (A.9) on making the assumption that D is independent of the concentration, or k is independent of the temperature. Fig. A2 shows a solid divided into layers of thickness Δx. In the finite difference heat transfer analysis each layer is assumed to be at a uniform temperature T_i. The temperature gradient between the ith and $(i+1)$th layer can be approximated by

$$\frac{\mathrm{d}T}{\mathrm{d}x} \simeq \frac{T_{i+1} - T_i}{\Delta x} \tag{A.10}$$

In a time interval Δt the increase in the thermal energy stored in layer i is the difference between the heat flows across the left and right hand boundaries. The calculation is made for unit cross sectional area, and yields in finite differences form the equation

$$\Delta x \rho c_p (T_i^* - T_i) = \Delta t \left[k \frac{(T_{i-1} - T_i)}{\Delta x} - k \frac{(T_i - T_{i+1})}{\Delta x} \right] \tag{A.11}$$

where ρ is the density and c_p is the specific heat capacity. T_i^* is the layer temperature at the end of the time interval. The equation can be rearranged to yield

$$\frac{T_i^* - T_i}{\Delta t} = \alpha \frac{(T_{i-1} + T_{i+1} - 2T_i)}{\Delta x^2} \tag{A.12}$$

where the thermal diffusivity α replaces the combination of constants $k/\rho c_p$. Equation (A.12) can be used as a recurrence relation for finite difference calculations, or it can be expressed in differential form by going to the limit as Δt and $\Delta x \to 0$.

$$\frac{dT}{dt} = \alpha \frac{d^2T}{dx^2} \tag{A.13}$$

SOLUTIONS TO THE DIFFERENTIAL EQUATIONS

Only solutions of wide general applicability having simple boundary conditions will be discussed. The steady state solution to Fick's first law is a constant concentration or temperature gradient, and it is only the non-steady state solutions that will be discussed.

(i) A constant surface concentration C_0 on a semi-infinite body

Equation (A.13) is a linear differential equation because only the first power of the differentials occur in it. Consequently any solution of it can be combined with any other solution to provide a further solution. Let us start with the equation (A.4) for an initially planar source of impurity atoms. This can be re-expressed in terms of the diffusion coefficient using equation (A.8)

$$C(x) = \frac{M}{(\pi 4Dt)^{1/2}} \exp\left(\frac{-x^2}{4Dt}\right) \tag{A.14}$$

The problem is modelled as an infinite body, in which there is initially a constant concentration $2C_0$ for $x < 0$. The initial impurity is split up into layers (planar sources) each of strength $2C_0\,d\zeta$ (Fig. A3). Impurity reaching x on the right hand side has diffused a distance of at least x from one of the planar sources, and the total concentration is given by summing the individual contributions as

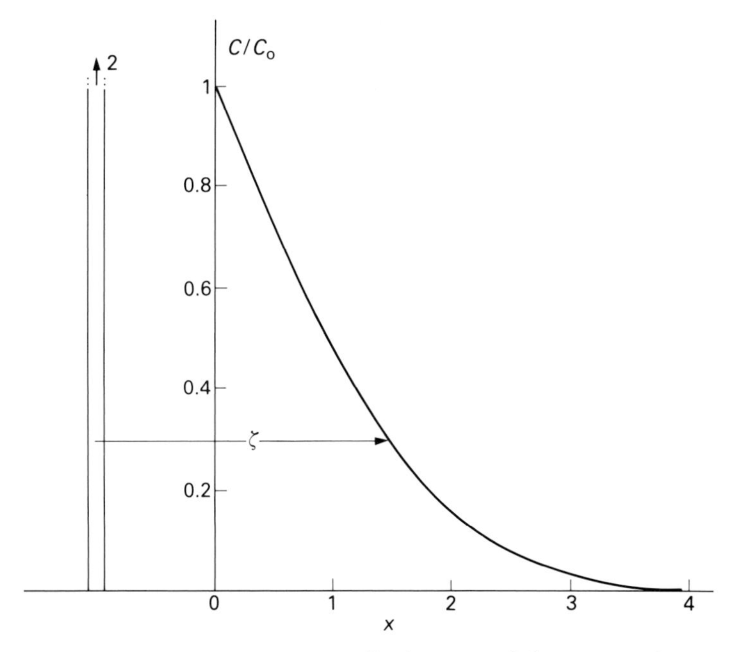

Fig. A3 Superposition of the concentration profiles from a set of planar sources from $-\infty < x < 0$ to produce the erfc profile in the semi-infinite body $x > 0$. ζ is the diffusion distance from a typical source, and x is in units of $2\sqrt{Dt}$

$$C(x) = \frac{2C_0}{(\pi 4Dt)^{1/2}} \int_x^\infty \exp\left(-\frac{\zeta^2}{4Dt}\right) d\zeta \tag{A.15}$$

The integral in equation (A.15) can be written in a standard form by the change of variable $\eta^2 = \zeta^2/4Dt$. The integral known as the error function, used in probability theory, is defined as

$$\operatorname{erf}(z) = \frac{2}{\pi^{1/2}} \int_0^z \exp(-\eta^2)\, d\eta \tag{A.16}$$

It is the area under the normalised gaussian function (Fig. A1) between the ordinates $-z$ and z. The error function complement $\operatorname{erfc}(z) = 1 - \operatorname{erf}(z)$ represents the remainder of the area under the gaussian curve. Equation (A.15) can be written as

$$C(x) = C_0 \operatorname{erfc} \frac{x}{2(Dt)^{1/2}} \tag{A.17}$$

and tables of erfc are available (erfc $0 = 1.0$, erfc $0.2 = 0.777$, erfc $0.4 = 0.572$, erfc $0.6 = 0.396\ldots$). This solution for an infinite body maintains $C = C_0$ at $x = 0$, and so is a solution to the semi-infinite body problem with a constant surface concentration.

(ii) Constant surface concentration(s) on a plane sheet

The simplest problem of this type is when a sheet of thickness $2L$ has an initially constant temperature T_0 and the surfaces $x = 0$, $2L$ are held a constant zero temperature. The problem can be solved analytically by assuming that the variables are separable, so that equation (A.13) has a solution

$$T = X(x)\theta(t) \tag{A.18}$$

On substituting and dividing by θX the equation has the form

$$\frac{1}{\theta}\frac{d\theta}{dt} = \frac{\alpha}{X}\frac{d^2X}{dx^2} \tag{A.19}$$

in which one variable occurs on one side and the other only on the other side. Both sides can then be equated to a constant ($-\alpha\lambda^2$ for convenience) and solved separately to give

$$T = (A \sin \lambda x + B \cos \lambda x) \exp(-\lambda^2 \alpha t) \tag{A.20}$$

The constants A and B are evaluated from the boundary conditions; that at $z = 0$ means that $B = 0$, and that at $x = 2L$ requires that

$$\sin \lambda 2L = 0 \qquad \text{so} \qquad 2\lambda L = \pi, 3\pi, 5\pi \ldots \tag{A.21}$$

The even terms are omitted because the temperature distribution has mirror symmetry about the midplane. Consequently the solution is

$$T = \sum_{m=1,3,5,\ldots} A_m \sin\left(\frac{m\pi x}{2L}\right) \exp\left(-\frac{m^2\pi^2\alpha t}{4L^2}\right) \tag{A.22}$$

The constants A_m are evaluated in the Fourier transform of the initial temperature distribution. The higher the value of m the more rapidly does the sine term die away, so that eventually the temperature distribution becomes a single half sine wave.

In order to implement the finite difference method a value for the time interval Δt must be chosen. The largest value that will give a stable solution is

$$\Delta t = 0.5\, \Delta x^2/\alpha \tag{A.23}$$

in which case equation (A.12) reduces to the simple recurrence relation

$$T_i^* = \tfrac{1}{2}(T_{i-1} + T_{i+1}) \tag{A.24}$$

which is convenient for computation. The results (Fig. A4) are usually displayed in graphical form with dimensionless axes. x/L is the dimensionless distance from the centreline, $(T - T_0)/(T_m - T_0)$ is the dimensionless temperature, and the curves are labelled with the dimensionless time or *Fourier number*.

$$F_0 = \frac{\alpha t}{L^2} \tag{A.25}$$

The reason for presenting the results in this dimensionless way is that they can

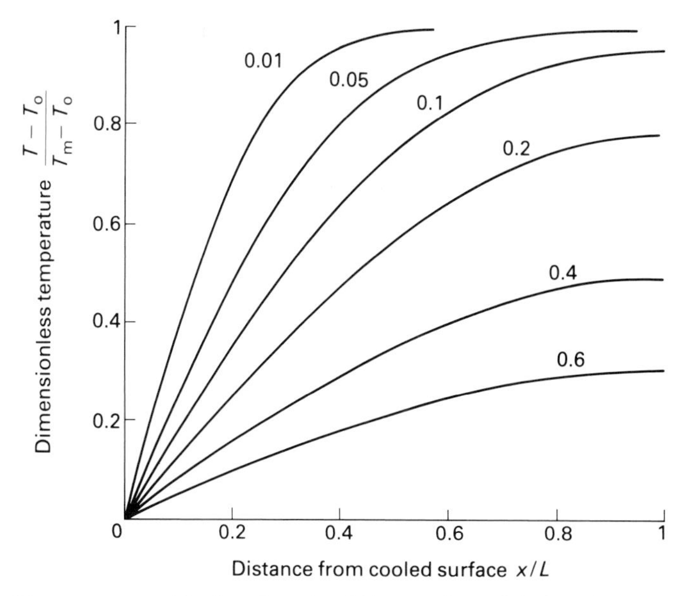

Fig. A4 Temperature profiles in a sheet for different values of the Fourier number. The initial uniform temperature is T_m, the surface at $x = 0$ is kept at T_0 and the surface at $x = L$ is a mirror plane, or is insulated

be applied to sheets of any thickness and any material. A uniform background temperature can also be added.

If the surface concentrations are maintained at constant but different levels, then as well as the transient material flows there will also eventually be a steady state flow. A problem of this type is met in Chapter 10.

Appendix B
Polymer melt flow analysis

We need to know how to convert the data from at least one type of melt rheometer into a flow curve, and then how to use such a curve to estimate the pressure drops in simple melt processing equipment. If the reader wishes to go beyond the elementary treatment given here, specialist texts* are available. The most general flows involve both shear strain rates and extensional strain rates, but there are some simple cases where there are only shear strains. Fig. B1 shows four types of flow in channels. These are assumed to be steady laminar flows; turbulence is rare in high viscosity polymer melts, and oscillatory flows are generally avoided because of the detrimental effect they have on product quality. The first task is to identify the strain rates in Fig. B1. Polymer melts adhere to metal surfaces so the melt velocity is zero at a stationary channel wall. The four cases are as follows.

(i) A rectangular slot of breadth $b \gg$ height h

With the xyz axes shown the only non zero strain rate is the shear strain rate $\dot{\gamma}_{yz}$ in the yz axes. Consequently we can drop the subscripts yz on this and on the shear stress τ_{yz} without causing any confusion. Near the sides of the slot there will be a strain rate $\dot{\gamma}_{xz}$, but these edge effects will be ignored. The analysis will also apply to an annular channel when the radius $r \gg h$, since the slot can be 'bent' until the sides meet and an annulus is formed.

(ii) A cylindrical channel

The only non zero strain rate is $\dot{\gamma}_{rz}$ in the cylindrical axes r, θ, z. Consequently the subscripts on this and the corresponding shear stress may again be dropped.

* Han, C. D., 'Rheology in Polymer Processing', Academic Press, New York, 1976.

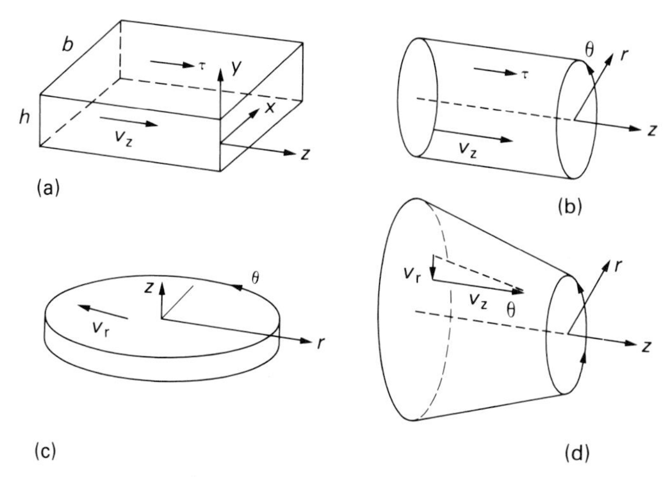

Fig. B1 Flows in channels used in polymer processing. (a) and (b) are shear flows in a rectangular slot and a cylindrical die respectively. (c) and (d) are combinations of shear and extensional flow in a spreading disc and a tapering cylinder. The non-zero velocity components are shown

(iii) A spreading disc flow between plates of separation *h*

This flow commonly occurs at a gate in an injection moulding. There is a similarity with case (i) in that there is a shear strain rate, this time denoted $\dot{\gamma}_{zr}$ in the cylindrical axes. There are also extensional strain rates in the $r\theta$ plane, with

$$\dot{e}_z = 0 \qquad \text{and} \qquad \dot{e}_\theta = -\dot{e}_r$$

because the melt extends in the hoop direction while contracting in the radial direction.

(iv) A tapering cylindrical channel

As well as the shear strain rate $\dot{\gamma}_{rz}$ as in (ii) there is a uniaxial extension strain rate, due to the melt extending in the z direction as it progresses, with

$$\dot{e}_r = \dot{e}_\theta = -\tfrac{1}{2}\dot{e}_z$$

flows; (iii) and (iv), although common in polymer processing, are not easy to instrument or analyse. Consequently we concentrate on the analysis of the *simple shear flows* (i) and (ii). Even with these the anisotropy of polymer melts flowing under high stresses (Chapter 2) produces cross effects; in this case tensile stresses arise in a shear flow as a result of the elasticity of the melt. There is no generally accepted molecular or phenomenological model for the behaviour of polymer melts under complex flows like cases (iii) and (iv), and consequently ad hoc analyses are often necessary.

SIMPLE SHEAR FLOWS IN A SLOT AND IN A CYLINDRICAL DIE

The analysis of the simple shear flows in Fig. B1 proceeds in three stages. The analysis will be given for the rectangular slot, with the similar result for the cylinder given in parentheses.

(a) Find how the shear stress varies with position. The section of channel shown is assumed to be remote from any sudden changes of cross section, so that there are no elastic entrance or exit effects. The pressure p can be assumed to vary linearly along the length, and to be constant across the cross section. We can write

$$\frac{dp}{dz} = -\frac{\Delta p}{\Delta L} \tag{B.1}$$

where Δp is the pressure drop between the entrance and exit of this section. Next we consider the forces on the slab of liquid between $\pm y$ in the slot. The liquid is not accelerating, thus the forces on it are in equilibrium and

$$2yb\Delta p + 2b\Delta L\tau = 0$$

hence

$$\tau = -y\frac{\Delta p}{\Delta L}$$

so from equation (B.1)

$$\tau = y\frac{dp}{dz}\left[= \frac{r}{2}\frac{dp}{dz}\right] \tag{B.2}$$

This linear variation of shear stress with position is a consequence of the linear pressure variation down the channel.

(b) Use the melt flow relationship to find the shear strain rate variation. The relationship between the shear stress and the shear strain rate is usually shown graphically and referred to as the *flow curve*. We need to assume a form for the flow curve before we can analyse the data from a pressure flow rheometer. Luckily there are other types of rheometer—drag flow rheometers in which one surface of the channel moves relative to the other—for which this assumption is not necessary. The melt is usually assumed to be *power law fluid* for which

$$\tau = k\dot{\gamma}^n \tag{B.3}$$

where k is a constant that decreases with increasing temperature, and n is a constant that changes with the polymer and the width of the molecular mass distribution. If the pressure drops in the channel are less than 10 MN m^{-2} the flow is effectively isothermal and k in equation (B.3) is a constant. At very low strain rates n in equation (B.3) may tend to 1 so the behaviour reduces to that of a newtonian fluid with

$$\tau = \eta\dot{\gamma} \tag{B.4}$$

the constant η being the viscosity.

In order to combine equations (B.2) and (B.3) we introduce the value of the shear stress at the channel wall

$$\tau_w = -\frac{h}{2}\frac{\Delta p}{\Delta L}\left[= -\frac{R}{2}\frac{\Delta p}{\Delta L}\right] \tag{B.5}$$

so equation (B.2) becomes

$$\frac{\tau}{\tau_w} = \frac{2y}{h}$$

then substitute this in equation (B.3) to obtain

$$\frac{\dot{\gamma}}{\dot{\gamma}_w} = \left(\frac{2y}{h}\right)^{1/n}\left[= \left(\frac{r}{R}\right)^{1/n}\right] \tag{B.6}$$

(c) Integrate to find the velocities and the flow rate. The definition of the shear strain rate in a simple shear flow is

$$\dot{\gamma} = \frac{dV_z}{dy}\left[= \frac{dV_z}{dr}\right] \tag{B.7}$$

where V_z is the z component of velocity.

The volume flow rate Q in the channel is given by the integral

$$Q = 2b\int_0^{h/2} V_z\, dy$$

$$= 2b[V_z y]_0^{h/2} - 2b\int_0^{h/2} y\frac{dV_z}{dy}\, dy$$

The first term in the integration by parts is zero since $V_z = 0$ when $y = h/2$ at the channel wall, so using equation (B.7) in the second term we have

$$Q = -2b\int_0^{h/2} y\dot{\gamma}\, dy \tag{B.8}$$

Equation (B.6) for the variation of the strain rate can now be substituted to give

$$Q = -\frac{2b\dot{\gamma}_w}{(h/2)^{1/n}}\int_0^{h/2} y^{1+1/n}\, dy = -\frac{b\dot{\gamma}_w h^2}{2[2+(1/n)]} \tag{B.9}$$

Equation (B.9) is more useful in its inverted form

$$\dot{\gamma}_w = -\frac{2[2+(1/n)]Q}{bh^2}\left[= \frac{-[3+(1/n)]Q}{\pi r^3}\right] \tag{B.10}$$

Fig. B2 shows the variation of shear stress and shear strain rate across a rectangular channel, for a power law fluid for which $n = 0.5$. The shear strain rate has the same sign as the shear stress but varies in a non-linear manner. The velocity variation is shown alongside. For a newtonian fluid the velocity variation would be parabolic, but here the variation is increased near the channel walls.

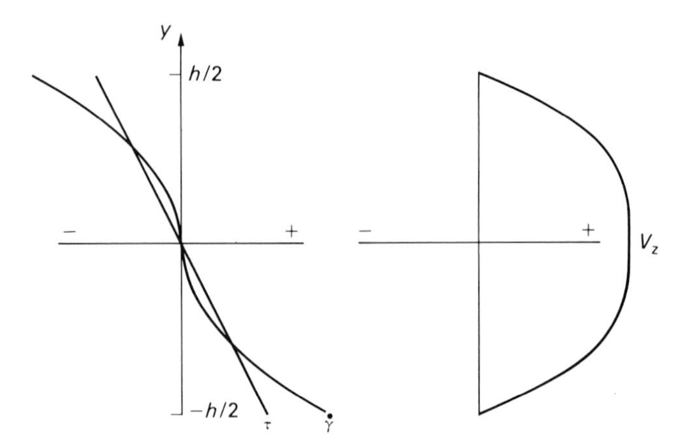

Fig. B2 Variation of the shear stress, shear strain rate and velocity V_z for the flow in Fig. B1 (a) with the coordinate y

PRESENTATION OF MELT FLOW DATA

Melt rheometers either impose a fixed flow rate and measure the pressure drop along a die of constant cross section, or, as in the Melt Flow Indexer, impose a fixed pressure and measure the flow rate. Equation (B.5) can be used to find the shear stress, but equation (B.10) requires a knowledge of n before the shear strain rate can be calculated. Therefore it is conventional to plot the data against the *apparent shear rate* $\dot{\gamma}_{wa}$, calculated assuming newtonian behaviour, with $n = 1$ in equation (B.10). This 'error' is of no consequence when the flow curve is used to calculate the pressure drop in a cylindrical channel because the apparent shear rate can be calculated from the flow rate Q, then the wall shear stress can be read from the flow curve at this apparent shear rate, and the pressure drop calculated from equation (B.5). However, there will be a slight error if a flow curve determined with a cylindrical die is used to predict the pressure drop in a rectangular channel.

The apparent shear rate at the channel wall can be calculated from the mean velocity in the channel $\bar{V}$ instead of the volume flow rate Q, using

$$\dot{\gamma}_{wa} = -\frac{6\bar{V}}{h}\left[= -\frac{4\bar{V}}{R}\right] \tag{B.11}$$

which can be derived from equation (B.10).

It is common to refer to the *apparent viscosity* of a melt, which is defined as

$$\eta = \frac{\tau W}{\dot{\gamma} WA} \tag{B.12}$$

The apparent shear viscosity decreases as the shear strain rate increases, so the use of the term viscosity here does not imply a constant quantity.

Further reading

CHAPTER 1

Albright, L. F., *Processes for Major Addition-Type Plastics and Their Monomers*, McGraw-Hill, New York, 1974.
Brighton, C. A., Pritchard, G. and Skinner, G. A., *Styrene Polymers: Technology and Environmental Aspects*, Elsevier Applied Science, Barking, 1979.
Burgess, R. H., Ed., *Manufacture and Processing of PVC*, Elsevier Applied Science, Barking, 1982.
Frank, H. P., *Polypropylene*, Macdonald, London, 1969.
Kresser, J. O. J., *Polyolefin Plastics*, Van Nostrand, New York, 1969.
Mascia, L., *The Role of Additives in Plastics*, Edward Arnold, London, 1974.

CHAPTER 2

Bassett, D. C., *Principles of Polymer Morphology*, Cambridge University Press, Cambridge, 1981.
Campbell, D. and White, J. R., *Polymer Characterisation: Physical Techniques*, Chapman and Hall, London, 1989.
Hemsley, D. A., *The Light Microscopy of Synthetic Polymers*, Oxford University Press, Oxford, 1985.
Hemsley, D. A., Ed., *Applied Polymer Light Microscopy*, Elsevier Applied Science, Barking, 1989.
Kelly, A. and Groves, G. W., *Crystallography and Crystal Defects*, Longman, London, 1970.
Samuels, R. J., *Structured Polymer Properties: The Identification, Interpretation and Application of Crystalline Polymer Structure*, Wiley, New York, 1974.
Schultz, J. M., *Polymer Materials Science*, Prentice Hall, Englewood Cliffs, NJ, 1974.
Treloar, L. R. G., *Physics of Rubber Elasticity*, 3rd Ed., Oxford University Press, Oxford, 1975.

CHAPTER 3

Allport, D. C. and Janes, W. H., Eds., *Block Copolymers*, Applied Science, Barking, 1973.
Bucknell, C. B., *Toughened Plastics*, Elsevier Applied Science, Barking, 1977.
Freakley, P. K. and Payne, A. R., *Theory and Practice of Engineering with Rubber*, Elsevier Applied Science, Barking, 1978.
Gibson, L. J. and Ashby, M. F., *Cellular Solids: Structure and Properties*, Pergamon Press, Oxford, 1988.
Hilyard, N. C., Ed., *Mechanics of Cellular Plastics*, Applied Science, Barking, 1982.
Hull, D., *Introduction to Composite Materials*, Cambridge University Press, Cambridge, 1981.
Manson, J. A. and Sperling, L. H., *Polymer Blends and Composites*, Heyden, London, 1976.

CHAPTER 4

Bown, J., *Injection Moulding of Plastics Components*, McGraw-Hill, Maidenhead, 1979.
Fischer, E. G., *Blow Moulding of Plastics*, Iliffe, London, 1971.
Frados, J., Ed., *Plastics Engineering Handbook*, 4th Ed., Van Nostrand, New York, 1976.
Han, D. H., *Rheology in Polymer Processing*, Academic Press, New York, 1976.
Levy, S., *Plastics Extrusion Technology Handbook*, Industrial Press, New York, 1981.
Macosko, C. W., *RIM, Fundamentals of Reaction Injection Moulding*, Hanser, Munich, 1989.
Rubin, I. I., *Injection Moulding*, Wiley, New York, 1972.
Tadmor, Z. and Gogos, C. G., *Principles of Polymer Processing*, Wiley, New York, 1979.

CHAPTER 5

Injection Moulds, VDI-Verlag, Dusseldorf, 1980.
Butters, G., Ed., *Particulate Nature of PVC*, Elsevier Applied Science, Barking, 1982.
Ogorkiewicz, R. M., Ed., *Thermoplastics: Effects of Processing*, Iliffe, London, 1969.
Also see References from Chapter 4.

CHAPTER 6

Aklonis, J. J. and MacKnight, W. J., *Introduction to Polymer Viscoelasticity*, 2nd Ed., Wiley, New York, 1983.
Hertzberg, R. W. and Manson, J. A., *Fatigue of Engineering Plastics*, Academic Press, New York, 1980.
Ogorkiewicz, R. M., Ed., *Thermoplastics: Properties and Design*, Wiley, London, 1974.
Turner, S., *Mechanical Testing of Plastics*, 2nd Ed., George Godwin (Butterworths), London, 1983.
Williams, J. G., *Stress Analysis of Polymers*, 2nd Ed., Ellis Horwood, Chichester, 1980.

CHAPTER 7

Kausch, H. H., Ed., *Crazing in Polymers* (Vol. 52 of *Advances in Polymer Science*), Springer, Berlin, 1983.
Powell, P. C., *Engineering with Polymers*, Chapman and Hall, London, 1983.
Ward, I. M., *Mechanical Properties of Solid Polymers*, 2nd Ed., Wiley, Chichester, 1983.

CHAPTER 8

Dugdale, D. S., Yielding of Steel Sheets Containing Slits, *Journal of Mechanics and Physics of Solids* **8**, 100, 1960.
Engel, L., Klingele, G. W. and Schaper, H., *An Atlas of Polymer Damage*, Wolfe, London, 1981.
Kausch, H. H., *Polymer Fracture*, Springer, Berlin, 1978.
Kinloch, A. J., Ed., *Fracture Behaviour in Polymers*, Applied Science, Barking, 1983.
Parker, A. P., *The Mechanics of Fracture and Fatigue*, Spon, London, 1981.
Williams, J. G., *Fracture Mechanics of Polymers*, Ellis Horwood, Chichester, 1984.

CHAPTER 9

Allen, N. S., Ed., *Degradation and Stabilisation of Polyolefins*, Applied Science, Barking, 1983.
Böcker, H. *et al.*, High Performance PE Provides Better Safety for Pipelines, Kunststoff, *German Plastics*, **82**, 8, 1992.
Cullis, C. F., *Fundamental Studies of Fire and Flammability*, Oxford University Press.
Davis, A. and Sims, D., *Weathering of Polymers*, Elsevier Applied Science, Barking, 1983.

Ewalds, H. L. and Wanhill, R. J. H., Fracture Mechanics, Edward Arnold, London, 1985.

Kramer, E. J., Environmental Cracking of Polymers, in *Developments in Polymer Fracture*, Ed. Andrews, E. H., Elsevier Applied Science, Barking, 1979.

Kelen, T., *Polymer Degradation*, Van Nostrand, New York, 1983.

CHAPTER 10

Comyn, J. M., Ed., *Polymer Permeability*, Elsevier Applied Science, Barking, 1985.

Gilmore, M., *Fibre Optic Cabling: Theory, Design and Installation Practice*, Newnes, Oxford, 1991.

Membranes in Gas Separation and Enrichment, Special Publication No. 62, Royal Society of Chemistry, London, 1986.

Mills, N. J., Optical Properties, in Vol. 10 of *Encyclopaedia of Polymer Science and Technology*, 2nd Ed. Wiley, New York, 1987.

Spilman, R. W., Economics of Gas Separation, *Chemical Engineering Progress*, **35**, 41, 1989.

CHAPTER 11

Blythe, A. R., *Electrical Properties of Polymers*, Cambridge University Press, Cambridge, 1979.

Bowden, M. J. and Turner, S. R., Eds., *Electronic and Photonic Applications of Polymers*, Advances in Chemistry Series No. 218, American Chemical Society, Washington, DC, 1988.

Margolis, J. M., Ed., *Conductive Polymers and Plastics*, Chapman and Hall, New York, 1989.

Salanek, W. R., Clark, D. T. and Samuelson, E. J., Eds., *Science and Applications of Conducting Polymers*, Adam Hilger, Bristol, 1991.

CHAPTER 12

MacDermott, C. P., *Selecting Thermoplastics for Engineering Applications*, Marcel Dekker, New York, 1984.

Rosato, D. V., Di Mattia, D. P. and Rosato, D. V., *Designing with Plastics and Composites: A Handbook*, Van Nostrand Reinhold, New York, 1991.

Atkins, R. T., Ed., *Information Sources in Polymers and Plastics*, Bonker-Saur (Butterworths), London, 1989.

CHAPTER 13

Boaiwhuis, G., Ed., *Principles of Optical Disc Systems*, Adam Hilger, Bristol, 1985.
Chow, W. W. C., *Cost Reduction in Product Design*, Van Nostrand, New York, 1978.
Ehrenstein, G. W. and Erhard, G. E., *Designing with Plastics*, Hanser, Munich, 1984.
Mills, N. J., Residual Stresses in Plastics, *Journal of Material Science*, **17**, 558, 1982.
Mills N. J. and Gilchrist, A., The Effectiveness of Foams in Bicycle and Motorcycle Helmets, *Accident Analysis and Prevention*, **23**, 153, 1991.
Morton-Jones, D. H. and Ellis, J. W., *Polymer Products: Design, Materials and Manufacturing*, Chapman and Hall, London, 1986.
Siebourg, W., Polycarbonate—A Material for Optical Storage Media, *Kunststoff German Plastics*, **76**, 61, 1986.
Wimberger-Friedl, R., Analysis of the Birefringence Distributions in Compact Discs, *Polymer Engineering Science*, **30**, 813, 1990.

Questions

CHAPTER 1

1. Examine the data in Table 1.2 for a correlation between a high T_g and presence of in-chain benzene rings or large side groups. Can you explain such a correlation in terms of chain flexibility?

2. Verify equation (1.8) for the molecular weight averages of a step growth polymerisation. Sum the series obtained when equation (1.7) is substituted in equations (1.4) and (1.6).

3. Model the stereoregularity of a PVC molecule with a 0.71 probability of a racemic conformation *r*. Generate 100 random numbers R with a calculator, and assume a racemic conformation occurs if $0.71 > R > 0$ and a meso if $1 > R > 0.71$. How many of the 100 units are contained in *rrrr* . . . sequences of length $\geqslant 10$?

4. Plot a graph of the mole fraction F_1 of acrylonitrile in an acrylonitrile styrene copolymer, against the mole fraction $f_1 = [A]/([A]+[S])$ in the monomer mixture, using Table 1.5. If the initial $f_1 = 0.25$ in a batch polymerisation, what is the initial F_1, and how does it vary as the monomer is used up?

5. Discuss the reasons for the relatively low cost of polyethylene compared with step-growth polymers.

6. What are the main two molecular parameters that are controlled in the polymerisation of ethylene and what quality control tests on the polyethylene are used to assess these parameters?

7. PVC is unique among the commodity plastics in that 50% is sold in a plasticised form. In what way does its microstructure differ from the other commodity plastics to make this possible?

CHAPTER 2

1. Discuss the differences between the conformation of a molecule in a

polymer crystal, and in a rubber. Show how the conformations can be modelled by using different sequences of C—C bond rotations.

2. Describe how the disorder arises in a polymer glass, and discuss whether there is any evidence for the presence of micro-crystals of size 10 nm.

3. Contrast the mechanisms behind the elastic behaviour of a rubber and glassy polymer, explaining the range of shear moduli that are feasible for rubbers.

4. Explain how the 2*3/1 helical conformation of polypropylene molecules spaces out the methyl side groups along the helix. Give reasons why the polypropylene crystal has a lower modulus in the **c** direction than the polyethylene crystal.

5. How do entanglements between two polymer chains arise in a polymer melt, and what consequences do they have on the melt flow properties?

6. Semi-crystalline polymers are isotropic on a scale larger than 100 μm but anisotropic on a scale less than 1 μm. Discuss this statement.

CHAPTER 3

1. Design a laminated steel and rubber spring for use as a bridge bearing. The compressive load on the bearing is 10^6 N and it is mounted on concrete that has a compressive design stress of 8 MN m^{-2}. A shear deflection of 7 mm due to the thermal expansion of the bridge deck should not produce a shear force larger than 10^4 N. The compressive load should not produce a vertical displacement of more than 3 mm. The rubber used has a shear modulus of 0.9 MN m^{-2} and is in the form of square sheets.

2. A sandwich panel (as in Fig. 3.4a) is made from a rigid polyurethane form core, of Young's modulus 40 MN m^{-2} and density 92 kg m^{-3}, with equal thickness skins of glass fibre reinforced plastic of modulus 18 GN m^{-2} and density 1900 kg m^{-3}. A one metre width of this panel must have a bending stiffness MR of 2000 N m^2. Consider core thicknesses of 5, 10 and 20 mm and see which gives the lightest panel. What will happen to the surface strains at a given load as the panel thickness is increased?

3. Discuss how the size and shape of the rubber particles in a rubber toughened glassy polymer influence the toughening mechanism, for a tensile stress applied to the material in any direction.

4. Is the mean length of the glass fibre reinforcement in injection moulded polymers long enough to give optimal stress transfer to the fibres? Explain how the stiffness anisotropy in such a moulding arises.

5. What new mechanism of energy absorption arises in a thermoplastic when it is foamed, and how can the compressive yield stress be controlled over a range of values?

6. Explain how an increase in the volume fraction crystallinity of

polyethylene changes the moduli of a stack of lamellar crystals with amorphous interlayers, and hence changes the macroscopic Young's modulus of the polymer.

CHAPTER 4

1. How has the low value of the thermal conductivity of polyethylene influenced the design of the equipment for producing polyethylene sheet?

2. A reduction in the molecular mass of a polymer reduces the melt viscosity. Why does this change not lead to a greater output rate from an extruder/die combination? What disadvantages might occur if blown film was being produced?

3. Polystyrene sheet is being extruded at $T_m = 200$ °C on to a pair of rolls of the type shown in Fig. 4.10, which are at $T_0 = 20$ °C. Calculate the Fourier number required for the whole of the sheet to be cooled below 80 °C at the end of the contact with the rolls. If the sheet is 1.25 mm thick and the roll surfaces move at 0.12 m s^{-1} what diameter should the rolls be? The thermal diffusivity of polystyrene is 0.9×10^{-7} m^2 s^{-1}.

4. Take a blow moulded polyethylene bottle for detergent or fabric softener and section it vertically and horizontally with a razor blade. Examine the weld line at the base and look for roughness on the inner surface.

5. Consider the thermoforming of a domestic bath from 5 mm thick PMMA sheet. Why is it preferable to preheat the blanks in an oven rather than apply radiant heating to the sheet when it is over the mould? What are the stresses in the sheet when it is cylindrical in shape, radius 0.5 m and thickness 4 mm, when the pressure difference across it is 0.5 bar (50 kN m^{-2})? What is a typical draw ratio?

6. A 508 mm diameter cycle wheel, having 5 spokes, is injection moulded using a toughened nylon. It is gated at the hub and is 4 mm thick. Comment on the maximum flow lengths, and whether the positions of the weld lines may cause problems.

7. Contrast the materials needed and the complexity of the moulds for conventional injection moulding, with those for RIM. What difference in cycle time would you expect for a 3 mm thick part?

CHAPTER 5

1. A circular lid for a box is injection moulded in polystyrene. If the melt is injected at 600 bar pressure at 200 °C, what would be the volume shrinkage when the lid is at 1 bar and 20 °C, assuming no feeding of the mould. Explain how feeding can partially compensate for this shrinkage, and why the diameter of the lid has a lower % shrinkage than the lid thickness.

2. Explain the design guidelines, that injection mouldings should have a uniform thickness and that there should be a 1° to 2° draft angle on the inside walls of boxes, in terms of the way in which the moulding solidifies, and the volume shrinkage of plastics on cooling. Does the structural foam process (Chapter 3) overcome these design limitations?

3. Explain how the external water cooling of extruded pipes leads to tensile residual hoop stresses at the bore of the pipe. To see whether a similar effect exists in a blow moulded bottle, cut off the base and top, then cut down one side with a razor blade. If it curls up to a smaller diameter there are residual compressive stresses on the outer surface.

4. Discuss how increases in the mould wall temperature and the melt injection temperature would decrease the orientation in an injection moulding. What adverse effects would these changes have on the production costs?

5. The only way in which to remove the orientation from an injection moulding is to anneal it at a temperature above its melting point. Explain why this is not a feasible process, whereas metal components can be recrystallised at temperatures below their melting points.

6. Why does a particle structure sometimes occur in PVC products but rarely in other polymers?

CHAPTER 6

1. A single Voigt model has parameters $E = 500\ \mathrm{MN\,m^{-2}}$ and $\eta = 2 \times 10^{12}\ \mathrm{N\,s\,m^{-2}}$. What is the retardation time? What creep strain is predicted for a creep stress of $10\ \mathrm{MN\,m^{-2}}$ applied for 500 s?

2. A cantilever beam of length 400 mm and thickness 20 mm has a second moment of area of $2.0 \times 10^{-8}\ \mathrm{m^4}$. A load of 6 N is applied to the free end. Use the data for HDPE in Fig. 6.6 to calculate the deflection of the free end after 2000 hours. Suggest a cross sectional shape for the beam that would minimise its mass if it were to be (a) injection moulded (b) extruded.

3. A linearly viscoelastic polymer has a creep compliance $J = 5 \times 10^{-9}\ t^{0.1}\ \mathrm{m^2\,N^{-1}}$, where the time t is in hours. If it is subjected to an intermittent tensile stress of $5\ \mathrm{MN\,m^{-2}}$ for 6 hours on, 6 hours off, starting at $t = 0$, use Boltzmann's superposition principle to calculate the strain after 29 hours. How does this compare with the strain for a constant $5\ \mathrm{MN\,m^{-2}}$ stress applied for 17 hours?

4. A tensile fatigue test is carried out on a polyethylene, with the strain varying from 0 to 1% at 300 Hz. Calculate the rate of energy dissipation per unit volume if $E' = 1.0\ \mathrm{GN\,m^{-2}}$, $E'' = 20\ \mathrm{MN\,m^{-2}}$, and hence the initial heating rate if the product of density and specific heat is $1.2 \times 10^6\ \mathrm{J\,m^{-3}\,K^{-1}}$.

5. Calculate the complex compliance at a frequency of 0.01 Hz of the generalised Voigt model of Fig. 6.1(b).

6. Two aluminium panels each of thickness 1 mm sandwich a layer of polyurethane of thickness 2.5 mm. What level of shear modulus should the polyurethane have for it to be effective as a vibration damper? The vibration mode has wavelengh of 200 mm so $L = 100$ mm in equation (6.29). Take Γ as 0.5, R as 2 m and the Young's modulus of aluminium as 70 GN m^{-2}.

CHAPTER 7

1. A polyethylene is chosen for blow moulding of a 5 litre liquid container. When a tensile test is performed at an extension rate of 50 mm min^{-1} a neck propagates down the specimen in a stable manner, yet when a prototype container is dropped through 3 m on to a hard surface the tensile neck that forms immediately fractures. Explain.

2. Injection mouldings are usually thin walled to minimise the cooling part of the cycle time. Explain why they are more likely to fail under a compressive load by viscoelastic buckling, than by compressive yielding.

3. Increasing the thickness of a plastics product does not necessarily make it stronger. Discuss this statement with respect to surface indentation by a sharp object, and the transition from plane stress to plane strain fracture (Chapter 8).

4. Use the data in Fig. 7.14 to estimate the creep rupture time at a stress of 4 MN m^{-2} at 20 °C for the particular polyethylene illustrated.

5. How does crazing differ from crack initiation in other solids, in respect of (i) the orientation with respect to the principal stresses (ii) the energetics of growth (iii) the likelihood of multiple craze initiation.

CHAPTER 8

1. Make a collection of broken plastics products, then assess from the fracture appearance (i) whether fracture was due to yielding, fatigue or environmental stress cracking (ii) whether the major load system was tension, bending or torsion (iii) whether there were processing features involved such as weld lines and orientation (iv) whether there were design faults that could have been remedied?

2. A circular hole in a region of a product that is under tensile stress can lead to crack initiation. Contrast the mechanisms in a brittle plastic like polystyrene and a ductile plastic like polycarbonate (when the hole diameter is much less than the part thickness). How can the design be modified to avoid such crack initiation?

3. Differentiate clearly between a stress concentration factor and a stress intensity factor.

4. A surface craze in some PMMA breaks down and becomes a crack. Treat

it as an edge crack of length $a = 0.5$ mm in a body of width $w \gg a$. What tensile stress σ acting perpendicular to the crack will eventually cause the crack to grow? Use the data in Table 8.1 and the stress intensity factor $K_I = 1.12\,\sigma(\pi a)^{0.5}$ where 1.12 is the correction for a surface crack compared with a central crack in a wide sheet.

5. A cracked specimen of transparent PVC is being tensile tested. Is it realistic to expect to see the craze at the crack tip with the naked eye? The K_{IC} value is 2.0 MN m$^{-1.5}$ and the craze stress $\sigma_0 = 60$ MN m^{-2}.

6. Discuss whether the notched Izod impact strengths of 3.2 mm thick plastic sheet are of any value for design purposes.

CHAPTER 9

1. Why does a PVC melt discolour and become more viscous when it is overheated, whereas PP melt becomes less viscous?

2. Explain why step growth polymers such as PETP and PC have to be rigorously dried to prevent molecular mass reductions during melt processing, whereas this is not a problem with addition polymers.

3. Why is oxidation more of a problem for the outdoor use of polyethylene film than it is for moulded or extruded products that are sveral mm thick?

4. Cracking in an ABS product was suspected to be caused by the stapling of a plasticised PVC coated cloth to the inner surface. What tests would you perform to check this hypothesis?

5. The data for the slow crack growth of polyethylene in detergent at 19 °C in Fig. 9.18 has the equation

$$da/dt = 9 \times 10^{-8}\,K^{4.0}$$

when the velocity is in m s^{-1} and K in MN m$^{-1.5}$. Calculate how long it would take for an initial edge crack length $a_i = 0.1$ mm to grow to 5 mm length if the applied tensile stress $\sigma = 5$ MN m^{-2}. The stress intensity factor can be calculated from $K = 1.12\,\sigma(\pi a)^{0.5}$.

CHAPTER 10

1. How many grams of O_2 will pass through a 20 μm thick PVDC film in 1 day at 25 °C? The area of the film is 0.1 m^2 and the pressure differential across it is 1.0×10^5 N m^{-2}. Hence estimate the effectiveness of such a film for wrapping food.

2. Calculate the solubility of CO_2 in a MDPE at 25 °C with a volume fraction crystallinity of 0.59 using the data in Table 10.1. Compare this with the solubility of CO_2 for HDPE calculated from the data in Table 10.2. The data in the 2 tables comes from different sources.

3. Discuss why it is acceptable to use HDPE containers for fluids when there is a finite permeability i.e. for petrol through the wall of a petrol tank, or water vapour through the wall of a brake fluid container.

4. Investigate the orientation in a transparent injection moulding by examining it between polarising filters. Find out whether there is a residual stress effect by making a thin saw cut through the moulding to see if the fringe pattern alters.

5. Estimate the mass of polyurethane foam needed in 100 m^2 of house wall to provide a 'U' value of 0.6 $W\,m^{-2}\,K^{-1}$. How much energy does this save in 3 months compared with a cavity brick wall of $U = 2.0\,W\,m^{-2}\,K^{-1}$ if the average temperature differential is 10 degrees.

CHAPTER 11

1. Explain why the design electric field in a high voltage cable insulated with polyethylene is only 3 $MV\,m^{-1}$ when in laboratory tests the electric strength is measured as 800 $MV\,m^{-1}$.

2. Why is plasticised PVC suitable for insulating domestic mains cable, but not as the dielectric in a TV aerial?

3. A 2 mm wide ribbon of 1.5 μm thick PETP tape is to be used to construct a 1 μF capacitor. Calculate the length of tape necessary, given that the dielectric constant is 3.23 at 1 kHz. Calculate the resistance of the film at 50 °C using the data in Fig. 11.1, treating the geometry as a parallel plate capacitor.

4. Which thermoplastics could be used as the body of printed circuit boards, given that a soldering operation must be carried out after the components are assembled on the board?

CHAPTER 12

1. A thin walled polymer tube must have a specified torsional stiffness. The tube has a circular cross section of specified diameter, and the length of the tube is fixed. Derive a figure of merit that selects the material which will minimise the mass of the tube.

2. Take the case of the cross ribbed plate of Fig. 12.5, where the solid plate has a thickness of 5 mm and the ribbed plate that replaces it has a thickness of 2 mm. If the ribbed plate is to have isotropic bending stiffness there must be 3 sets of ribs running at 60 °C to each other. Calculate the mass saving in this case, compared with the 45% weight saving for the crossribbed plate.

3. Make cardboard models of the ribbed beams shown in Fig. 12.10 and compare their torsional stiffness with those of a simple I beam (this method of constructing prototypes is useful to find the deformation mechanisms). Twist

the models in either direction and observe which of the ribs tends to buckle when the torque becomes excessive.

4. If you can get access to one of the plastics selection packages try the selection of
a) materials with tensile strength >60 MPa and price <1000 £/tonne b) for these materials construct a graph of HDT versus notched Izod impact strength at 23 °C, and see if any material has high values for both these parameters c) examine the way in which the package selects materials for bending stiffness-to-mass ratio, to see which function of modulus and density is used. Decide if this assumes that the beam has a rectangular cross section.

Index